INDUSTRIAL REAL ESTATE DESIGN MANUAL

产业地产设计手册

佳图文化 主编

中国林业出版社

图书在版编目（CIP）数据

产业地产设计手册 / 佳图文化主编 . -- 北京 : 中国林业出版社 , 2018.10

ISBN 978-7-5038-9796-2

Ⅰ . ①产… Ⅱ . ①佳… Ⅲ . ①工业园区一建筑设计一手册 Ⅳ . ① TU27-62

中国版本图书馆 CIP 数据核字 (2018) 第 239852 号

中国林业出版社
责任编辑：李 顺 薛瑞琦
出版咨询：（010）83143569

出 版：中国林业出版社（100009 北京西城区德内大街刘海胡同 7 号）
网 站：http://lycb.forestry.gov.cn/
印 刷：固安县京平诚乾印刷有限公司
发 行：中国林业出版社
电 话：（010）83143500
版 次：2019 年 9 月第 1 版
印 次：2019 年 9 月第 1 次
开 本：889mm×1194mm 1 ／ 16
印 张：20
字 数：200 千字
定 价：298.00 元

PREFACE
前言

The modern society is always changing, and those changes appear in all walks of life. The book *Headquarters & Sci-Tech Parks Ⅱ* has been published for about two years, and during this period, industries related to this kind of architecture have also changed, including the changes of planning and design ideas about corporate headquarters and sci-tech parks, as well as the innovation and upgrading in building materials and design approaches. With the development of headquarters and sci-tech economy, headquarters bases and sci-tech parks become more and more important, thus the planning and design for projects of this category should define their unique identities and highlight their core values. And, the concept of sustainable architecture and the researches on it have provided the architects and designers with more guidance and reference on the design of headquarters and sci-tech parks.

In order to provide the architects and designers with the latest knowledge and information in this field and make them keep pace with the times, the book *Headquarters & Sci-Tech Parks Ⅲ* has carefully selected about 30 excellent headquarters and sci-tech parks which were completed recently and well presented with plenty of high-resolution photos, renderings and professional drawings to provide different solutions in the architecture and landscape design for headquarters buildings and sci-tech parks. In addition with detail descriptions, it has well interpreted every project from different perspectives, enabling the readers to deeply understand the design ideas and improve their taste and design skills in headquarters and sci-tech parks.

现代社会日新月异，各行各业时刻发生着变化。《总部科技园Ⅱ》付梓已有两年，相关行业资讯也有所改变，其中既包括企业总部、科技园的规划设计思路的转变，亦包括建筑材料、设计手法等方面的创新改革——随着总部经济的发展，总部基地的重要性愈加显著，这就要求总部科技园在设计方面更多地体现出自身的核心地位；而可持续发展建筑概念及相关专业研究也为总部科技园的设计提供了更为丰富的指导与参考。

为了能让总部科技园规划设计相关从业人员紧跟时代发展步伐，了解更多总部科技园的设计技法与实操经验，《总部科技园Ⅲ》整理辑录了将近30个优秀总部科技园项目，精选近年来完成的设计实例，通过大量的园区实景图、设计效果图以及各种技术分析图纸，直观形象地展现了总部科技园建筑、景观设计的多种形态，并且配以专业规范的文字解读，让读者得以全方位多角度地深入了解每一个设计范本，丰富自身的设计案例库存，借以提高在总部科技园方面的设计鉴赏能力与专业水平。

CONTENTS 目录

Headquarters Office Buildings 总部办公大楼

Sci-tech Parks 科技产业园

Headquarters Office Buildings

总部办公大楼

Core Characteristics
核心特征

Image Display
形象展示

Group Owned
集团属性

Intelligent Building
智慧建筑

Wuqing Development Zone Business Headquarters Base

天津武清开发区创业总部基地企业总部区

Features 项目亮点

Design concept of “City of Nature” strengthened the planning framework, formed mesh texture by stretching and deforming grid and enriched skyline changes after upgrading volume.

设计以“自然之城”为理念，强化规划构架，对格状进行拉伸、变形处理，形成网状肌理，将体量提升后丰富天际线变化。

Keywords 关键词

City of Nature
自然之城

Mesh Texture
网状肌理

Three-dimensional Landscape
立体景观

Location: Wuqing District, Tianjin, China
Developer: Wuqing Development Zone Corporation
Architecture Design: DC Alliance
Land Area: 148,000 m^2
Gross Floor Area: 416,000 m^2
Design Date: 2010
Completion Date: 2013

项目地点：中国天津市武清区
开发商：武清开发区总公司
建筑设计：DC 国际
用地面积：14.8 万 m^2
总建筑面积：41.6 万 m^2
设计时间：2010 年
竣工时间：2013 年

Overview

Base is located in Wuqing Development Zone that was set up at the end of 1991, as the national economic and technological development zone and high-tech industrial park approved by the State Council. The Business Headquarters Base was set in the north of the zone, extending from Guangyuan Road in the north to Fuyuan Road in the south, from Tsuiheng Road in the west to Quanwang North Road in the east, adjacent to Swan Lake Resort by the east side.

项目概况

基地位于天津市武清开发区。武清开发区于 1991 年底设立，是经国务院批准的国家级经济技术开发区和国家级高新技术产业园区。开发区内北侧的创业总部基地，北起广源道，南至福源道，西起翠亨路，东至泉旺北路，东侧紧邻天鹅湖度假村。

Design Idea

Design concept of “City of Nature” strengthened the planning framework, formed mesh texture by stretching and deforming grid and enriched skyline changes after upgrading volume. The road entrance is open to the city, inserting the public system at the junction to form the interaction and infiltration with city; grid courtyard sinking down, implanted different themes courtyard, forming a three-dimensional landscape system.

设计理念

设计以“自然之城”为理念，强化规划构架，对格状进行拉伸、变形处理，形成网状肌理，将体量提升后丰富天际线变化。道路入口对城市开放，交接处插入公共系统，与城市形成互动与渗透；网格内庭院下沉，植入不同主题的庭院景观，形成立体景观系统。

Site Plan of Plot B-1,B-3,B-4

B-1,B-3,B-4 总平面图

B-1地块经济技术指标表	
规划总用地面积	34926m^2
总建筑面积	91622m^2
其中 地上总建筑面积	66940m^2
其中 1#楼建筑面积	18400m^2
2#楼建筑面积	2920m^2
3#楼建筑面积	7280m^2
4#楼建筑面积	5330m^2
5#楼建筑面积	7280m^2
6#楼建筑面积	2920m^2
7#楼建筑面积	12350m^2
8#楼建筑面积	10460m^2
地下总建筑面积	24682m^2
计算容积率建筑面积	66940m^2
建筑物基底总面积	6975m^2
建筑密度	19.9%
容积率	1.92
绿地率	15%
机动车停车位总面积	27804m^2
其中 地上机动车停车位总面积	3122m^2
地下汽车库建筑面积	24682m^2
机动车停车位数量	637辆
其中 地上机动车停车位数量	227辆
地下机动车停车位数量	410辆

B-3地块经济技术指标表	
规划总用地面积	82899m^2
总建筑面积	240645m^2
其中 地上总建筑面积	176310m^2
其中 1#楼建筑面积	29600m^2
2#楼建筑面积	8225m^2
3#楼建筑面积	5330m^2
4#楼建筑面积	5330m^2
5#楼建筑面积	2920m^2
6#楼建筑面积	5330m^2
7#楼建筑面积	2920m^2
8#楼建筑面积	8225m^2
9#楼建筑面积	10460m^2
10#楼建筑面积	12350m^2
11#楼建筑面积	32790m^2
12#楼建筑面积	32790m^2
13#楼建筑面积	20040m^2
地下总建筑面积	64335m^2
计算容积率建筑面积	176310m^2
建筑物基底总面积	15855m^2
建筑密度	19.1%
容积率	2.13
绿地率	15%
机动车停车位总面积	71499m^2
其中 地上机动车停车位总面积	7164m^2
地下汽车库建筑面积	64335m^2
机动车停车位数量	1621辆
其中 地上机动车停车位数量	521辆
地下机动车停车位数量	1100辆

Site Plan of Plot B-1,B-3

B-1,B-3 总平面图

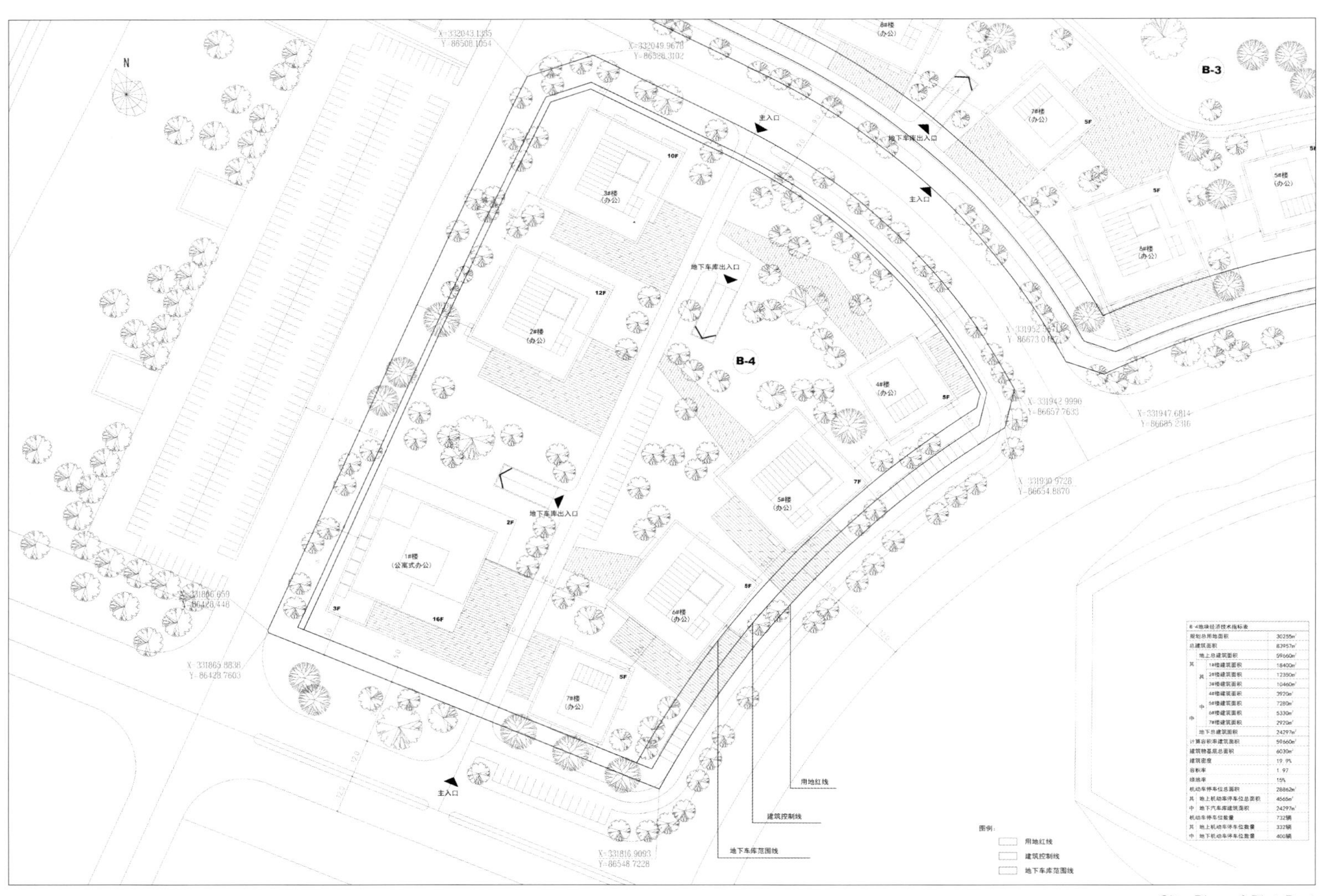

B-4地块经济技术指标表	
规划总用地面积	30255m^2
总建筑面积	83957m^2
其中 地上总建筑面积	59660m^2
其中 1#楼建筑面积	18400m^2
2#楼建筑面积	12350m^2
3#楼建筑面积	10460m^2
4#楼建筑面积	3920m^2
5#楼建筑面积	7280m^2
6#楼建筑面积	5330m^2
7#楼建筑面积	2920m^2
地下总建筑面积	24297m^2
计算容积率建筑面积	59660m^2
建筑物基底总面积	6030m^2
建筑密度	19.9%
容积率	1.97
绿地率	15%
机动车停车位总面积	28862m^2
其中 地上机动车停车位总面积	4565m^2
地下汽车库建筑面积	24297m^2
机动车停车位数量	732辆
其中 地上机动车停车位数量	332辆
地下机动车停车位数量	400辆

Site Plan of Plot B-4

B-4 总平面图

Functional Orientation

Project is positioned as a business headquarters base primarily serving entrepreneurial small businesses with independent office, supplemented by large- and medium-sized enterprises' centralized office. To meet the needs of different companies at different stages of development, the zone offers a variety of products from high-level units to single-family offices, especially single-family ecological offices of low density, low capacity and high green rate ushered in development climax of corporate office demands.

功能定位

项目定位为以服务创业型小型企业独立办公为主，大中型企业集中式办公为辅的城市创业型总部基地。为满足不同企业不同发展阶段的使用需求，这里提供从高层单元到独栋办公的多种产品，尤其是低密度、低容量、高绿地率的独栋生态办公迎来了企业办公诉求上的发展高潮。

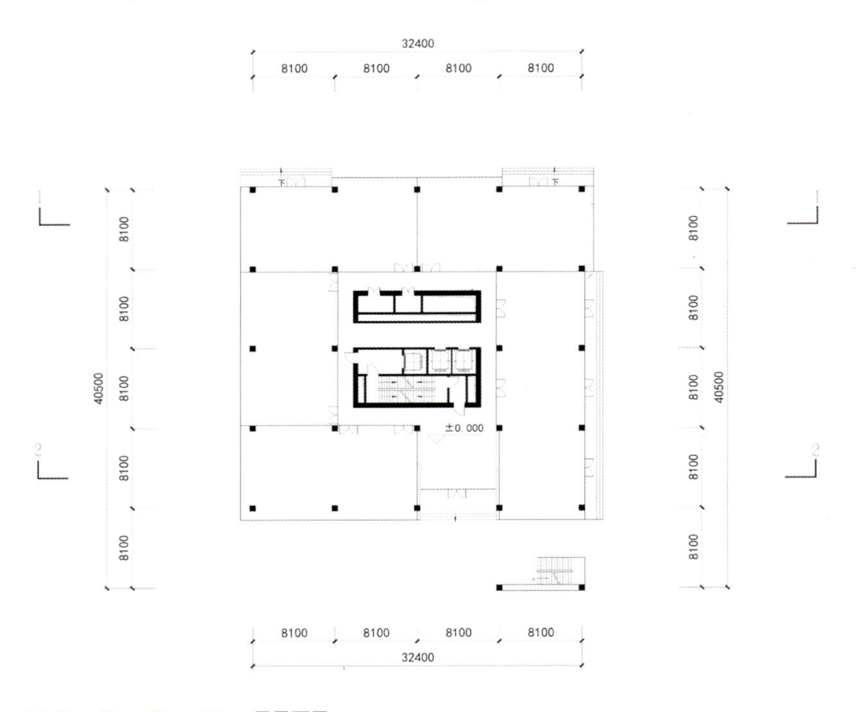

1# First Floor Plan 1# 一层平面图

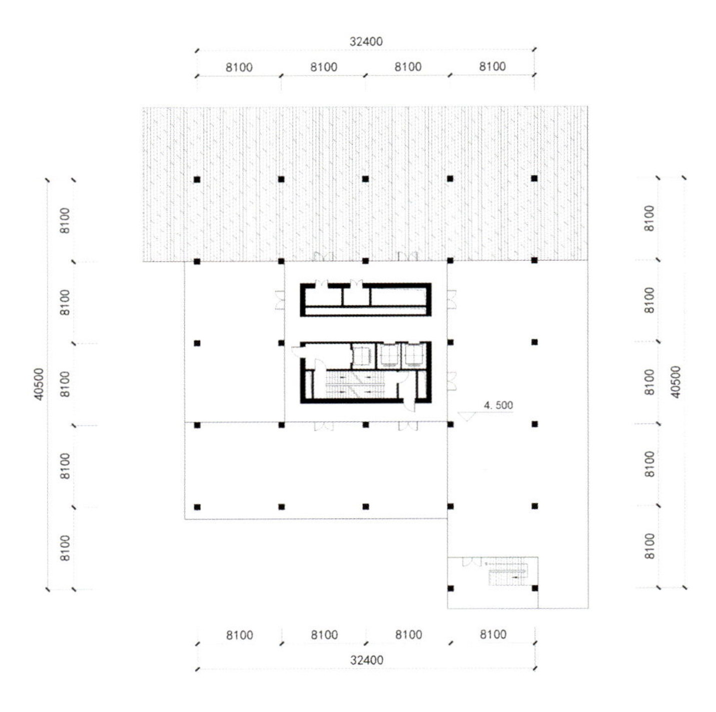

1# Second Floor Plan 1# 二层平面图

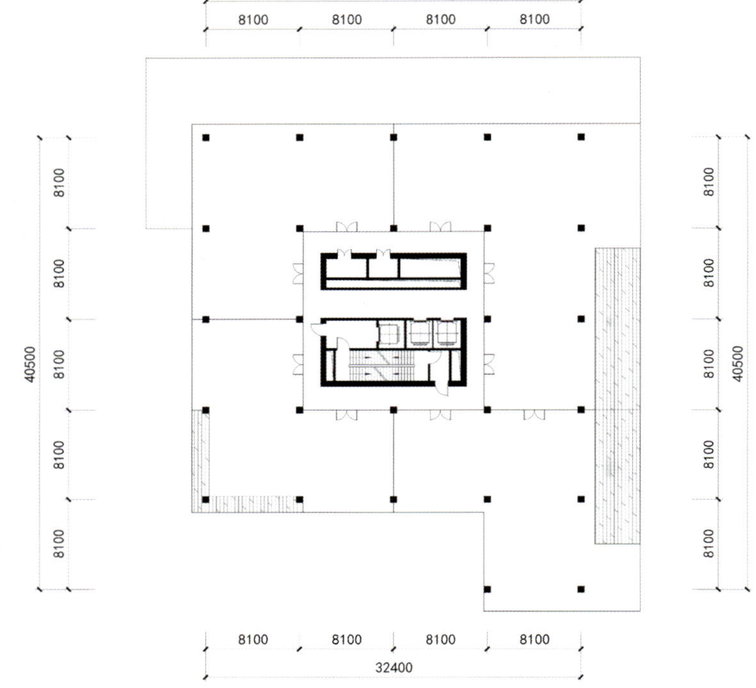

1# Third Floor Plan 1# 三层平面图

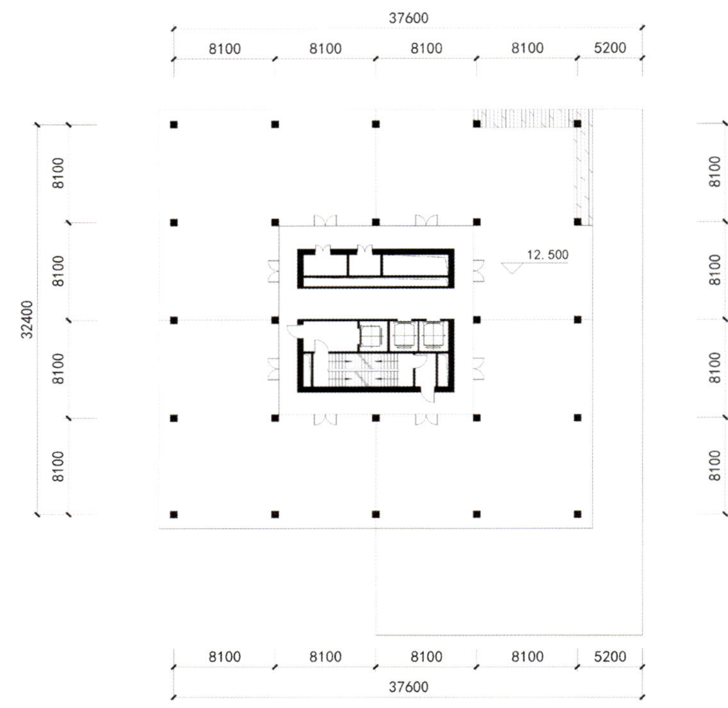

1# Fourth Floor Plan 1# 四层平面图

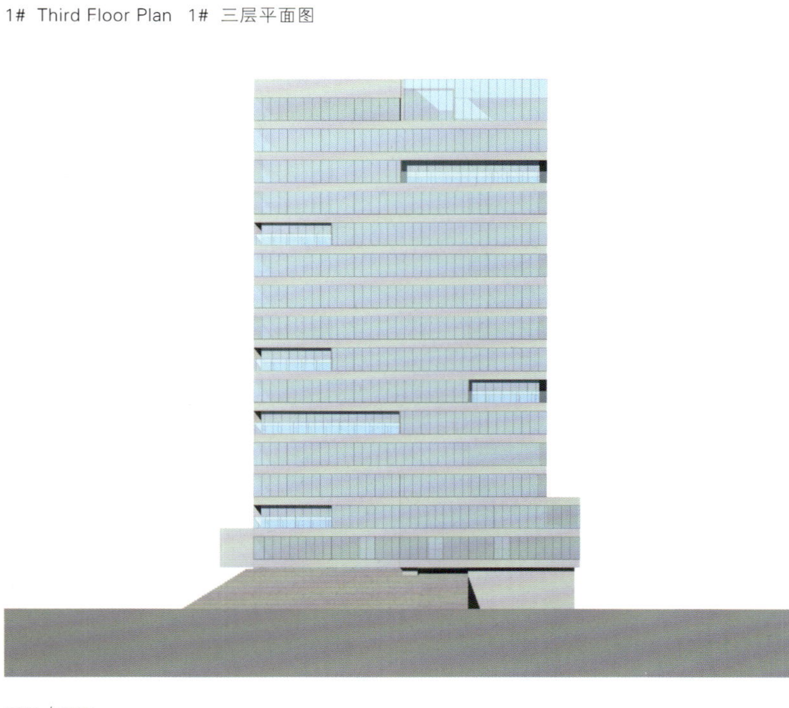

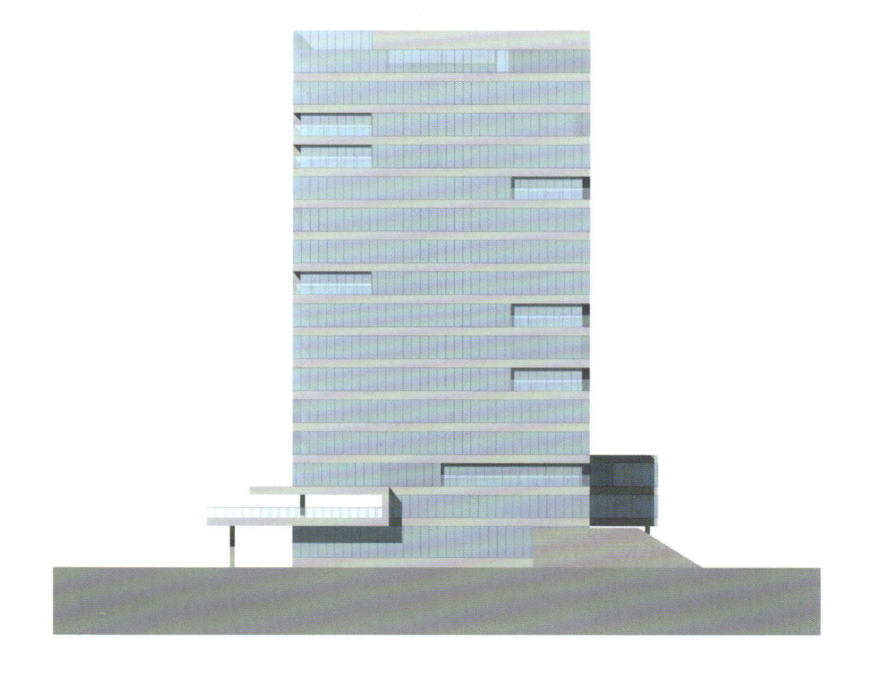

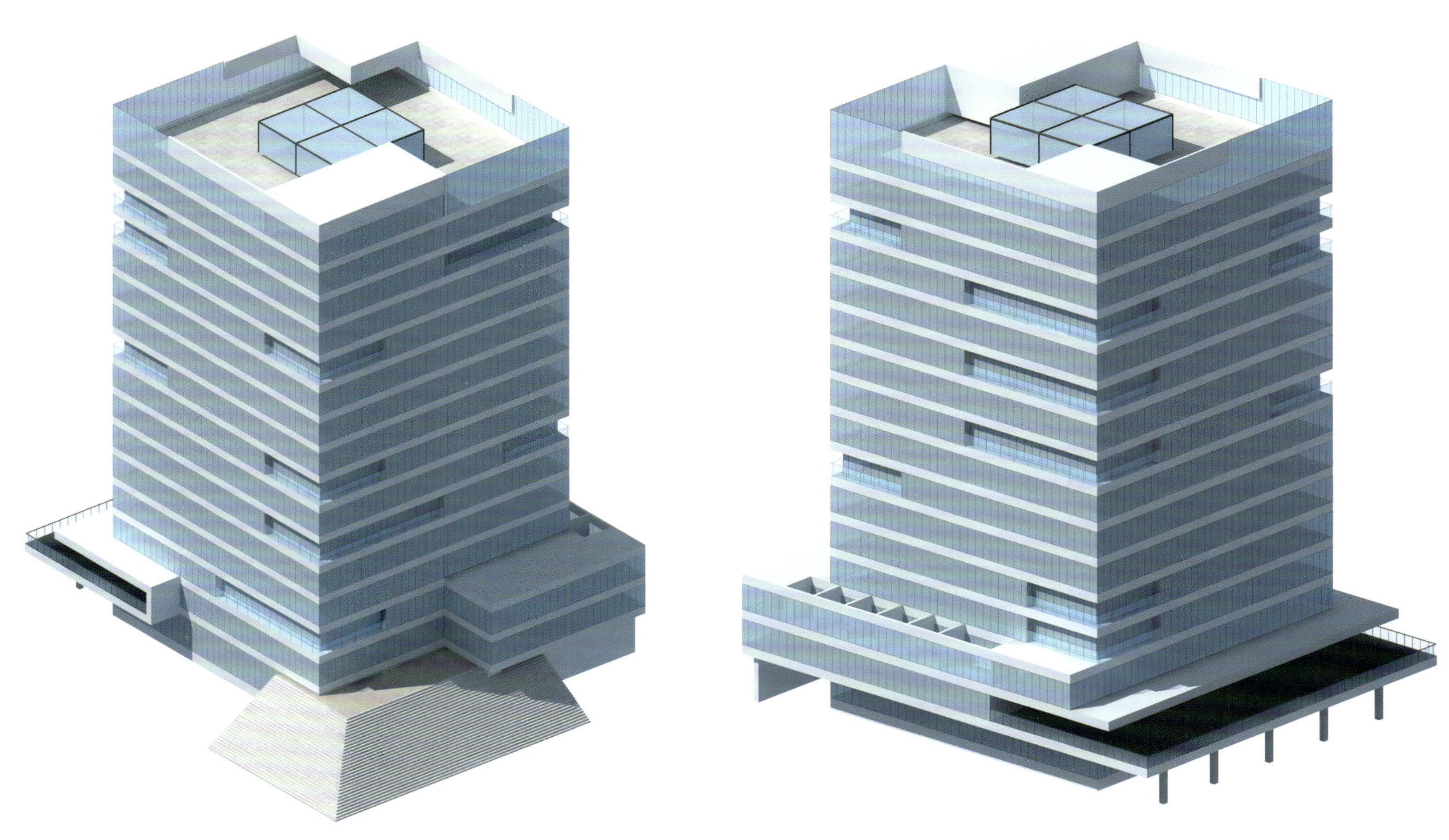

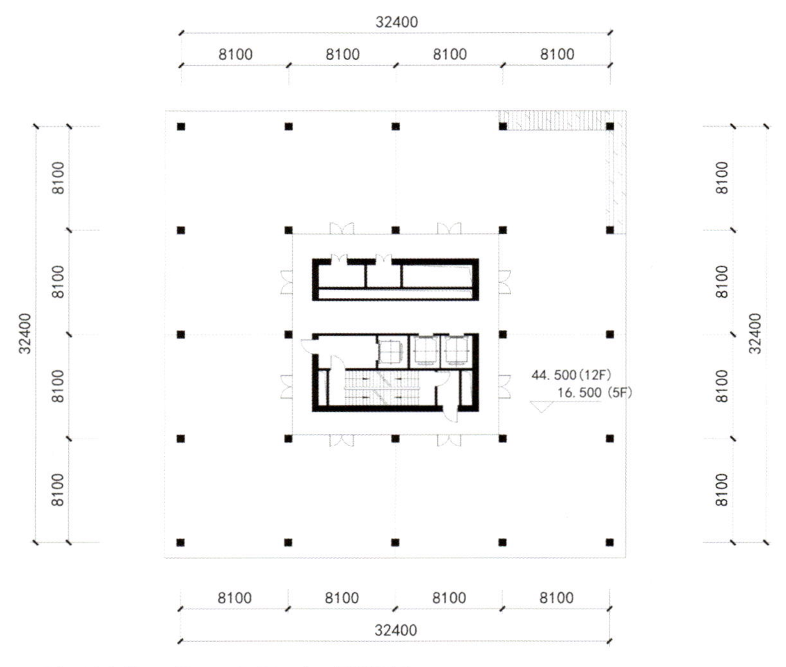

1# 5th, 12th Floor Plan 1# 五、十二层平面图

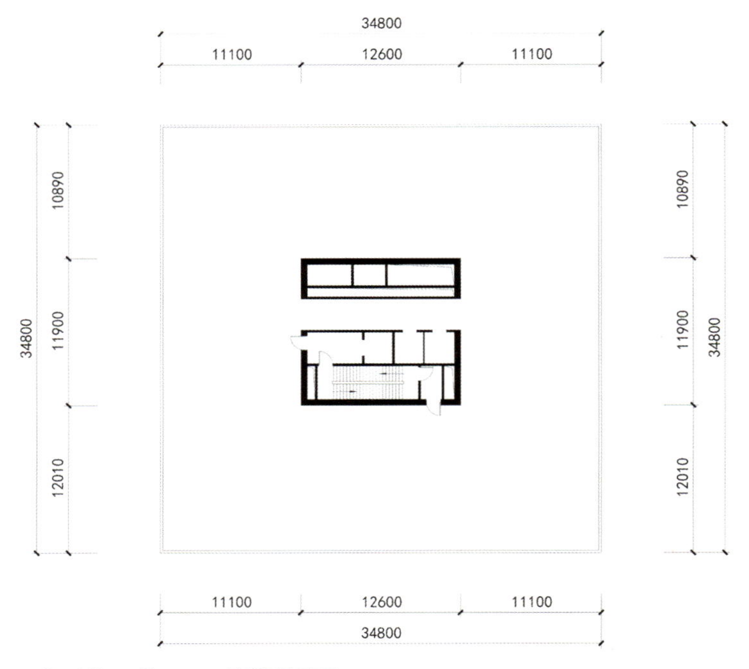

1# Roof Floor Plan 1# 屋顶层平面图

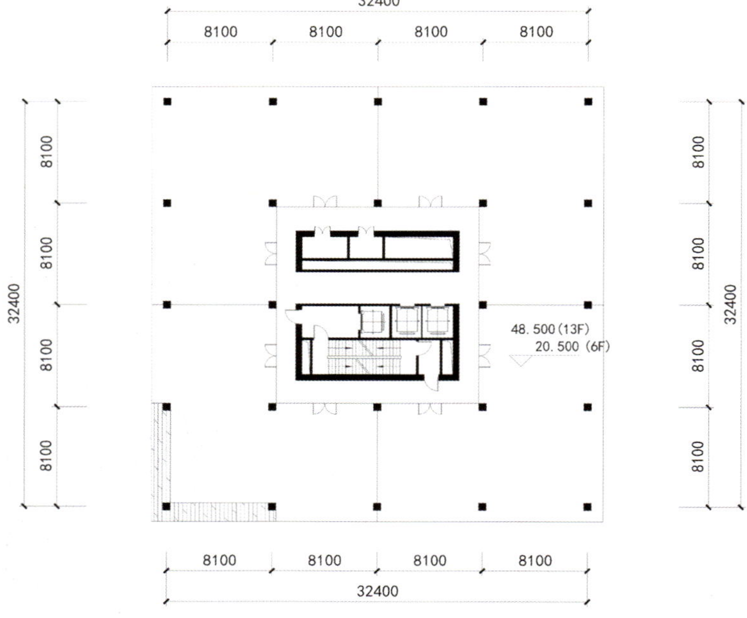

1# 6th,13th Floor Plan 1# 六、十三层平面图

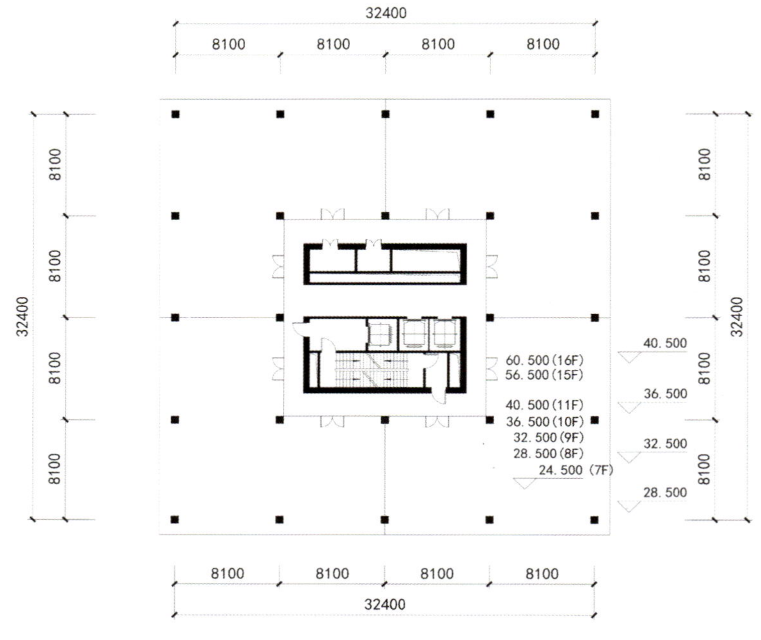

1# 7th~11th,15th,16th Floor Plan 1# 七～十一、十五、十六层平面图

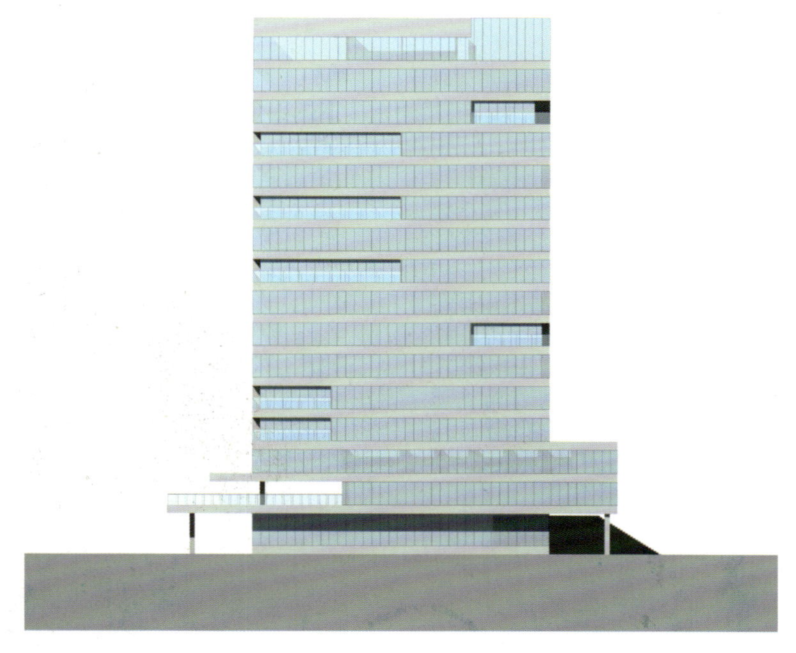

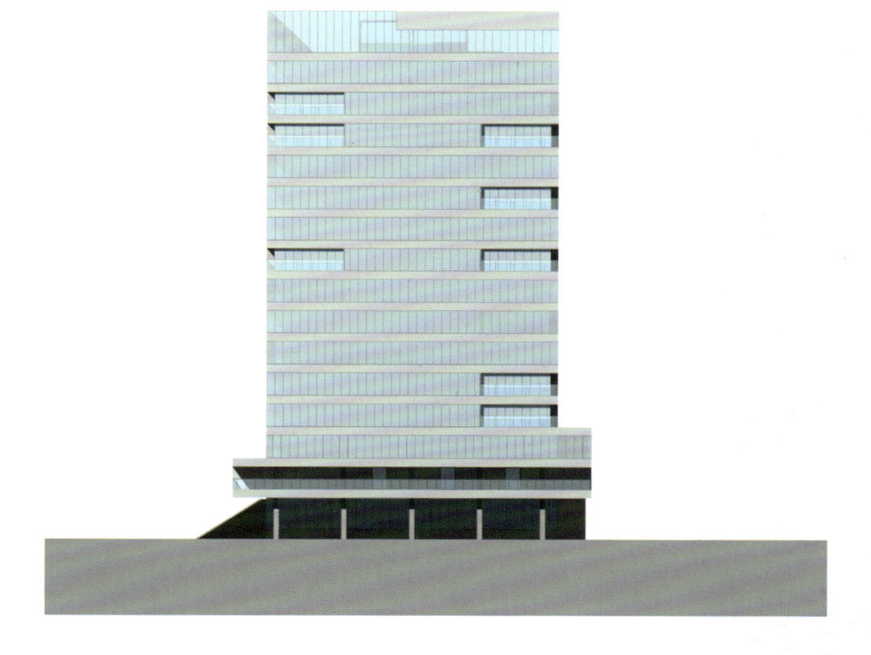

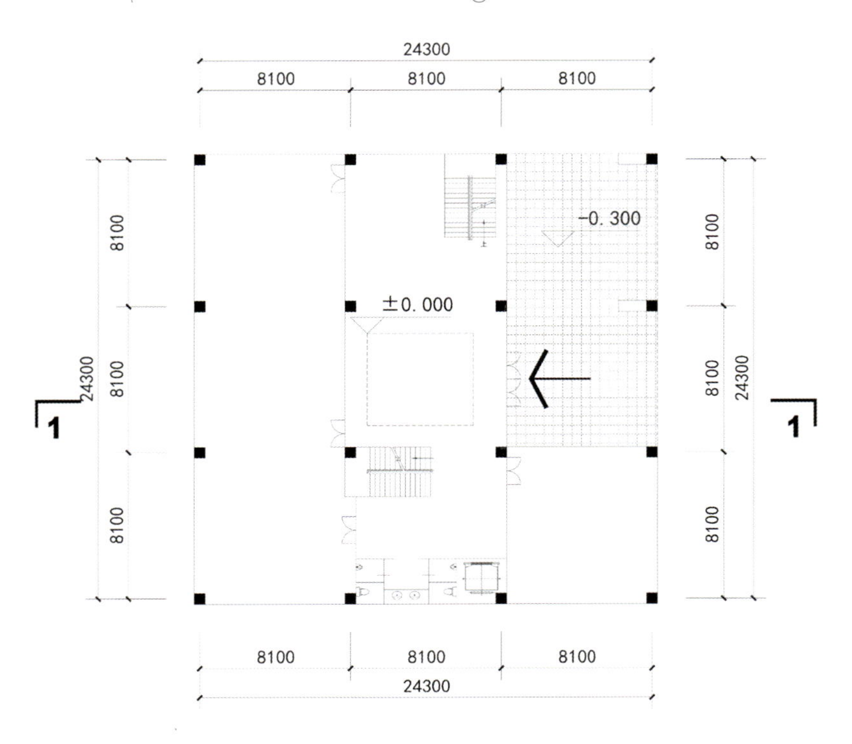

2# 6# First Floor Plan 2# 6# 一层平面图

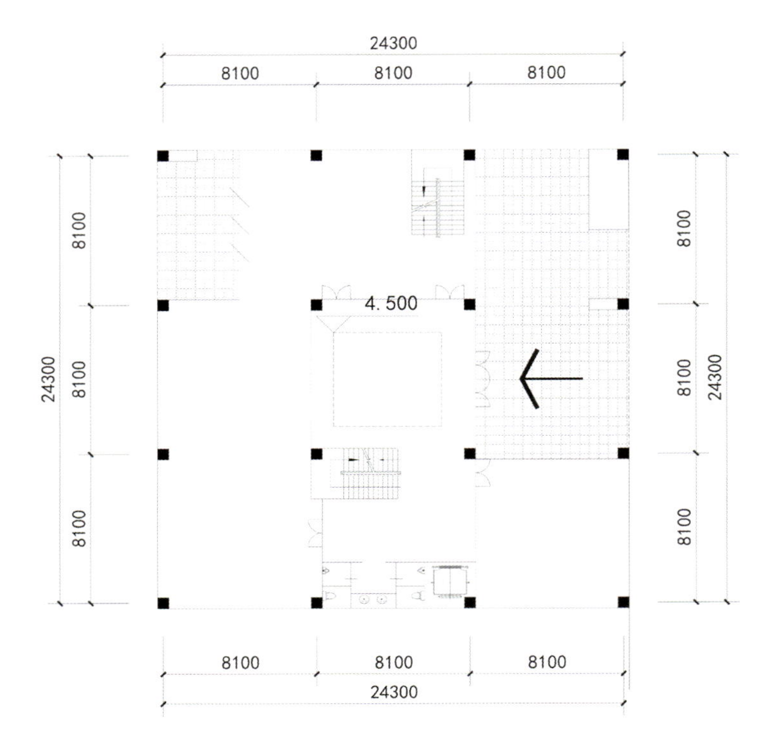

2# 6# Second Floor Plan 2# 6# 二层平面图

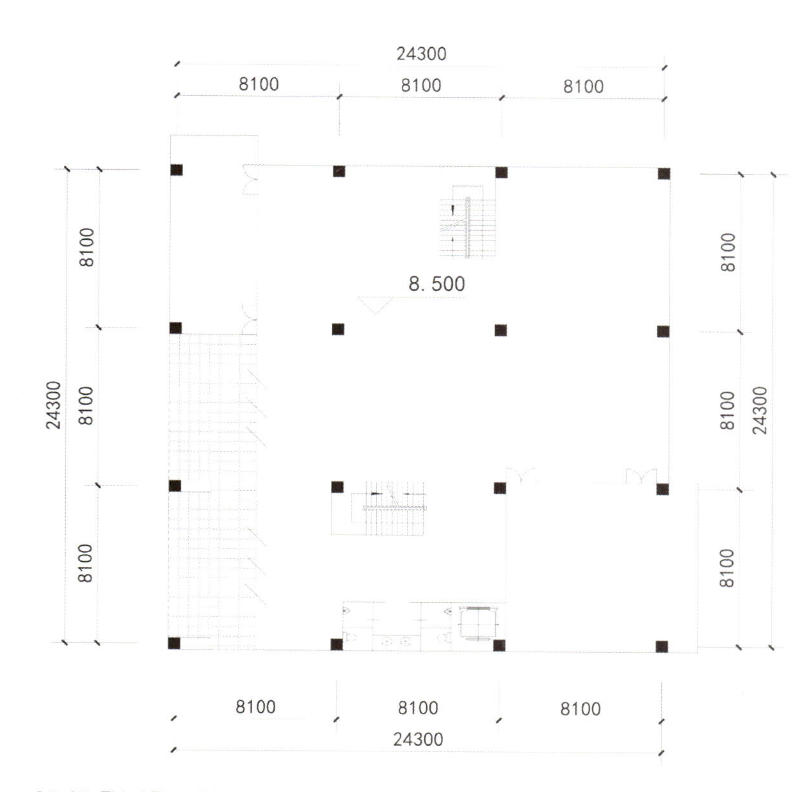

2# 6# Third Floor Plan 2# 6# 三层平面图

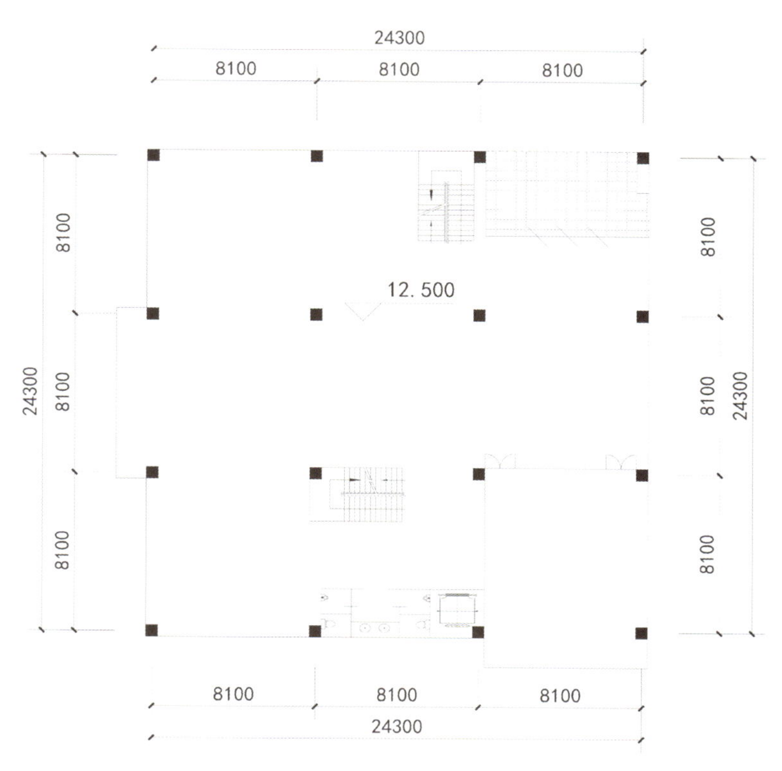

2# 6# Fourth Floor Plan 2# 6# 四层平面图

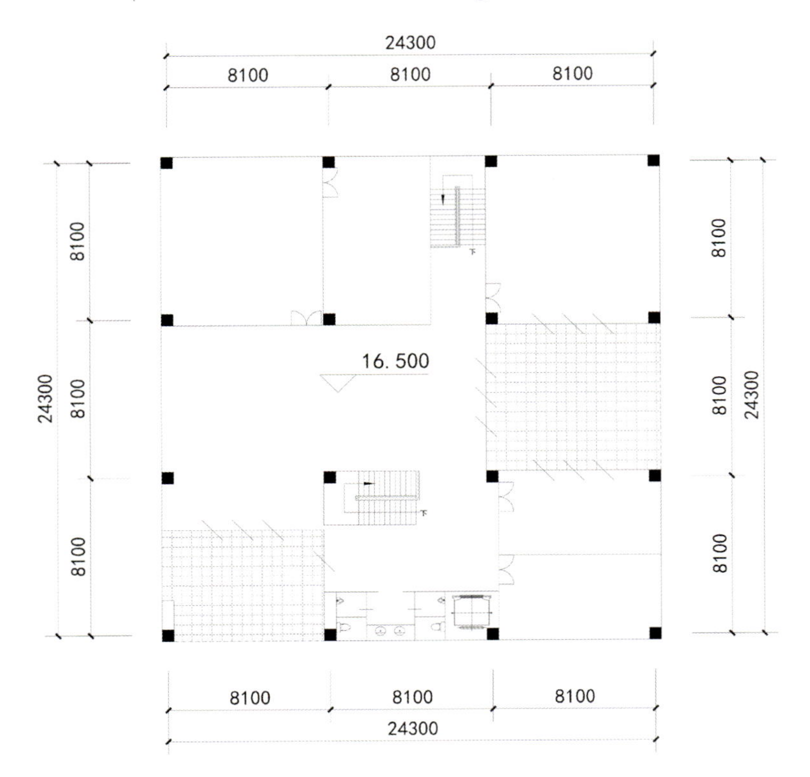

2# 6# Fifth Floor Plan 2# 6# 五层平面图

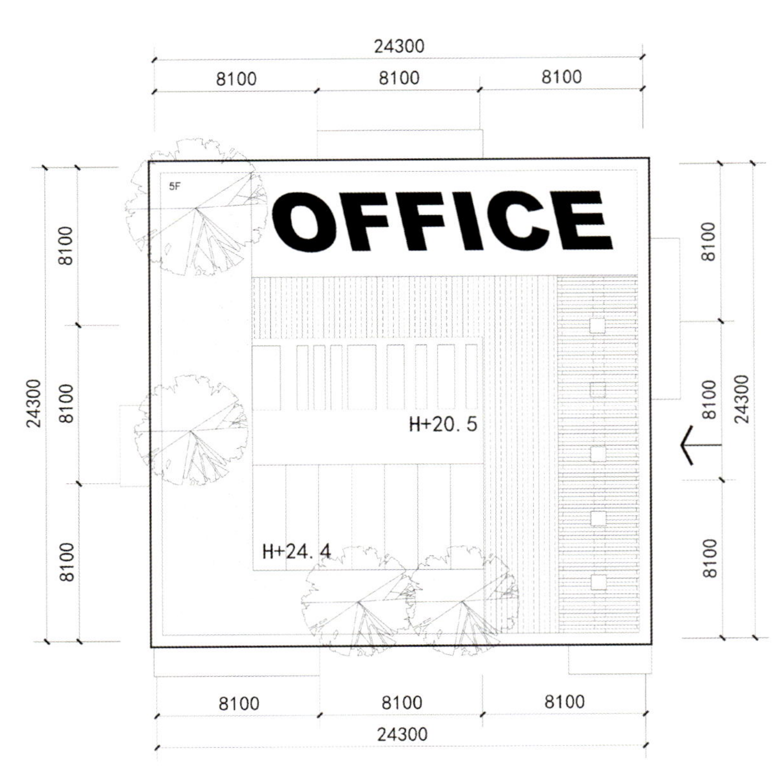

2# 6# Sixth Floor Plan 2# 6# 六层平面图

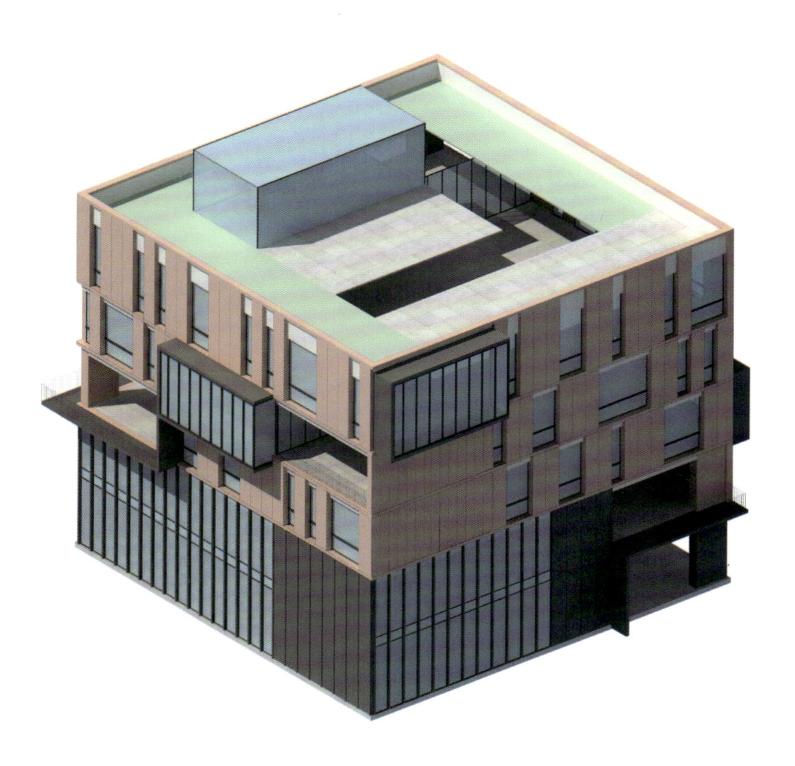

B06

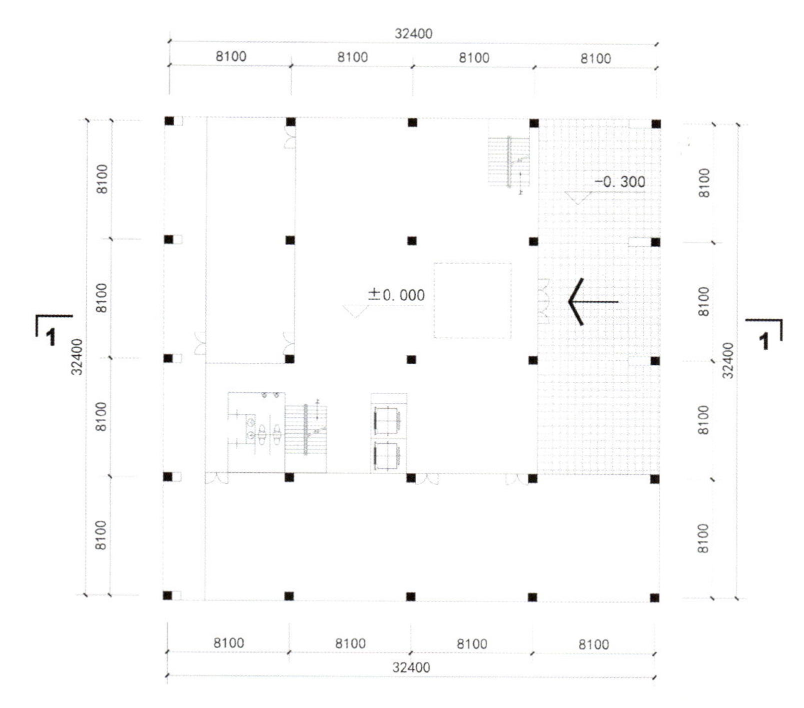

3# 5# First Floor Plan 3# 5# 一层平面图

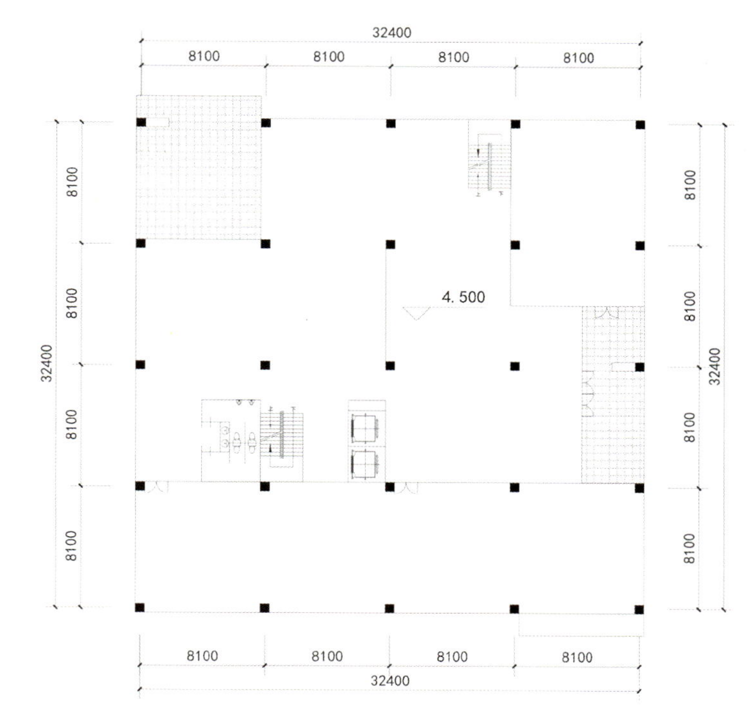

3# 5# Second Floor Plan 3# 5# 二层平面图

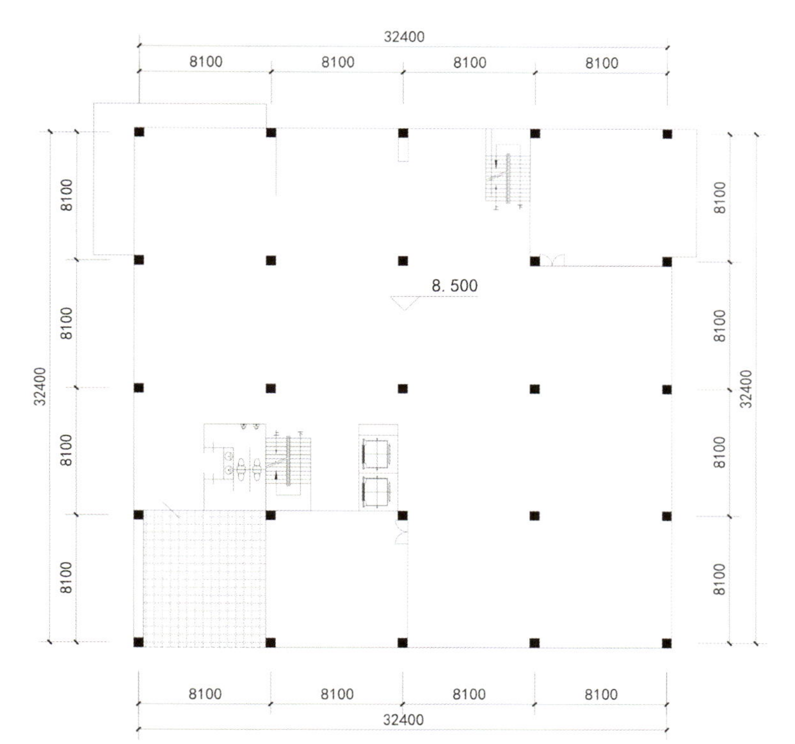

3# 5# Third Floor Plan 3# 5# 三层平面图

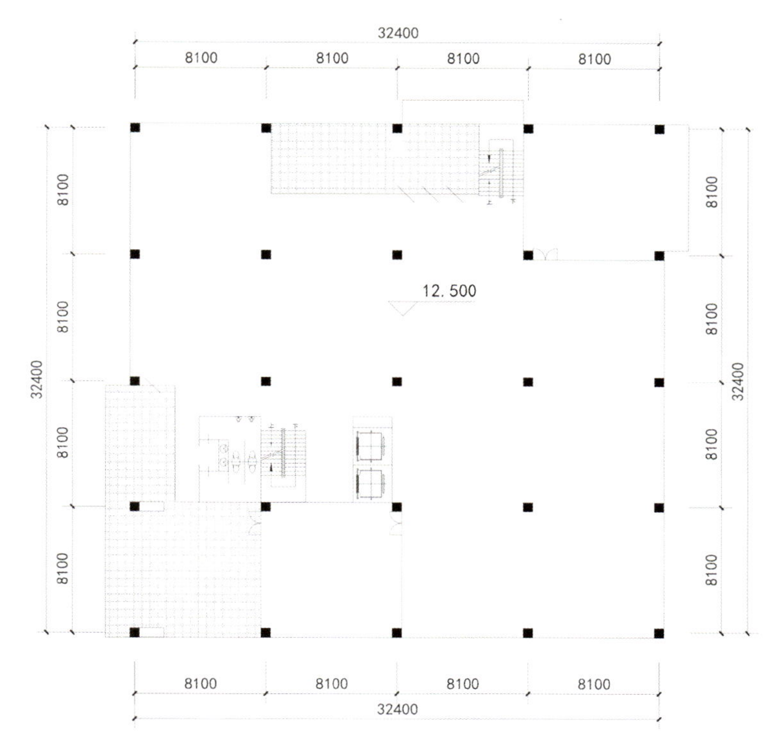

3# 5# Fourth Floor Plan 3# 5# 四层平面图

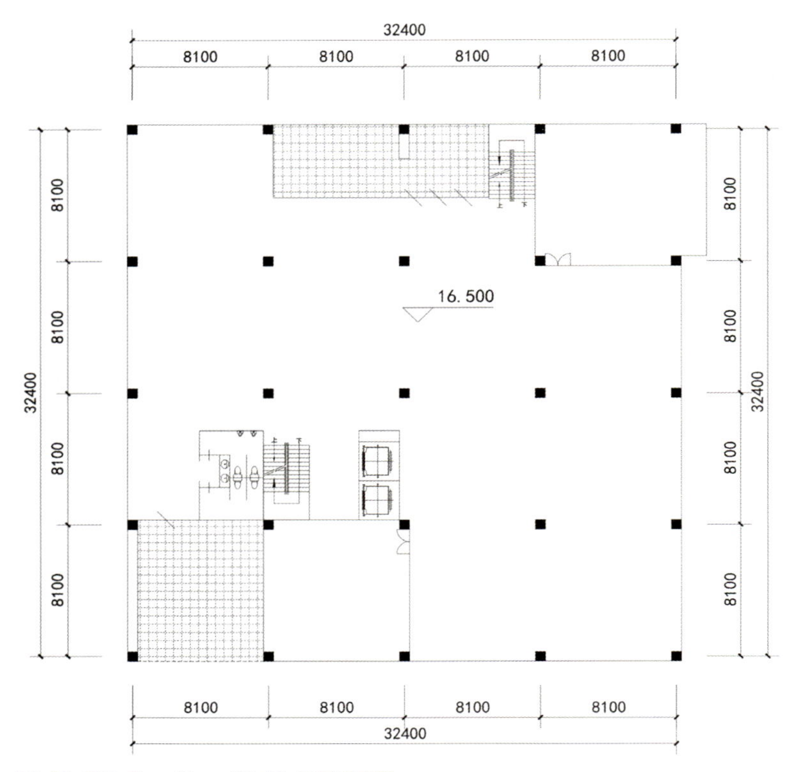

3# 5# Fifth Floor Plan 3# 5# 五层平面图

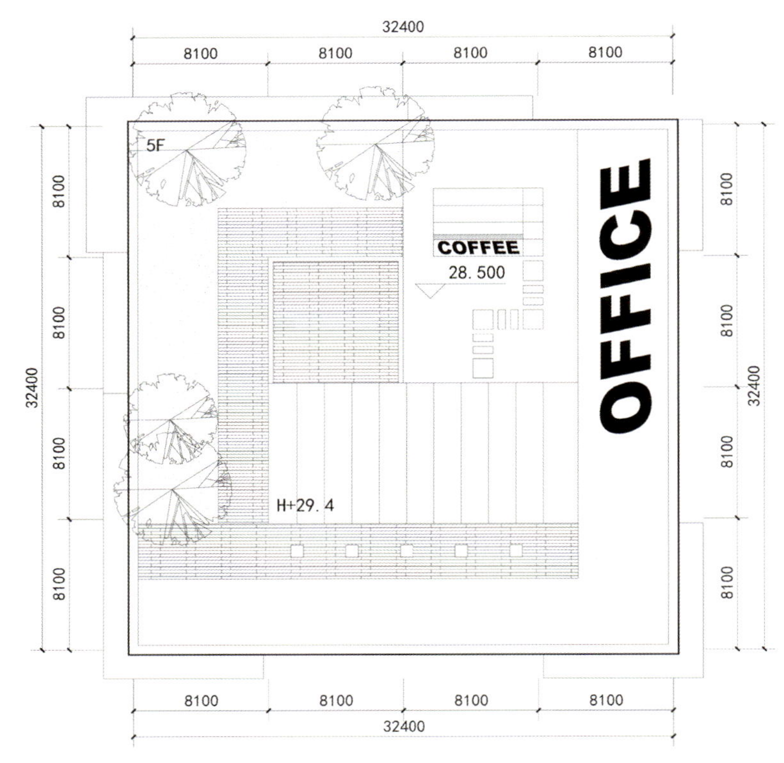

3# 5# Roof Floor Plan 3# 5# 屋顶层平面图

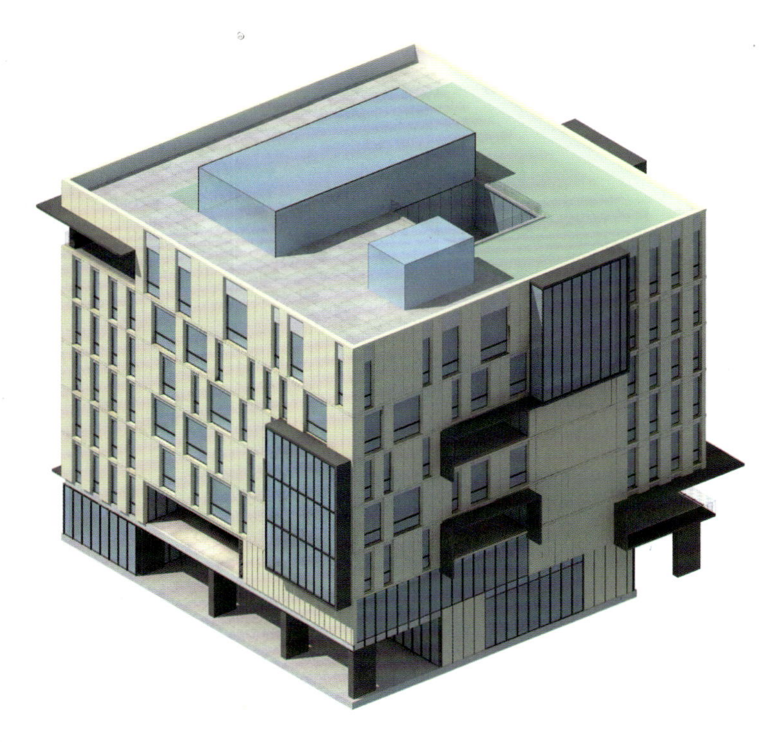

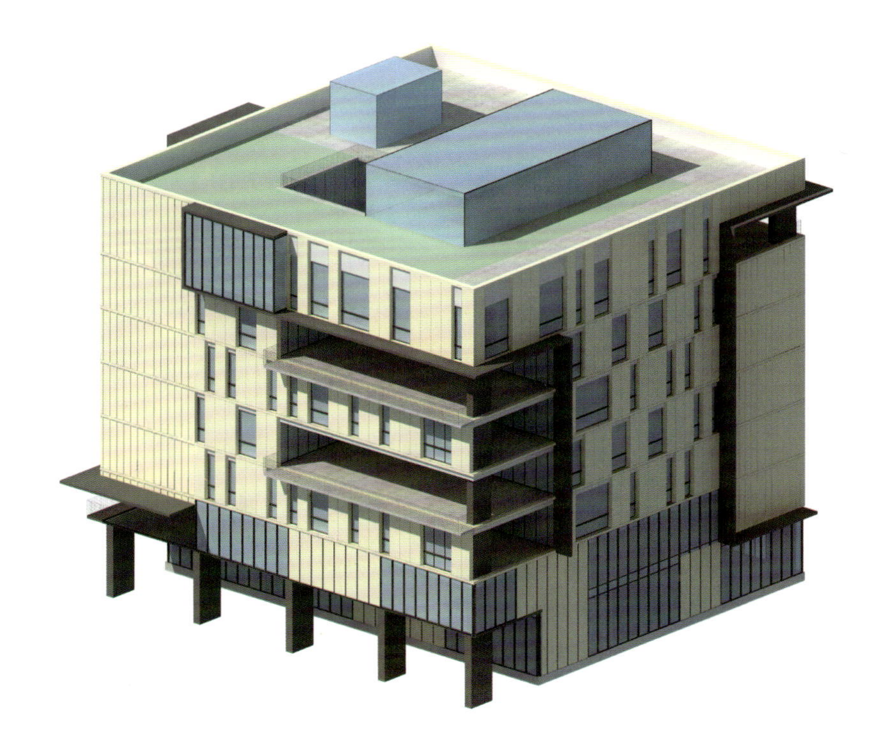

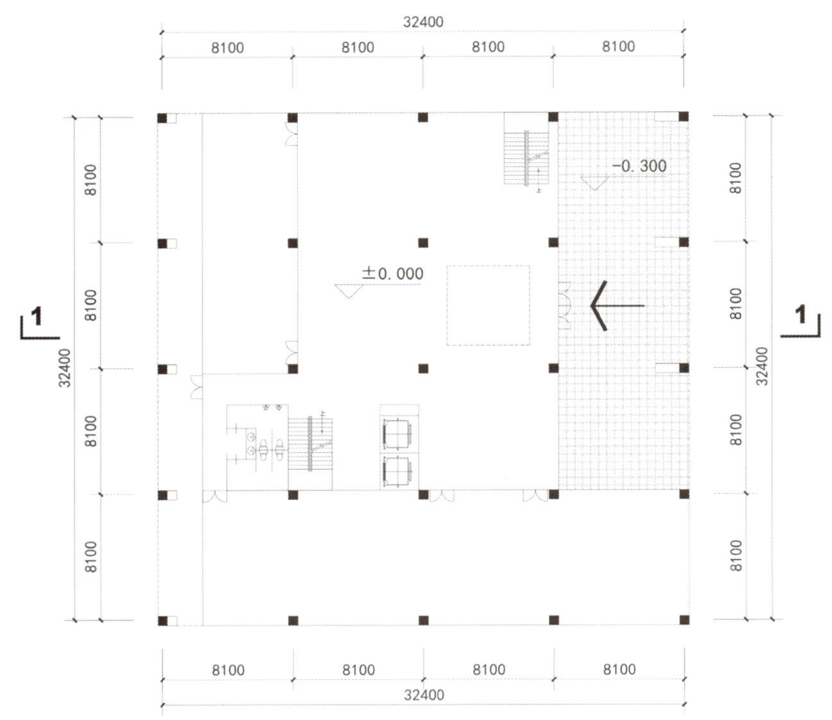

4# First Floor Plan 4# 一层平面图

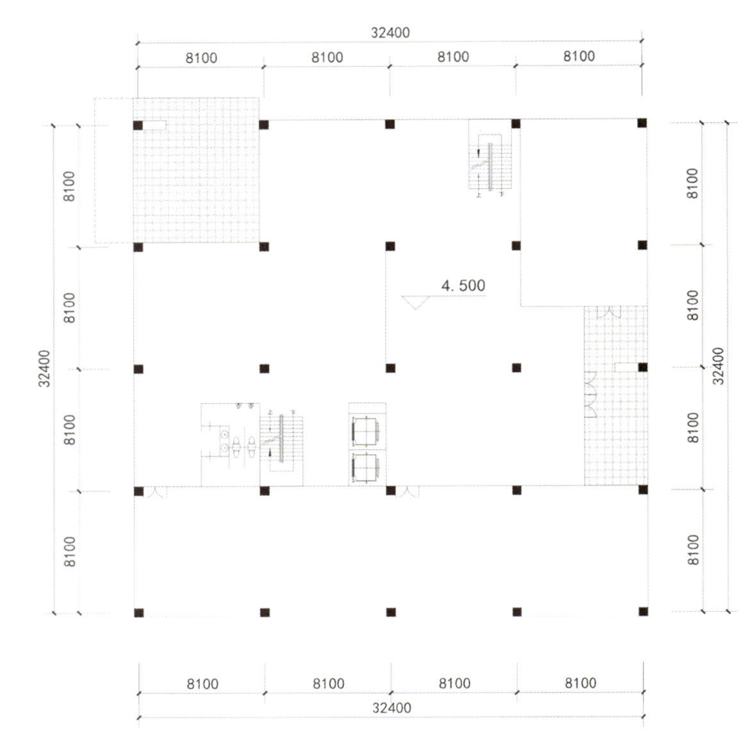

4# Second Floor Plan 4# 二层平面图

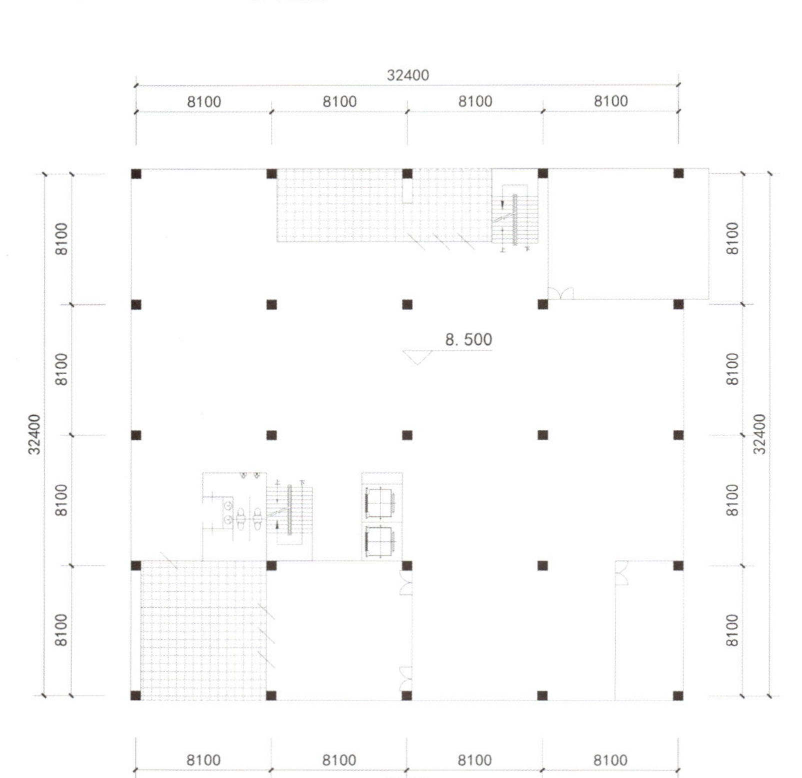

4# Third Floor Plan 4# 三层平面图

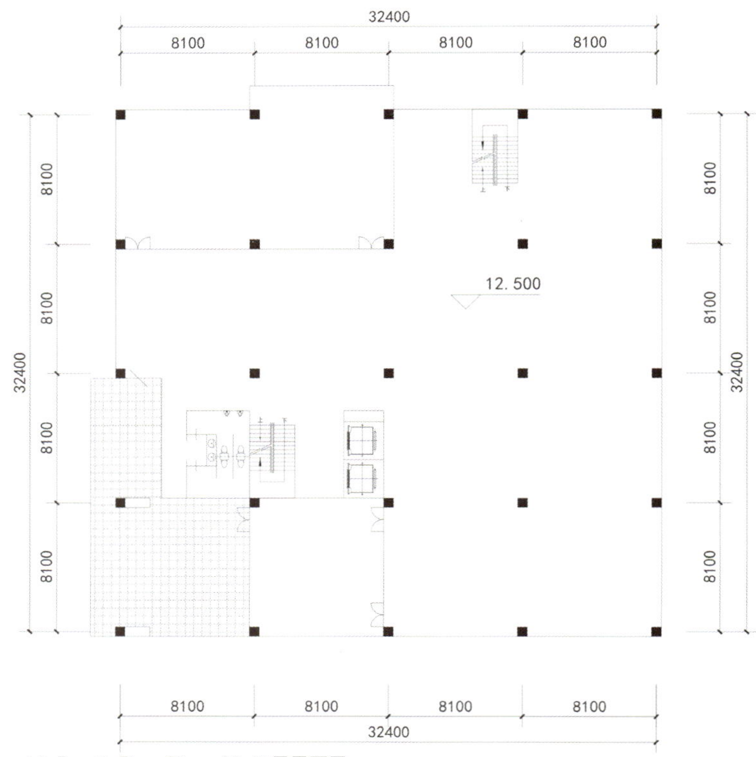

4# Fourth Floor Plan 4# 四层平面图

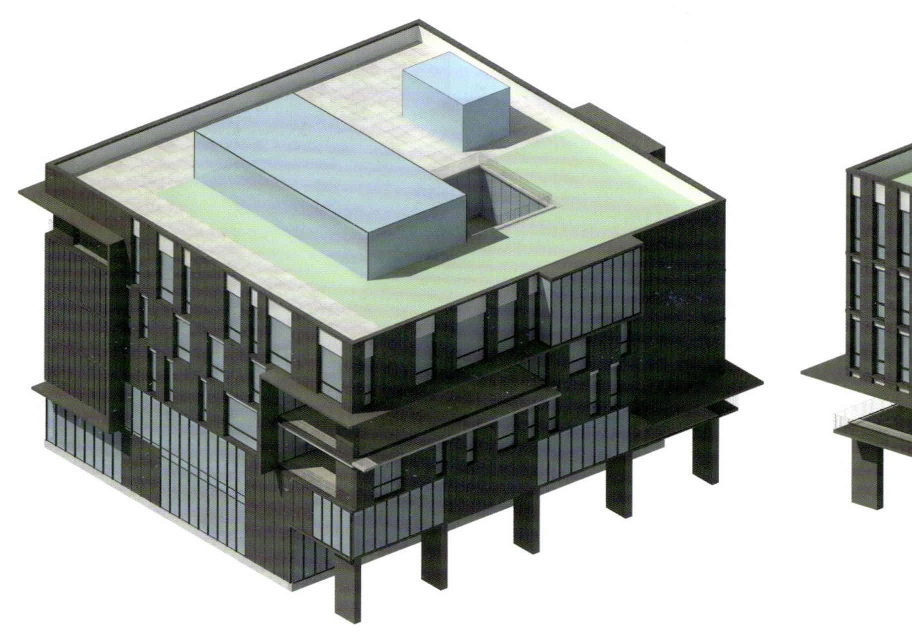

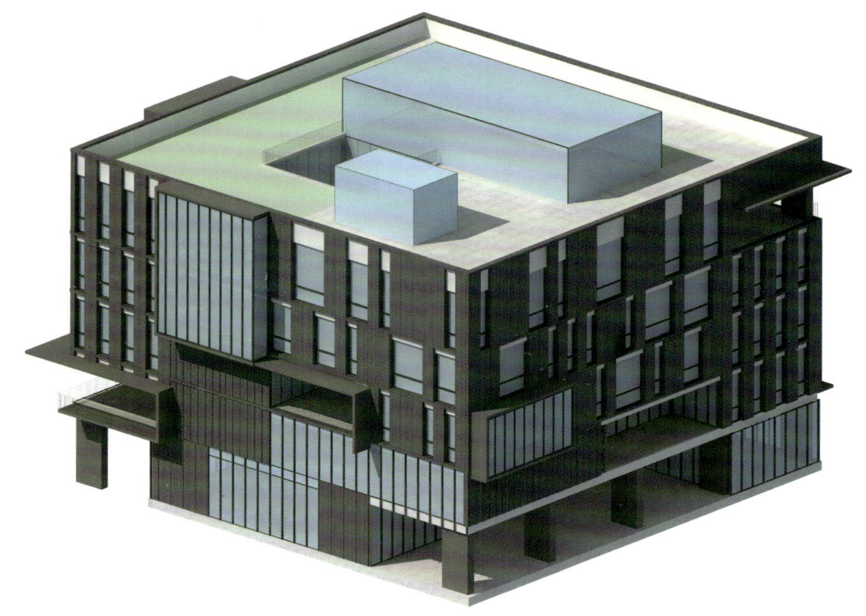

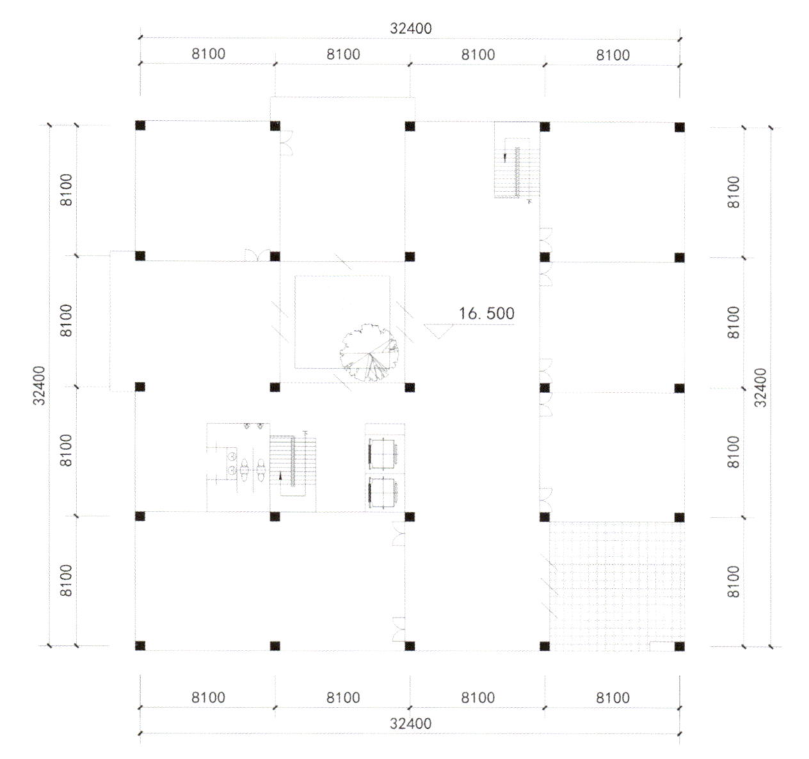

4# Fifth Floor Plan 4# 五层平面图

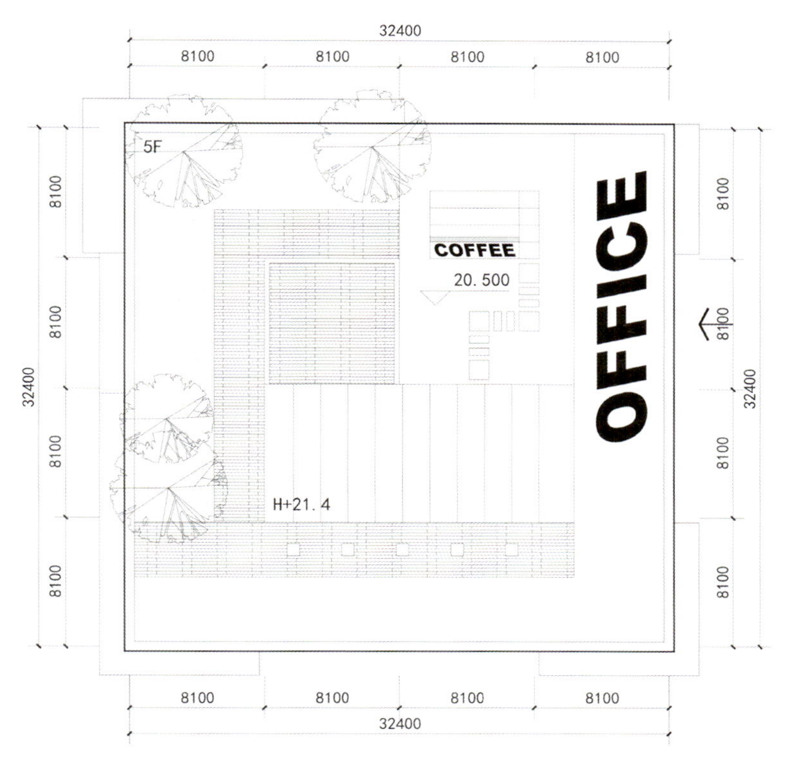

4# Roof Floor Plan 4# 屋顶层平面图

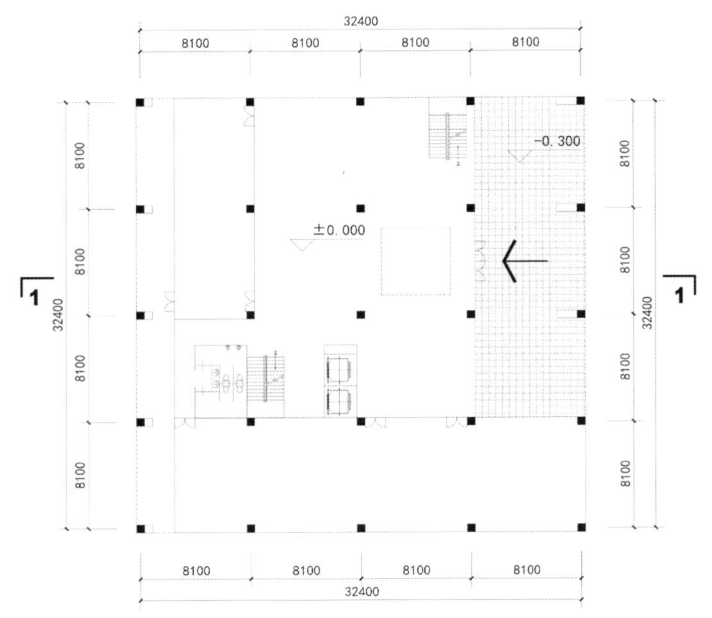

7# First Floor Plan 7# 一层平面图

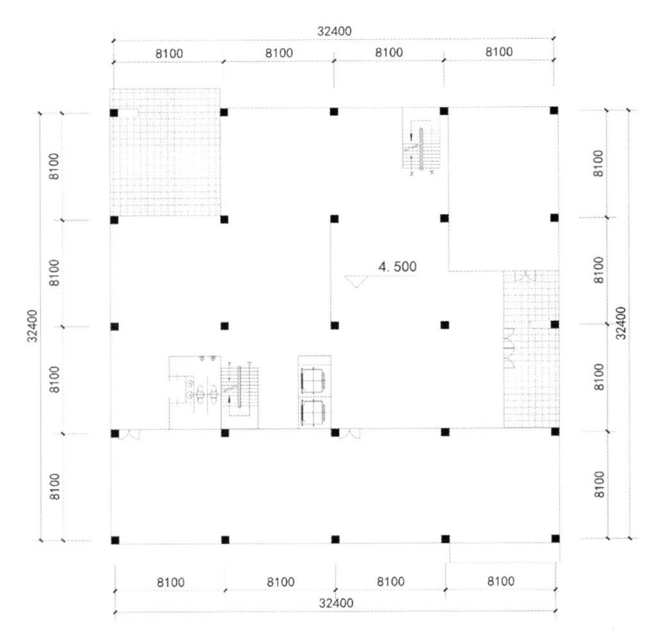

7# Second Floor Plan 7# 二层平面图

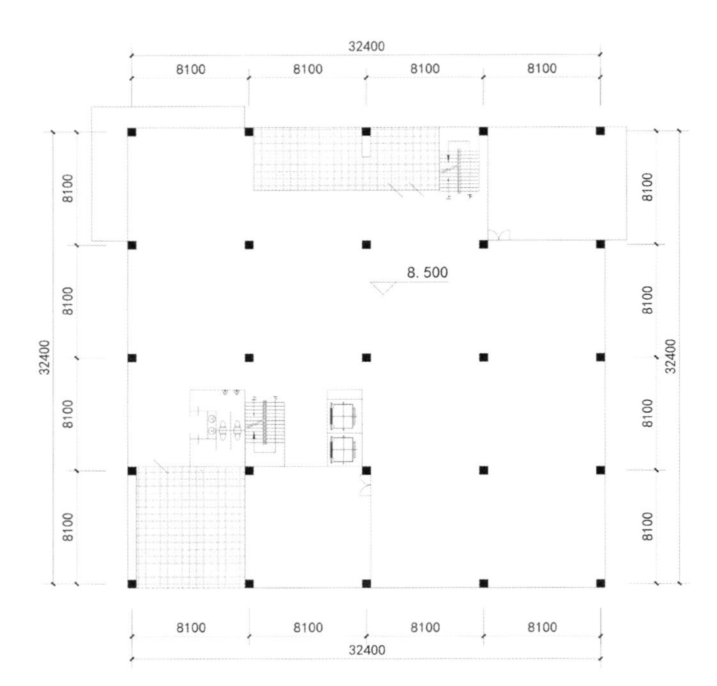

7# Third Floor Plan 7# 三层平面图

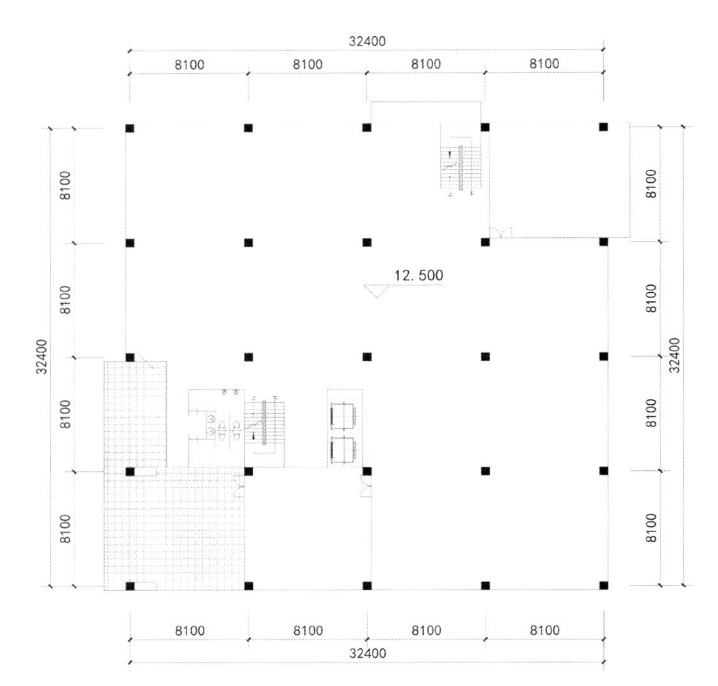

7# Fourth Floor Plan 7# 四层平面图

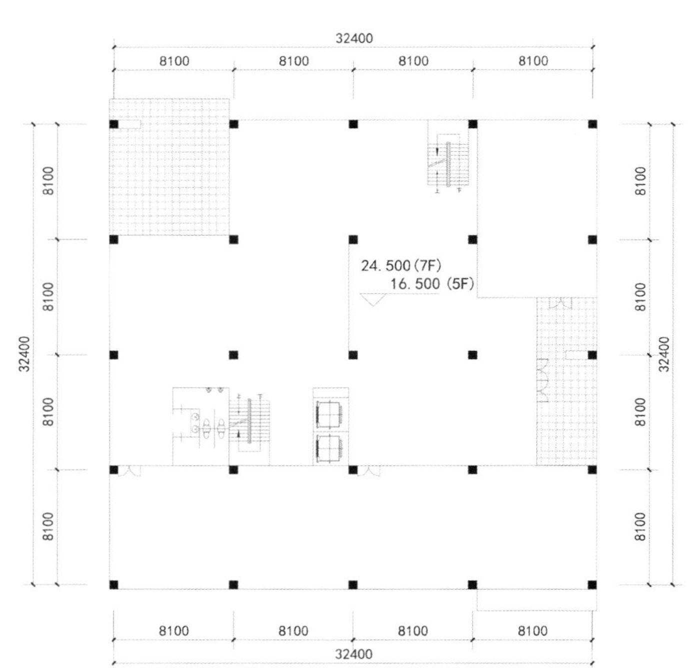

7# 5th,7th Floor Plan 7# 五、七层平面图

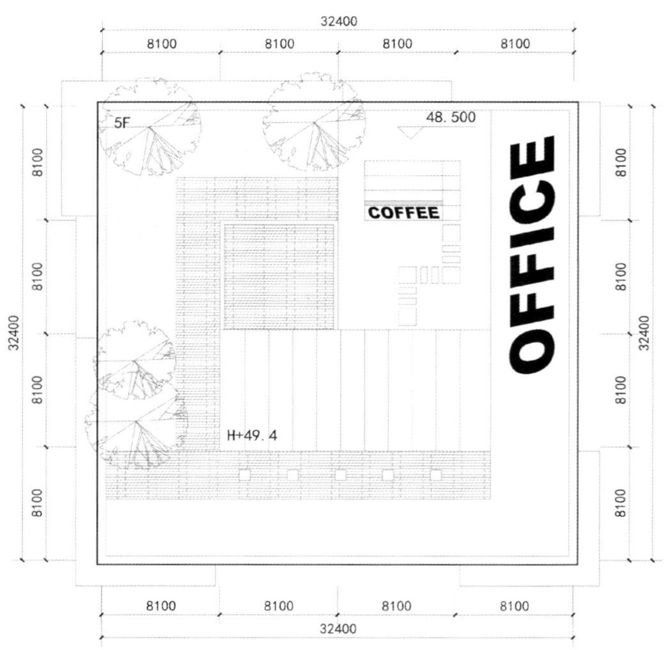

7# Roof Floor Plan 7# 屋顶层平面图

Architectural Design

By using split facades and walls, multi-level roof terrace, basement of partly overhead and other architectural elements, taking layer as a unit, design forms collage facade by rotation and superposition that is suitable for the requirements of different layers. Brick red and matt white aluminum plate mixed with black glass curtain wall, matching with black aluminum ceiling and wrapping. Single aluminum plate's size, taking story height as unit, emphasizes on the vertical lines in order to maintain the façade's unity and wholeness.

建筑设计

设计采用分割的立面和墙壁、多层级的屋顶平台、部分架空的底层等建筑元素，以层为单元，通过旋转、叠加形成了具有拼贴意味的立面，并适用于不同的层数要求。采用砖红色和亚光白的铝板混合玻璃幕墙，配以黑色铝板吊顶和包边。单块铝板尺度以层高为单位，强调竖向线条，以保持立面的统一和整体性。

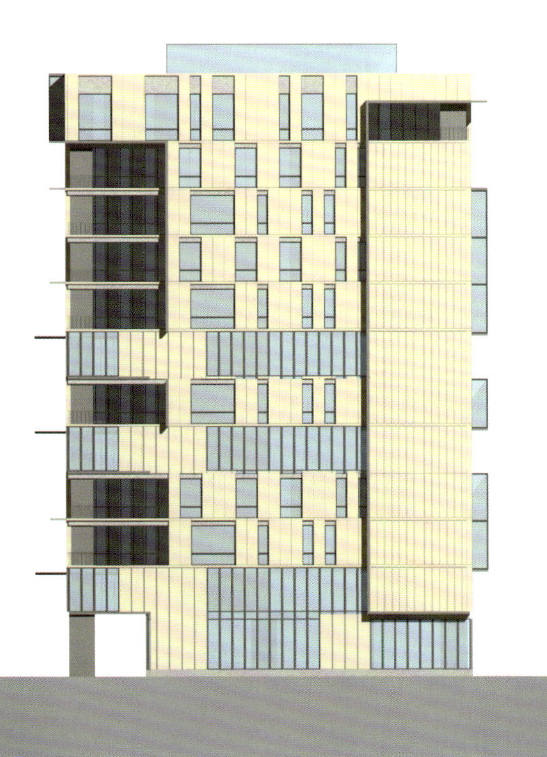

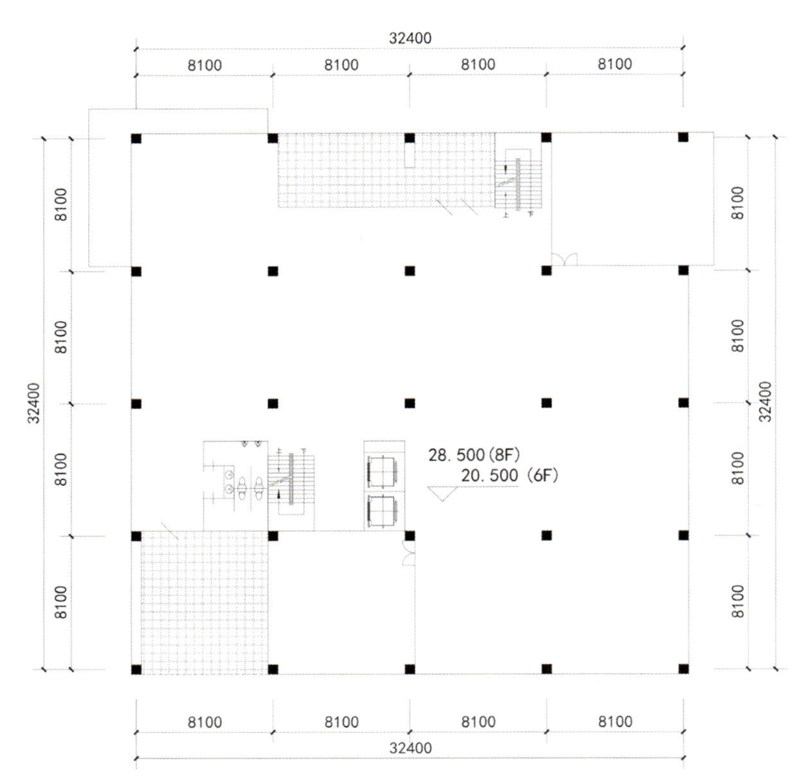

7# 6th,8th Floor Plan 7# 六、八层平面图

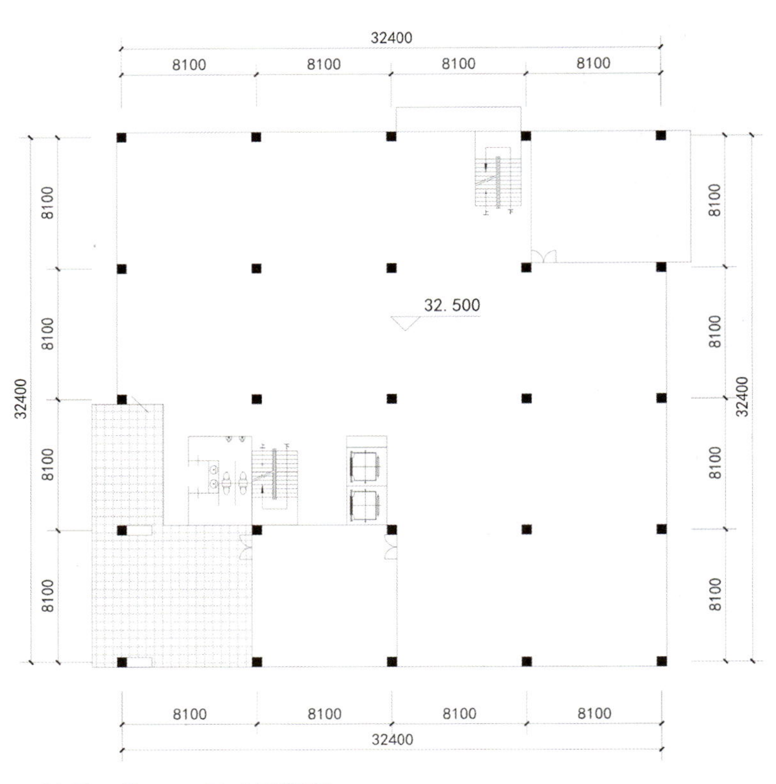

7# 9th Floor Plan 7# 九层平面图

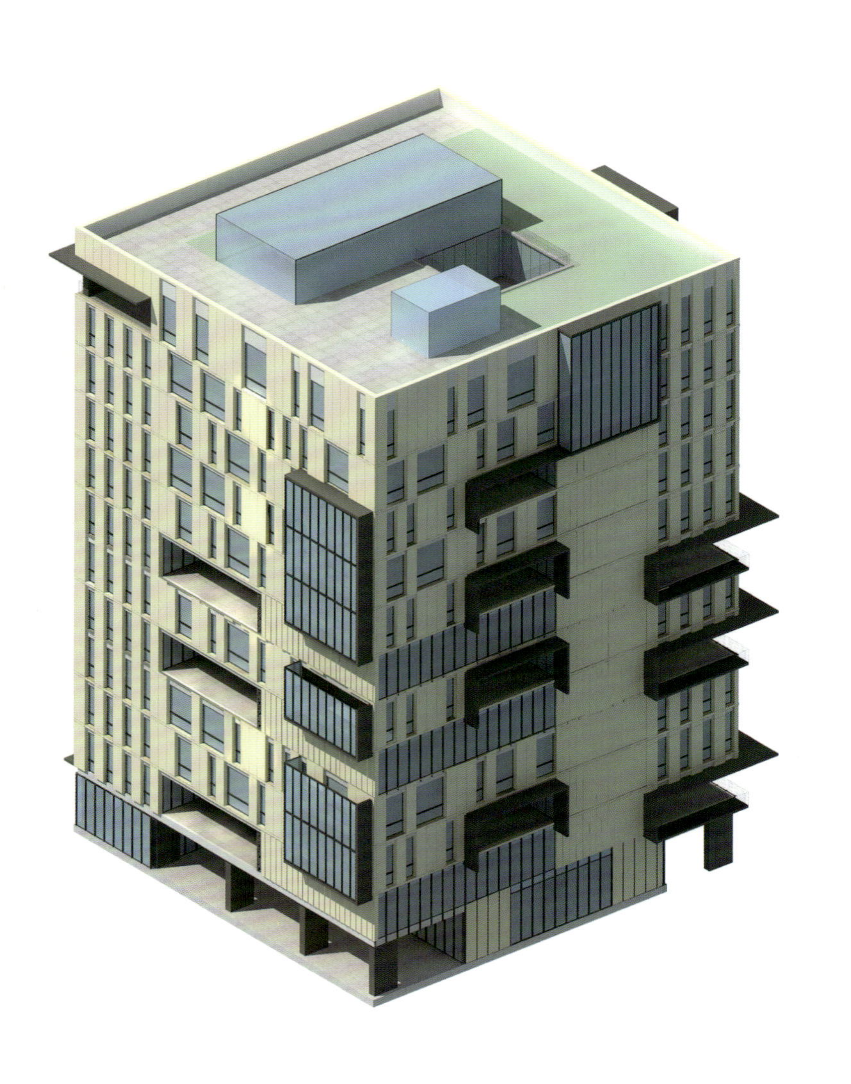

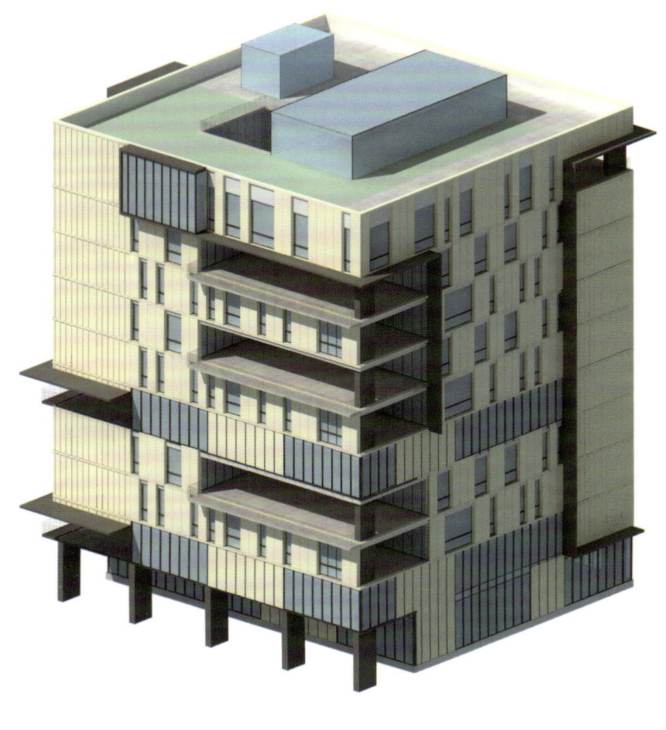

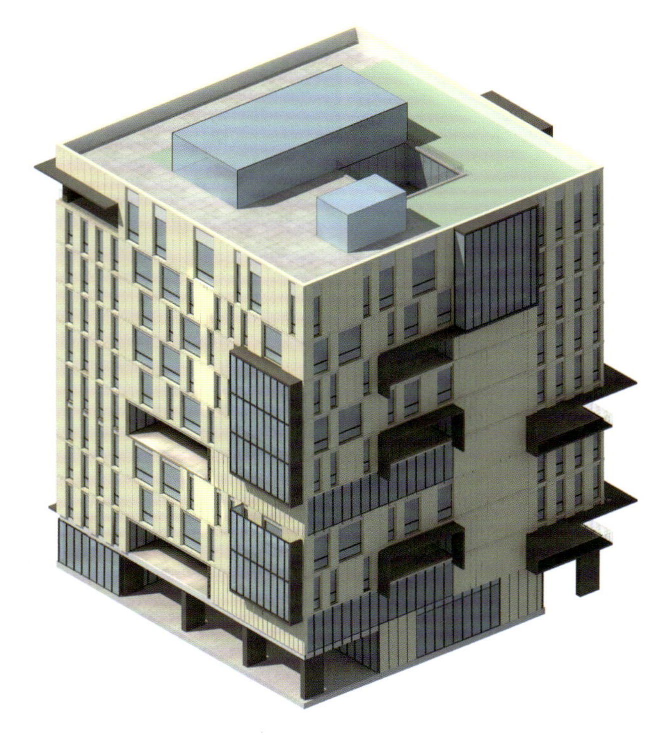

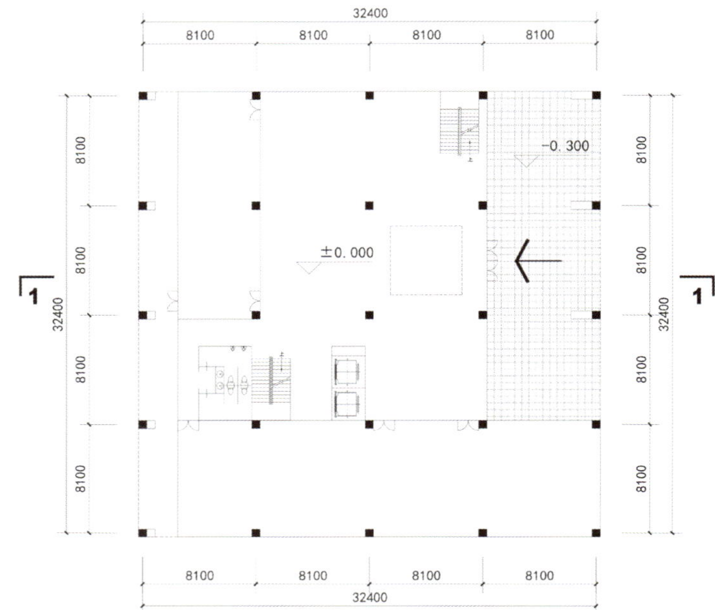

8# First Floor Plan 8# 一层平面图

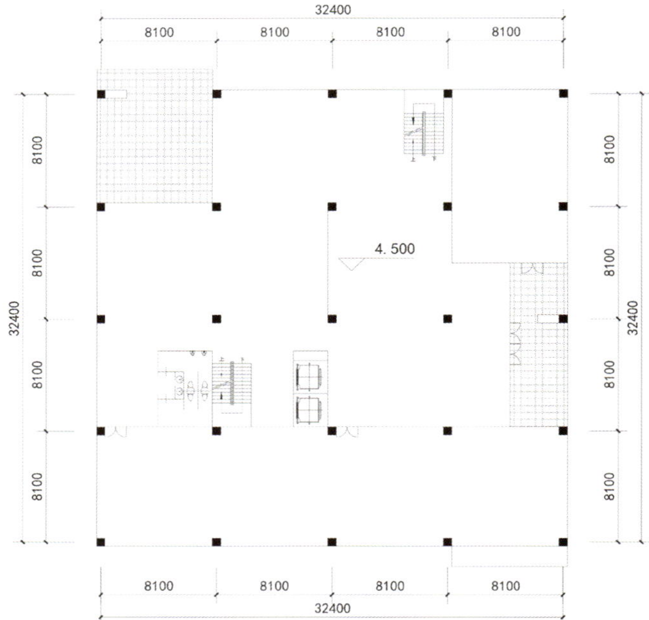

8# Second Floor Plan 8# 二层平面图

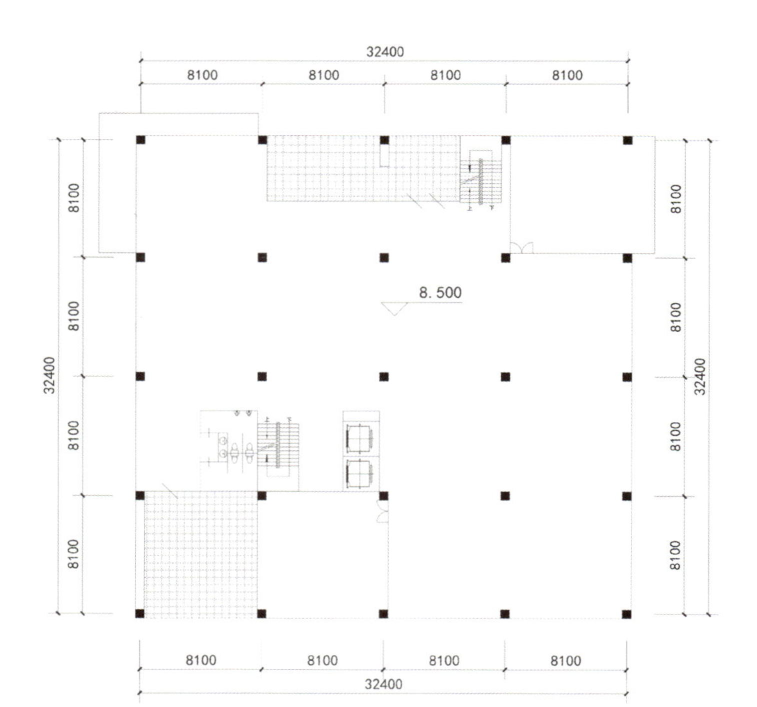

8# Third Floor Plan 8# 三层平面图

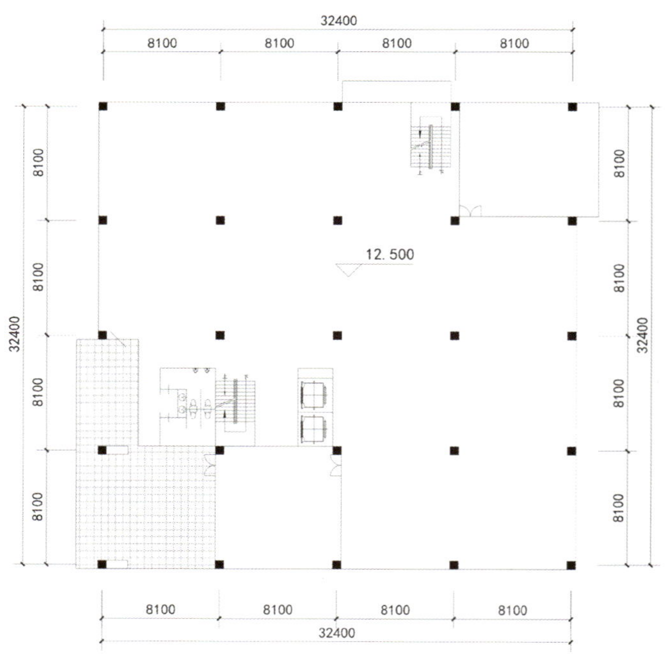

8# Fourth Floor Plan 8# 四层平面图

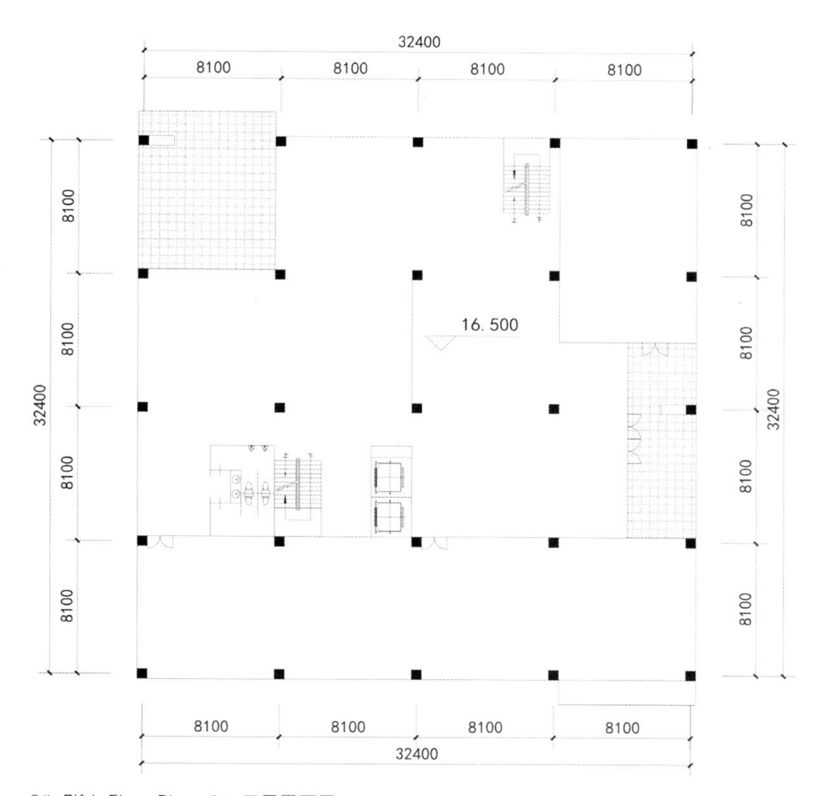

8# Fifth Floor Plan 8# 五层平面图

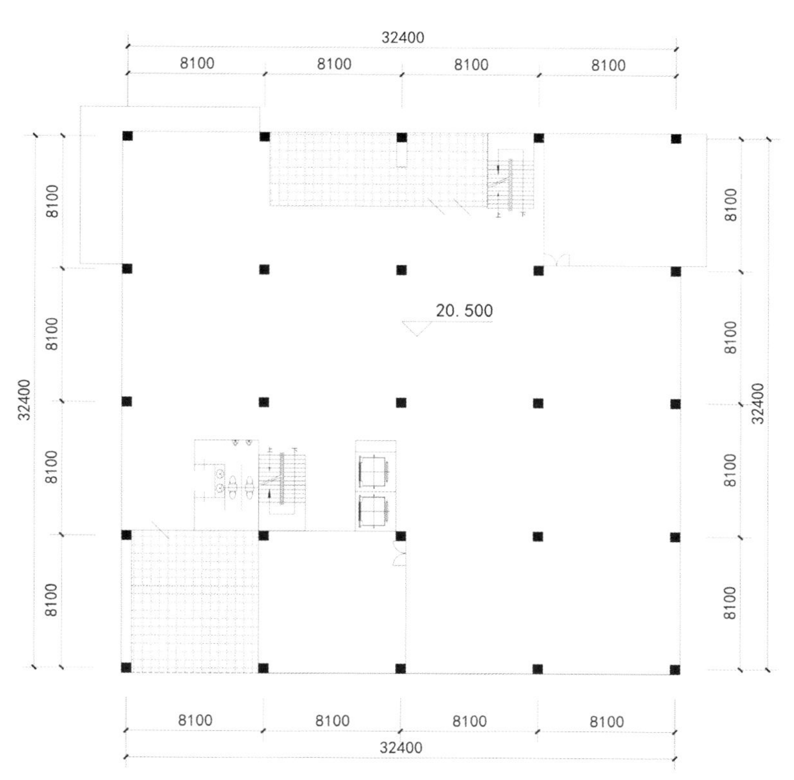

8# Sixth Floor Plan 8# 六层平面图

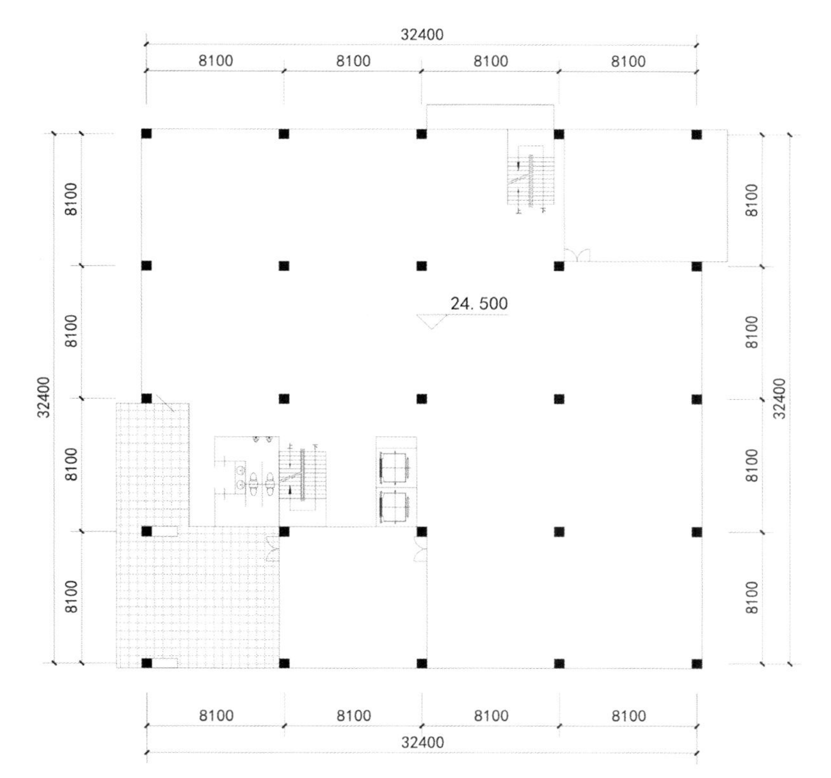

8# Seventh Floor Plan 8# 七层平面图

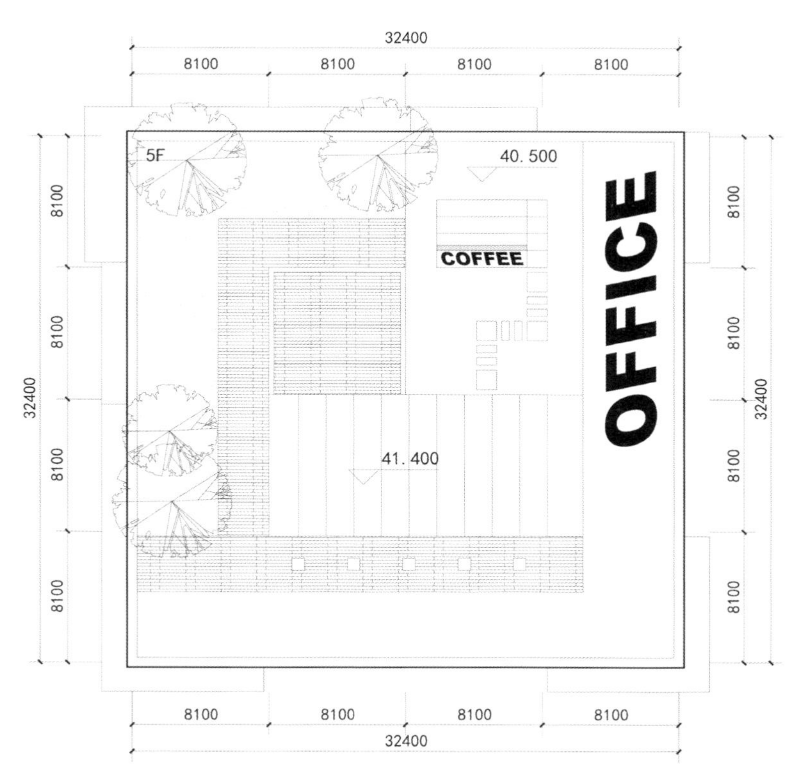

8# Roof Floor Plan 8# 屋顶层平面图

B09

Office Building in SSTT for Jiangsu Center of the Patent Office

苏州科技城国家知识产权局苏州中心办公楼

Features 项目亮点

Two blocks with different forms and skins interweave and enclose to form a staggered building volume, which looks simple and elegant.

项目设计通过将两个不同形态、不同表皮的体块进行穿插围合，塑造出交错互动的建筑外部形态，简洁清晰而又不失大气。

Keywords 关键词

Concise and Rigorous
简洁严谨

Interwoven Blocks
体块穿插

Clear Sequence
富有序列

Location: Suzhou, Jiangsu, China
Architectural Design: Suzhou 9-Town Design Studio for Urban Architecture
Land Area: 28,778 m^2
Floor Area: 59,536 m^2
Design Date: 2012
Completion Date: 2014

项目地点：中国江苏省苏州市
建筑设计：九城都市建筑设计有限公司
用地面积：28 778 m^2
建筑面积：59 536 m^2
设计时间：2012 年
竣工时间：2014 年

Overview

The site is located in Suzhou National New & Hi-tech Industrial Development Zone (SND), to the south of Kejie Road and to the east of Guangqi Road. The building for Jiangsu Center of the Patent Office just sits in the center of the office area, which will be the highest building in this area.

项目概况

项目地块位于苏州市高新区，地处科杰路以南、光启路以东。知识产权局大楼处于整个规划办公区域中最中心的位置，同时也是区域内最高的一栋建筑。

Design Strategy

With emphasis on the volume of the building, the design tries to make this office building as the soul and landmark of the office area. Two blocks with different forms and skins interweave and enclose to form a entrance square, highlighting the tension of the architecture.

设计策略

设计的策略是先强调建筑的体量感，将办公大楼作为整个办公区的灵魂和标志进行设计，通过两个不同形态、不同表皮的体块穿插围合形成出入口的广场空间。简洁大气的体块表现出建筑本身具有的张力。

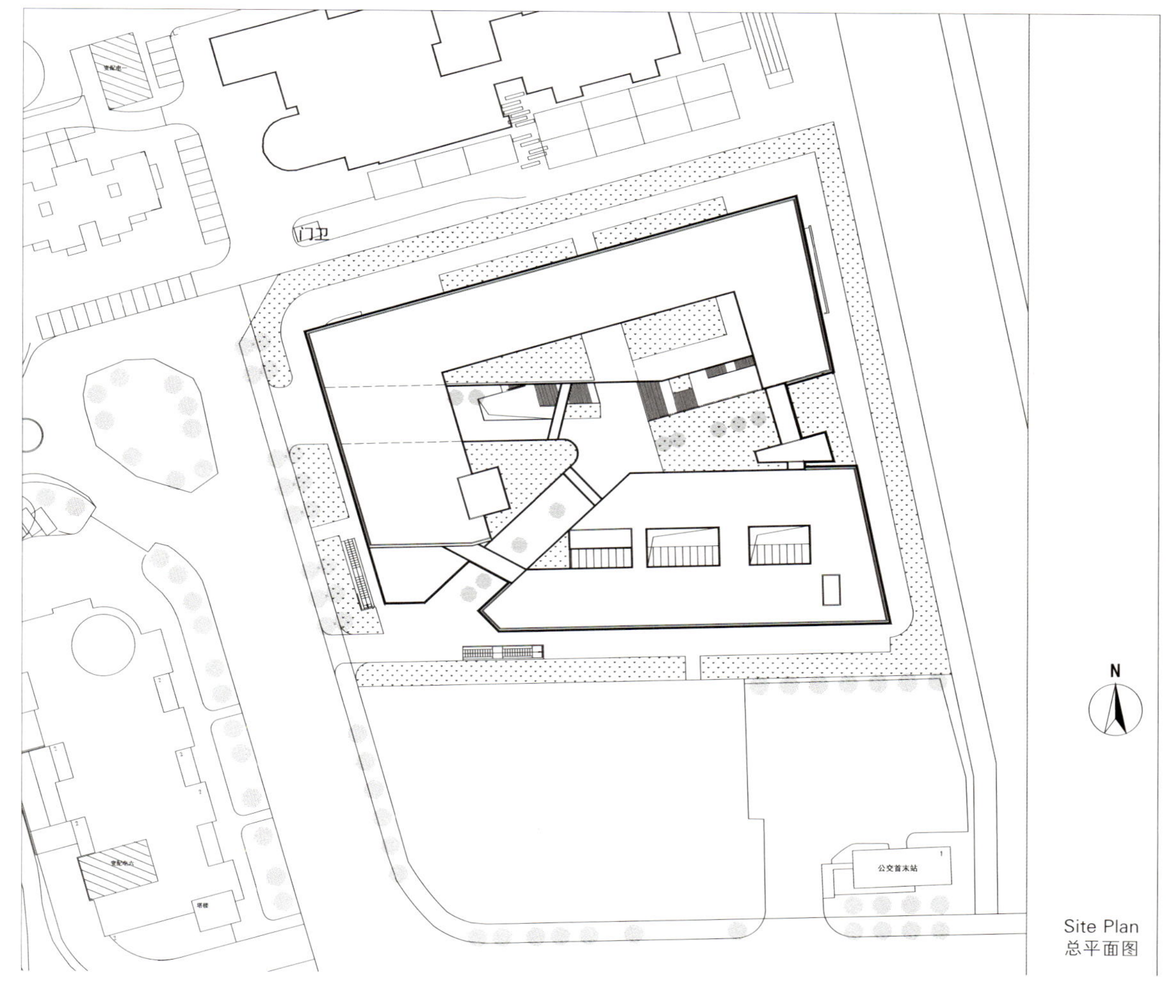

Site Plan
总平面图

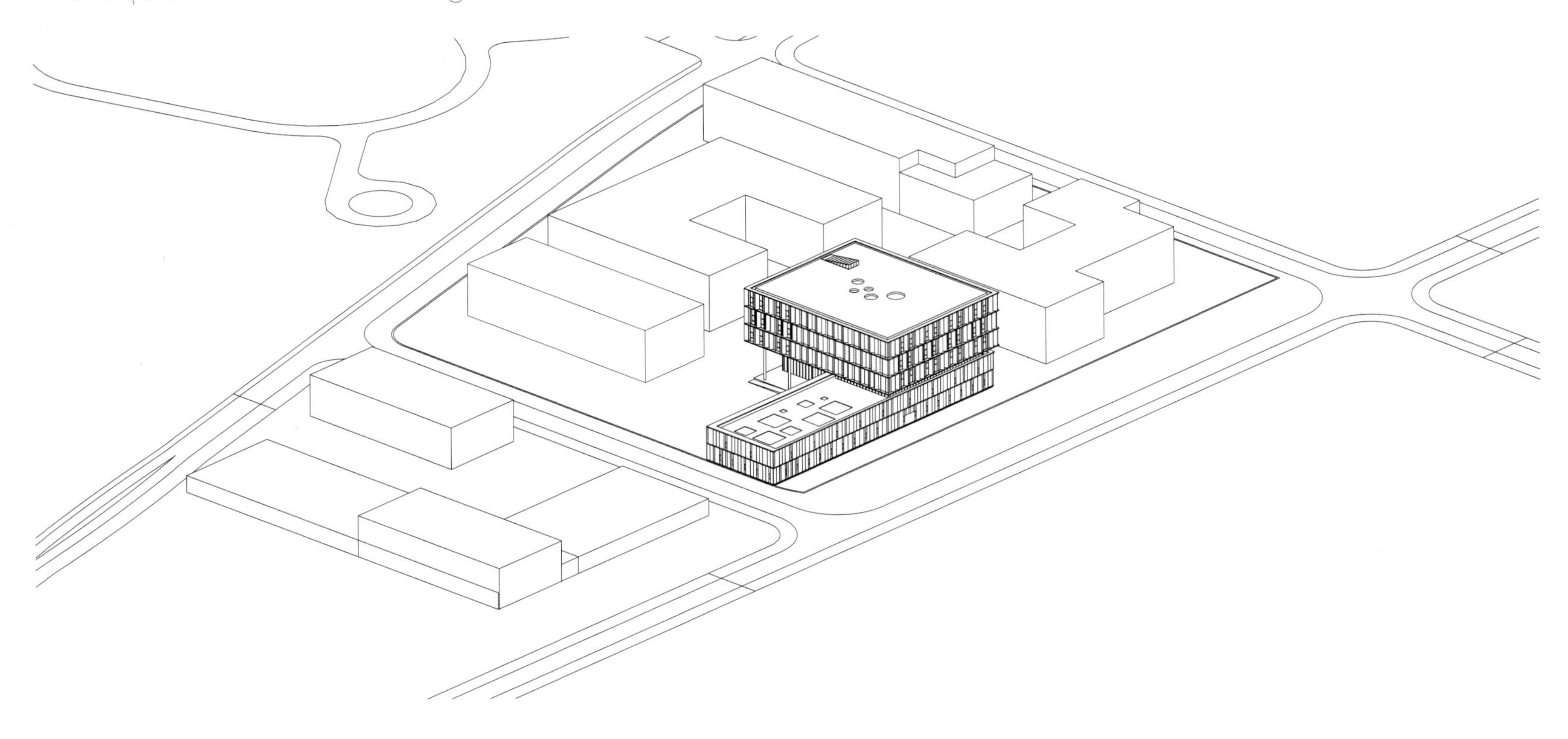

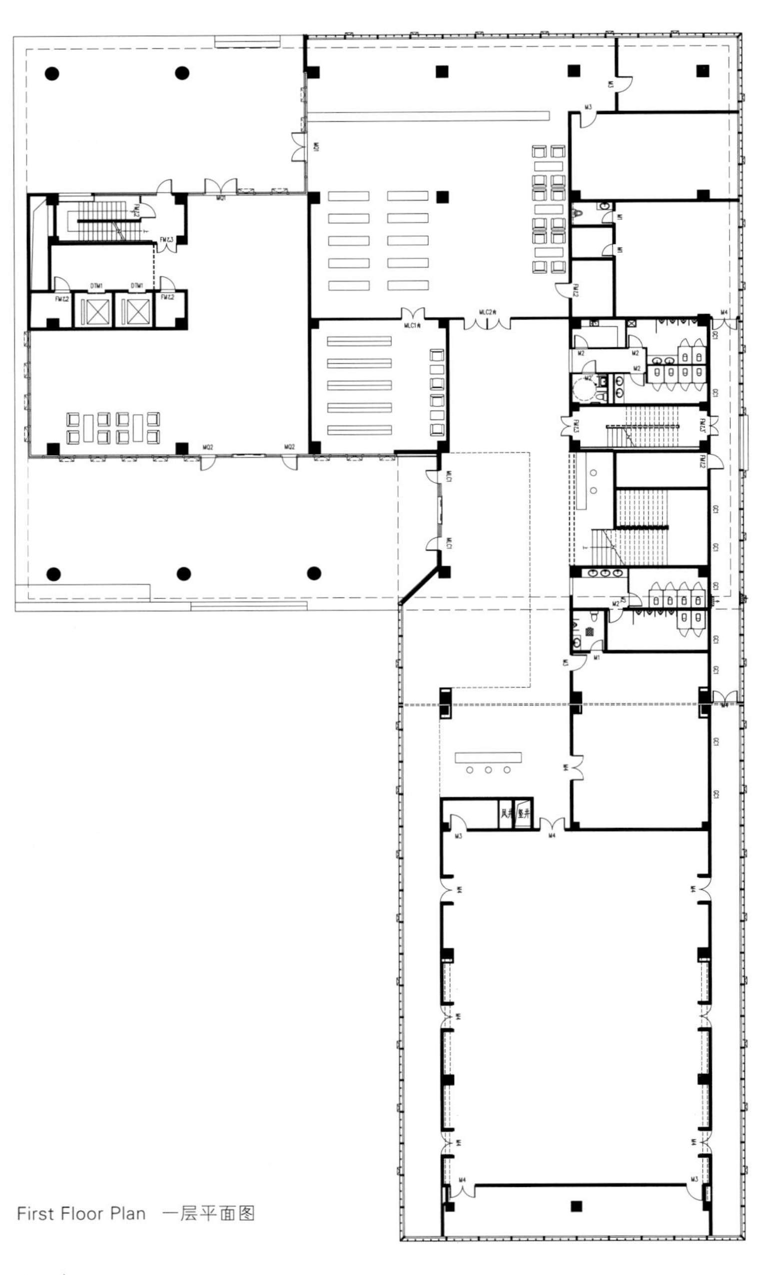

First Floor Plan 一层平面图

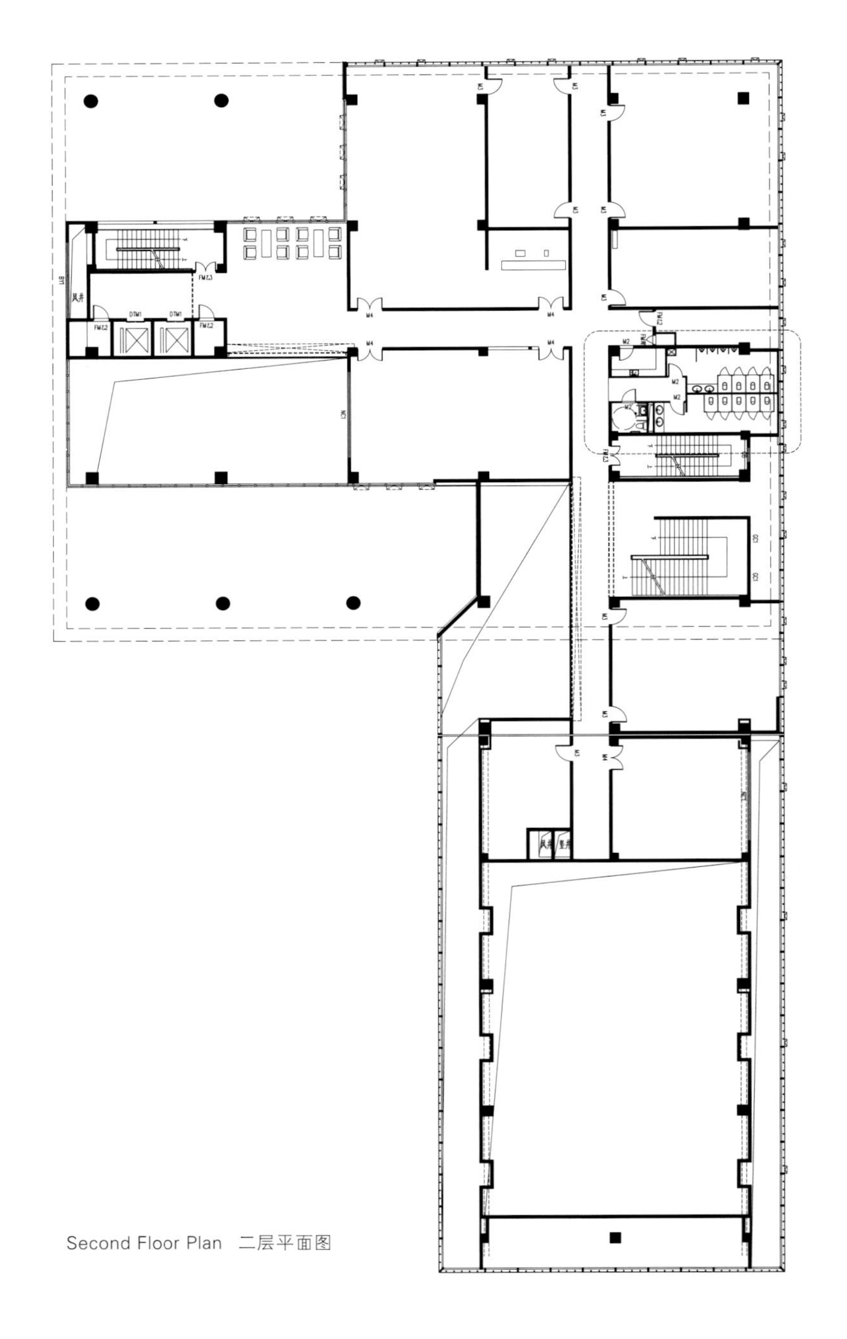

Second Floor Plan 二层平面图

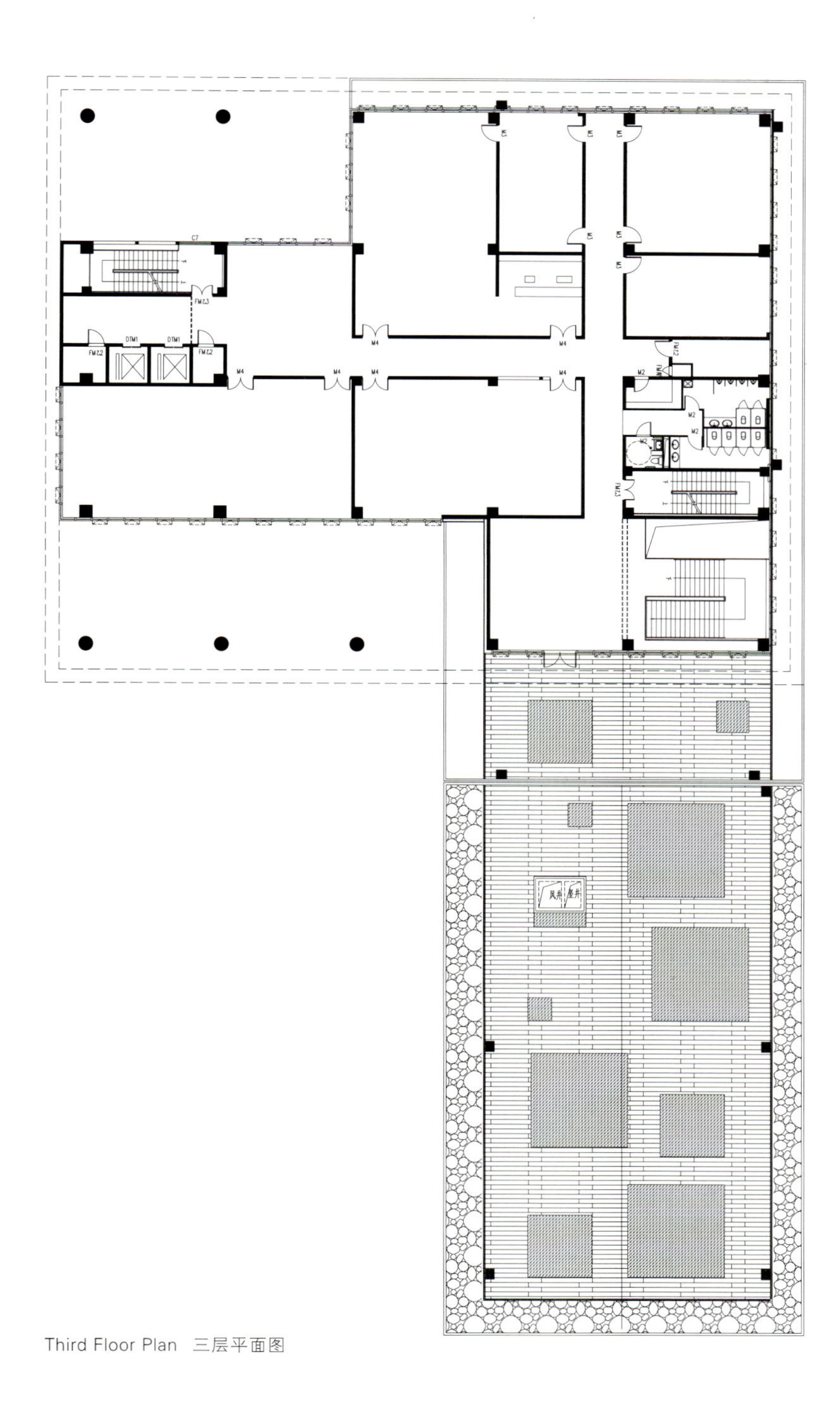
Third Floor Plan 三层平面图

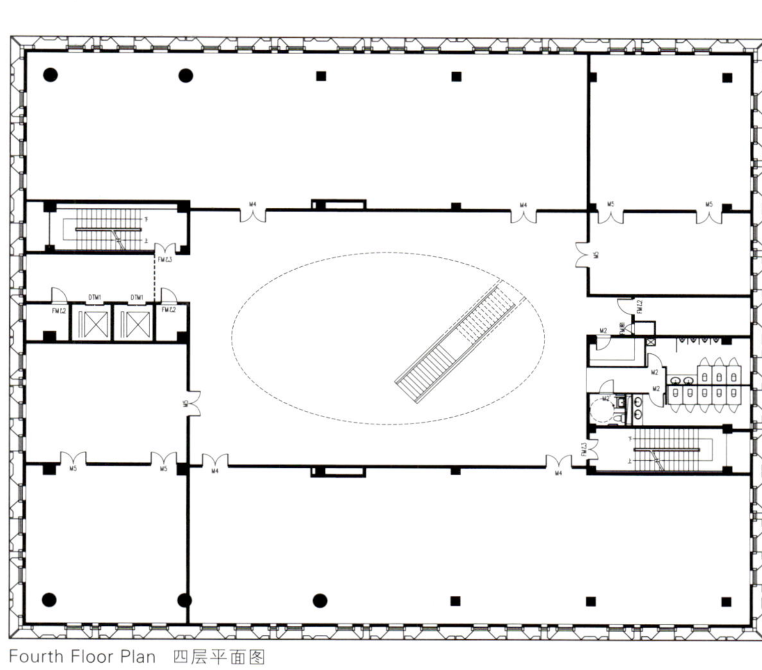
Fourth Floor Plan 四层平面图

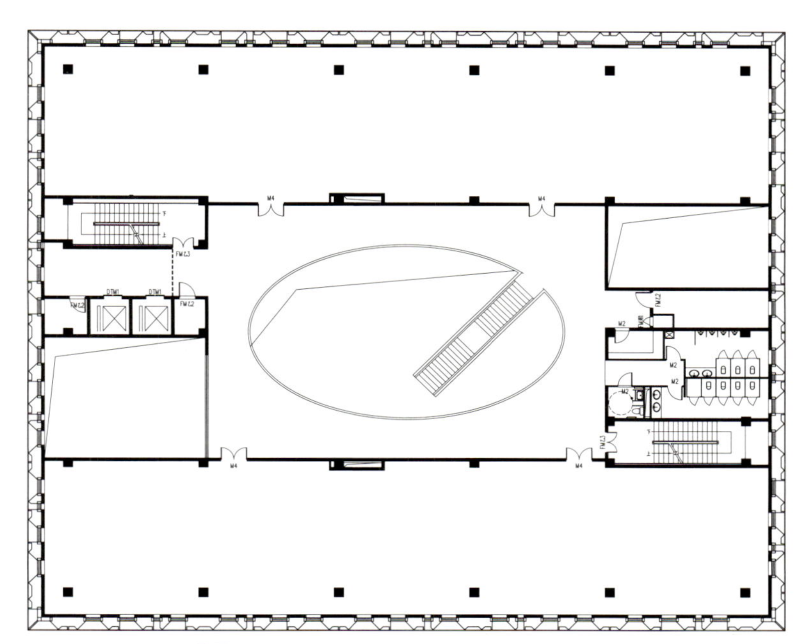
Fifth Floor Plan 五层平面图

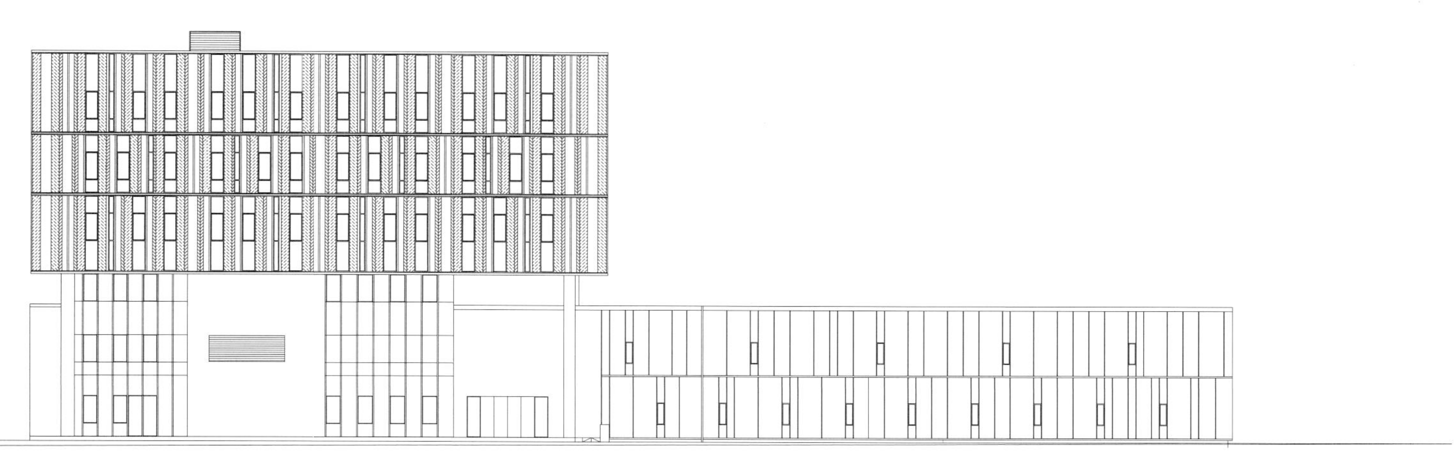
Elevation 1 立面图 1

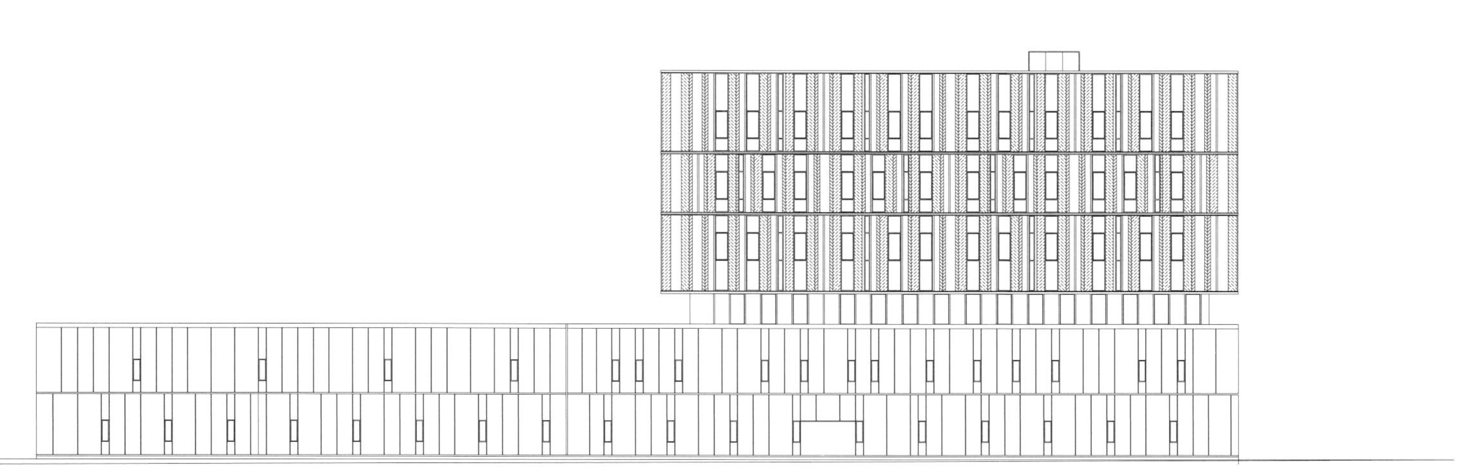
Elevation 2 立面图 2

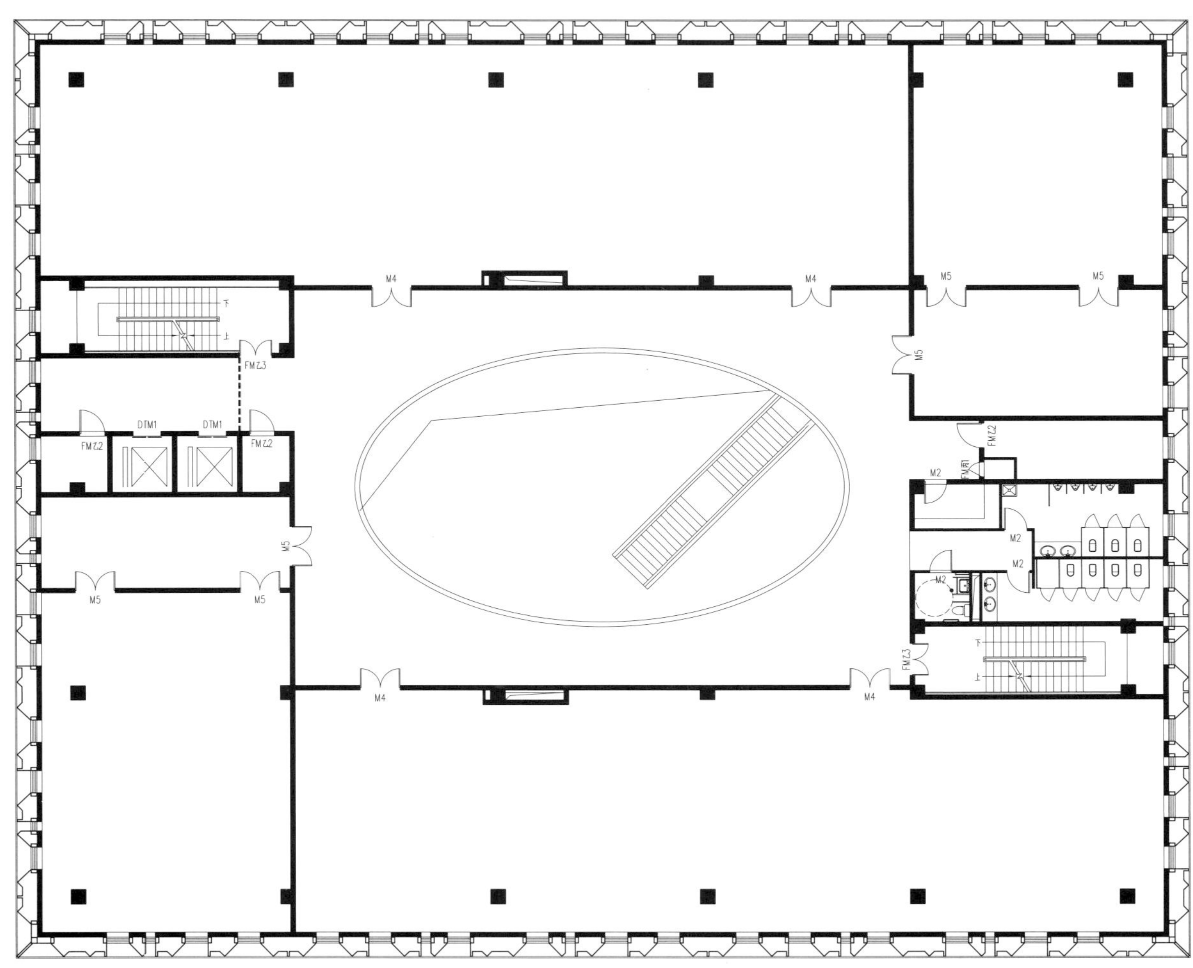

Sixth Floor Plan 六层平面图

Architectural Design

The two-storey podium designed with ultra clear glass skin is mainly used for the multi-functional lecture hall. Due to the requirement for large-scale space, the lecture hall is set beyond the office tower. And the main office tower is designed as a 6-storey quadrangle courtyard, using vertical white aluminum panels for sun shading. Thus the facade looks concise and rigorous with clear sequence to match its function as an office building.

The building is designed to be simple and elegant, reflecting the characteristics of modern science and technology. The beauty of the architecture firstly shows in its refreshing shape and staggered structure. And the quality of the architecture is embodied in the details and materials: glass facade combines with the metal curtain wall to shape unique style, highlighting the high quality of the Patent Office Building.

建筑设计

两层高的附属建筑体外表皮采用超白玻的体块，主要空间为多功能报告厅，由于需要大尺度空间，所以设置在主体建筑之外；而主体建筑为一个6层高的四合院，外侧采用竖向的白色铝板遮阳，立面形式清晰严谨而富有序列，与建筑本身的使用性质非常吻合。

建筑形象定位为体现时代科技发展的特征，简洁、严谨且干净、利落。建筑之美首先在于型体，体量简洁清晰，相互之间错动的关系也恰到好处。建筑的品质则在于细部和材质，玻璃与现代的金属幕墙的结合形成独特的感觉，体现了产权局大楼的特质。

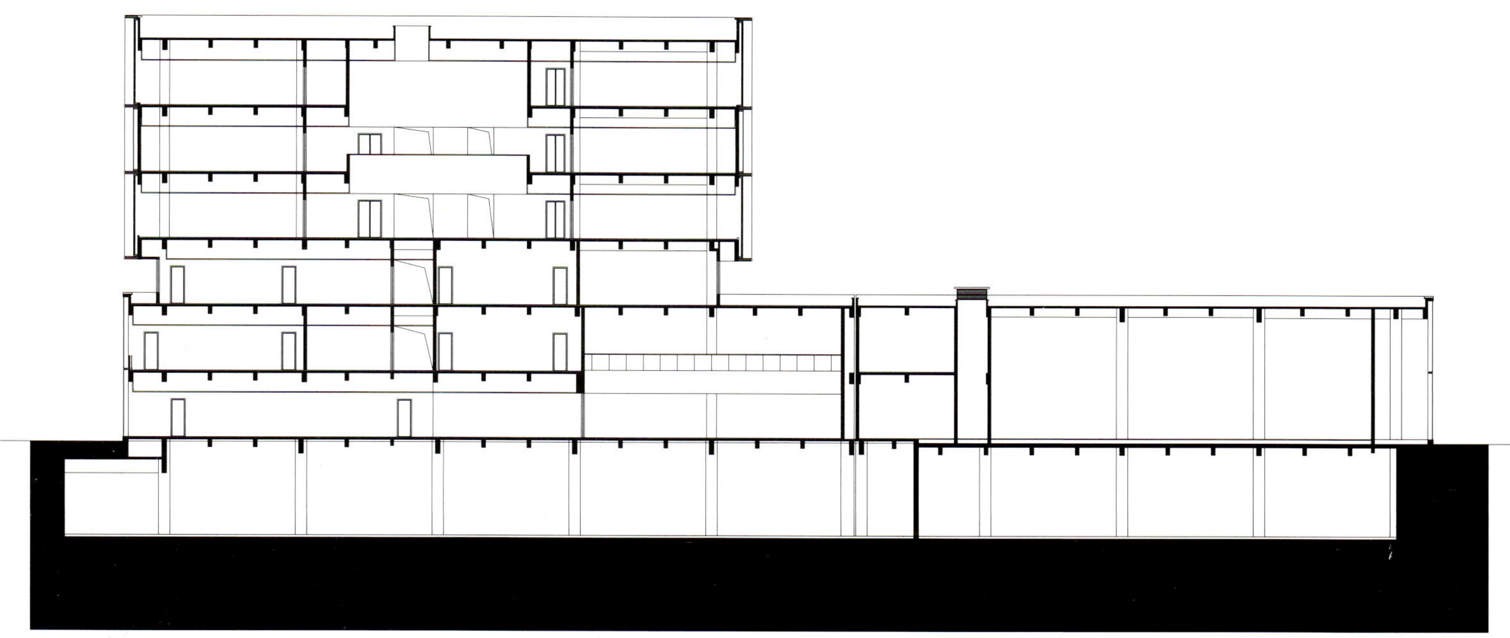

Sectional Drawing 剖面图

消火栓

Qingdao Hisense Research and Development Center

青岛海信研发中心

Keywords 关键词

Organic Growth
有机生长

Sci-tech and Humanities
科技人文

Skillful Landscaping
依势造景

Location: Laoshan District, Qingdao, Shandong, China
Architecture/Landscape Design: W&R Group
Developer: Qingdao Hisense Group
Land Area: 280,000 m²
Total Floor Area: 400,000 m²
Plot Ratio: 1.45
Design Time: 2012–2014

项目地点：中国山东省青岛市崂山区
建筑 / 景观设计：水石国际
开发商：青岛海信集团
占地面积：28 万 m²
总建筑面积：40 万 m²
容积率：1.45
设计时间：2012~2014 年

Features 项目亮点

By integrating the mountainous environment with the sci–tech requirements, it has shaped the identity of an international enterprise and provided a series of sustainable spaces for innovation, research and development with human–caring buildings to attract international talents and the magnificent mountain landscape that conforms with the urban environmental planning.

将山地自然环境与科技研发需求相结合，创造融入自然的国际大企业形象，提供永续发展的科技创新研发空间，塑造吸引国际人才的人文建筑风貌，打造符合城市环境规划的山林城市景观。

Overview

Hisense Research and Development Center, requested to be developed by 2 stages, is located in Laoshan District, Qingdao, Shandong and 20 km away from Qingdao urban area. Coastal Avenue is on its east and Tianshui Road on the south, planned Jianxi Road is on the north-west, it occupies a land area of 280,000 m² and over ground floor area of 400,000 m² in height of 24 m, with a plot ratio of 1.45.

项目概况

海信新研发中心位于山东省青岛市崂山区，距离市中心约 20 km，东临滨海大道，南临天水路，西北侧为规划中的涧西路，用地面积 28 万 m²，地上建筑面积约 40 万 m²，容积率 1.45，建筑高度 24 m，根据海信集团需求分两期建设。

Design Concept

A 3.5 hectare (35,000 m²) ecological park with lots of water features is requested by Hisense. The Logistical Support Centers and Academic Exchange Center are placed in the center of the plot and close to the park to provide services to the whole project. For the main entrance, it is set facing the Coastal Avenue to represent the grand and steady feelings of this international scientific and technological enterprise, while two sub-entrances are put in the north to meet the general requirements of transportation.

设计理念

园区需要一个面积约 35 000 m² 的以水体为中心的生态公园。园区的后勤保障中心和学术交流中心主要布置在地块中央临近公园的位置以便服务整个园区。沿滨海大道一侧设园区主入口，彰显国际化科技企业大气稳重的形象，北侧设置两处园区次入口满足日常交通需求。

Site Plan
总平面图

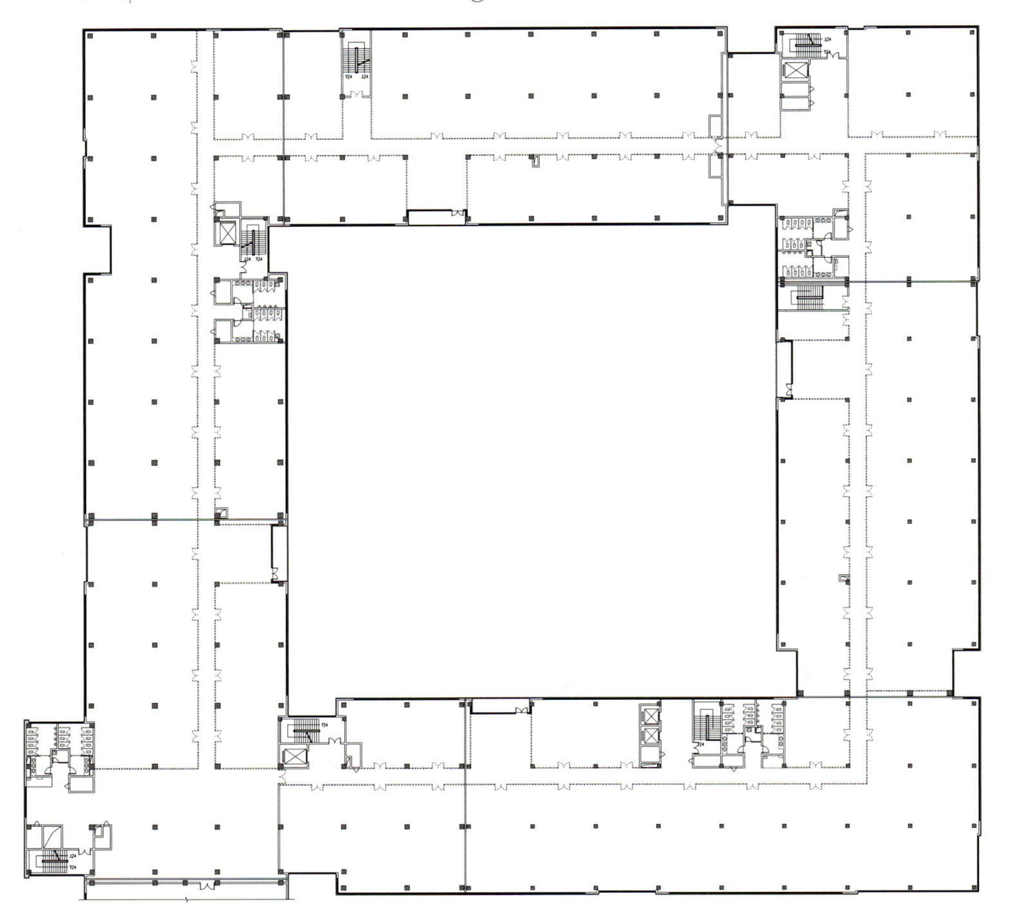

Floor Plan 平面图

Architectural Design

Research and development building groups are built as Zone A, B and C from west to east, with departments of brown goods, high and new technology and white goods respectively, and each zone is developed by two stages to meet the needs of next five and ten years. Before the completion of phase Ⅱ, it is essential to make the image of phase Ⅰ independent and completed.

Large-scale laboratories and garages that need less lighting are set at the ground floor with dark stones on the facades to integrate with mountains. Multi-storey offices with red face bricks facades are placed above the ground floor, while spaces for transportation and public communication are set between each single building that with glass curtain wall facades. The whole group has expressed the corporate image contains both humanity and high technology. Occupying the plot which is beside the central water feature, along with the fresh and unique shape, Academic Exchange Center features VIP reception, large meeting rooms and products display, it is the landmark and image of corporate spirit. Two Logistical Support Centers lie in the north with the functions of catering, fitness, healthcare and so on to provide staff with convenience.

建筑设计

研发建筑群从西至东分为A、B、C三区，分别设置黑色家电、高新技术、白色家电这三大类研发部门集群，每个集群各自分成一期和二期建设，分别应对未来五年和十年的需求。在二期整体形象尚未形成之时，还要考虑一期形象的相对完整。

建筑底层主要设置采光需求较弱的大型实验室和车库，立面采用深色石材形成基座与山地结合；基座上面是多层办公，立面采用红色面砖；建筑单元体之间是交通和公共交流空间，立面采用玻璃幕墙。建筑整体塑造了一种兼具人文感和科技感的企业形象。学术交流中心坐落于园区中央的水体旁边，功能包含贵宾接待、大型会议、产品展示，形象新颖独特，是园区标志建筑和企业精神象征。两个后勤保障中心位于园区北侧，功能包含餐饮、健身、医疗等，为员工生活提供便利的服务。

Landscape Design

Landscape design for the project also follows the principle of economy, beauty and application, learning from and also respecting nature, to keep the interests of nature and also create spaces for large-scale activities in key areas. Economical materials, exquisite plants, rich spatial designs, smooth functional organization, which have all cooperated together to make landscape integrate into the activities of the plot. The projects is situated on a slope that with various topography of Laoshan Mountain in Laoshan District, while meeting the requirements of high efficiency and convenience for the research and development center, it also makes the center integrate with natural landscape in the same time. By making use of the slope topography, designers create a park with ecological river system in the center to provide spaces for sorts of activities, besides, the park has also been a link between core landscape area and main nodes such as main entrance, Academic Exchange Center and visit roads.

景观设计

整个园区景观设计同样秉承经济、美观、适用的原则，师法自然，尊重环境，既保证整体环境的原始自然趣味性，又在重点区域打造符合大型功能需求的活动场地与空间。材料选择经济，植物配置考究，空间设计丰富，功能组织流畅，景观设计与园区活动形成有机的统一。项目基地位于青岛崂山区一块地形丰富的崂山余脉山坡地之上，景观在满足其研发园区使用功能的高效快捷的同时，也希望园区融入自然山体的优美风景之中。设计利用基地山体冲沟规划一个包含生态水系的园区内山体公园，既满足园区各种公共活动要求，也形成了整个园区的生态核心景观区，与主入口、交流中心、参观路径等重要节点相衔接。

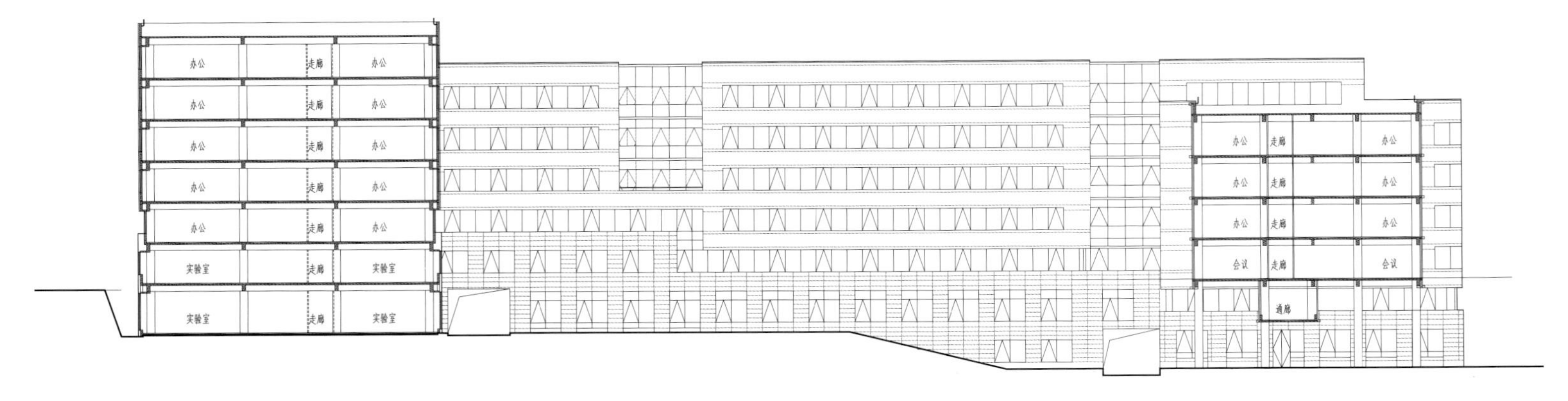

Side Elevation ②～⑯

②～⑯轴立面图

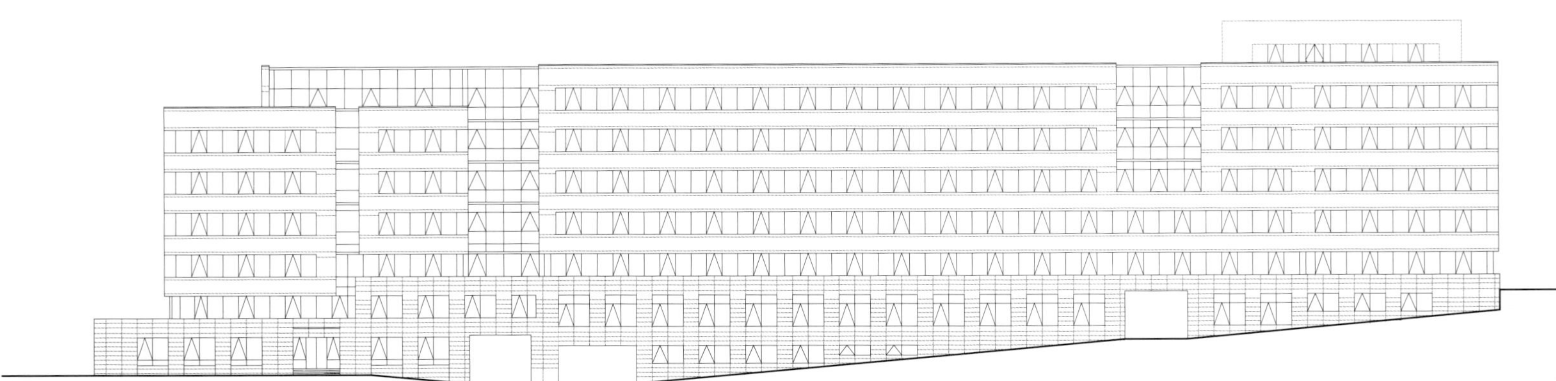

Side Elevation ⑯～②

⑯～②轴立面图

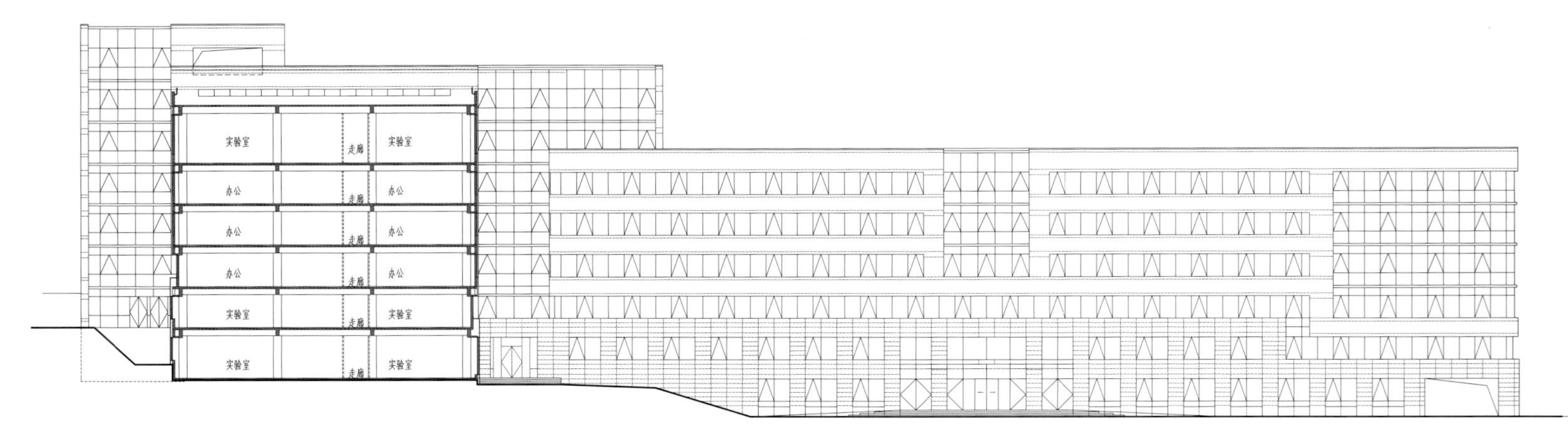

Side Section ②～⑯

②～⑯轴剖立面图

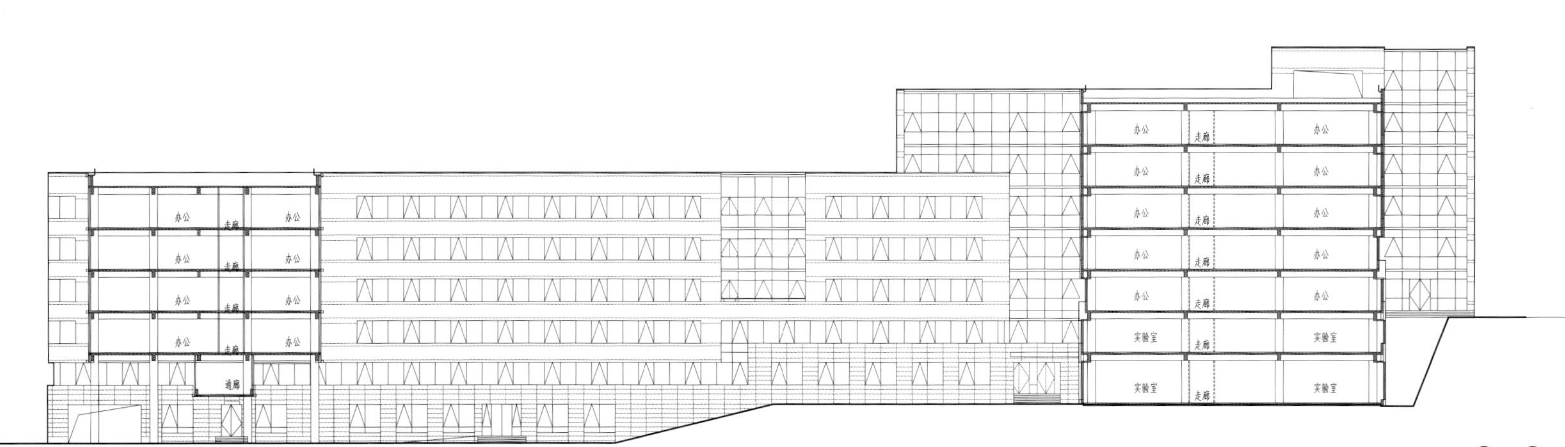

Side Section ⑯～②

⑯～②轴剖立面图

Hangzhou Huawei R&D Center

杭州华为研发中心

Features 项目亮点

The green interior courtyard of the main building continues the dynamism of the surrounding landscape terrain, extends to the stepped perceptual space and ingrates the office building into the landscape.

主楼绿色的内部庭院延续周围景观地形的动感，扩展到阶梯状的知觉空间，将办公室融入景观。

Keywords 关键词

Hierarchical Space
分级空间

Partially Overhead
局部架空

Coordinated Rhythm
节奏协调

Location: Hangzhou, Zhejiang, China
Developer: Huawei Technology Co., Ltd.
Architectural Design: HENN GmbH
Gross Floor Area: 310,000 m^2
Completion Date: 2013

项目地点：中国浙江省杭州市
开发商：华为技术有限公司
建筑设计：德国海茵建筑设计公司
总建筑面积：310 000 m^2
竣工时间：2013 年

Overview

Hangzhou is famous for art. The new Huawei R&D Park, located at the development area of Chinese eastern city Hangzhou, provides an office space for 8,000 staffs of a technology company. This building group is based on the independent buildings throughout the spacious landscape, and adopts the beautiful scenery of Hangzhou as the theme.

项目概况

杭州是一座以艺术而闻名的城市。新华为研发园位于中国东部城市杭州的开发区内，为该技术公司的 8 000 名员工提供办公空间。这一建筑群以宽敞的景观空间中的独立建筑为基础，以杭州秀美的风光作为主题。

Design Idea

The master planning concept of the new R&D Center, built for Huawei, a technology-based company, is to use the poetic scenery of artistic city Hangzhou as basic theme and organically integrate with each independent building unit. Along the water axis, there are six single building models of interior study room and office.

设计理念

技术型企业华为公司新建的研发中心的总体规划设想是，以艺术之城杭州诗情画意般的美景作为基本主题，与各个独立的建筑单元有机融合。沿着一条水轴线，两侧共分布有六个内设研究室和办公室的建筑单体模数。

Architectural Design

The green interior courtyard of the main building shows the terrain movements of the landscape park, forming a hierarchical perceptual space and integrating the office building into the landscape of the park. The partially overhead structure at the bottom of building makes the architectural art quite naturally transit to the beautiful scenery and form a harmonious unity with view of different angles in the center of the park. Two smaller entrances are set at the two sides of the western main entrance, and the main entrance connects with the canteen in the center of the park through the main landscape visual axis. Different tastes of buildings manifest along the road of the landscape park and integrate the nature and techniques into a unity with coordinated rhythm.

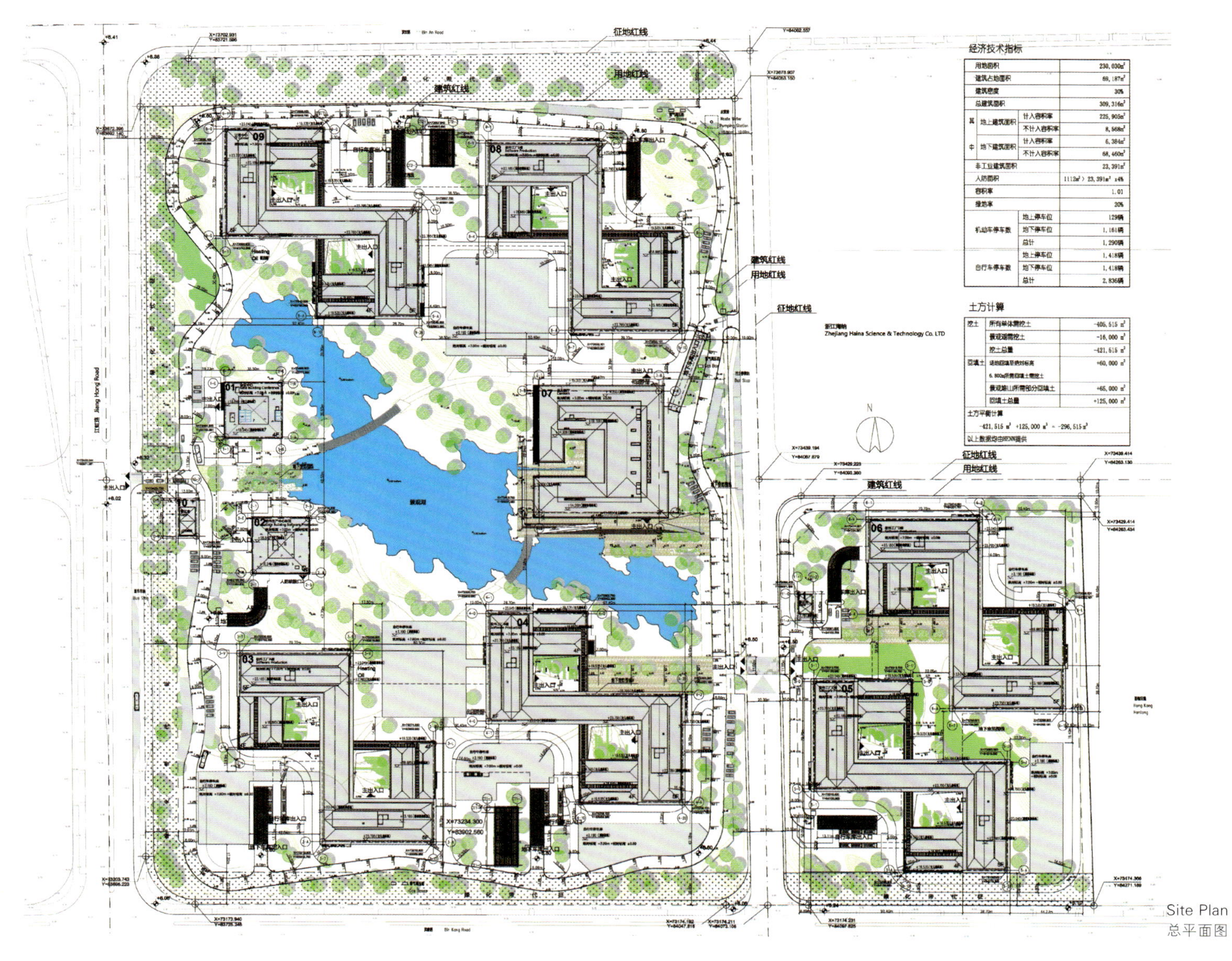

经济技术指标

用地面积			230,030㎡
建筑占地面积			69,187㎡
建筑密度			30%
总建筑面积			309,316㎡
其中	地上建筑面积	计入容积率	225,905㎡
		不计入容积率	8,568㎡
	地下建筑面积	计入容积率	6,384㎡
		不计入容积率	68,460㎡
非工业建筑面积			23,391㎡
人防面积			1112㎡ > 23,391㎡ x4%
容积率			1.01
绿地率			20%
机动车停车数		地上停车位	129辆
		地下停车位	1,161辆
		总计	1,290辆
自行车停车数		地上停车位	1,418辆
		地下停车位	1,418辆
		总计	2,836辆

土方计算

挖土	所有单体需挖土	-405,515 ㎥
	景观湖需挖土	-16,000 ㎥
	挖土总量	-421,515 ㎥
回填土	场地回填至绝对标高 6.800m所需回填土需挖土	+60,000 ㎥
	景观堆山所需部分回填土	+65,000 ㎥
	回填土总量	+125,000 ㎥
土方平衡计算 -421,515 ㎥ +125,000 ㎥ = -296,515 ㎥		
以上数据均由HENN提供		

Site Plan
总平面图

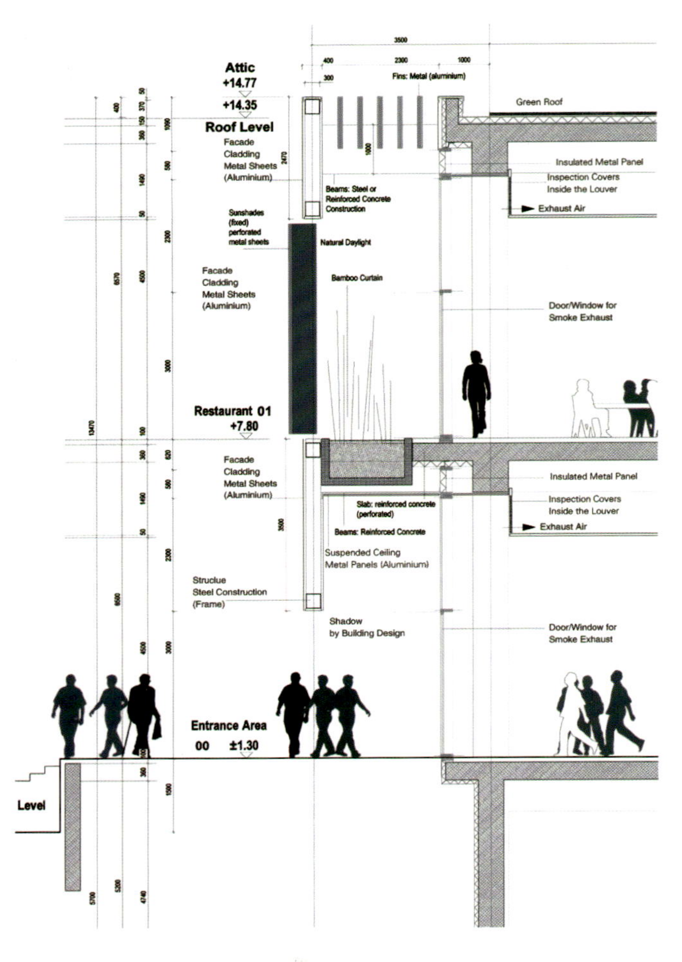

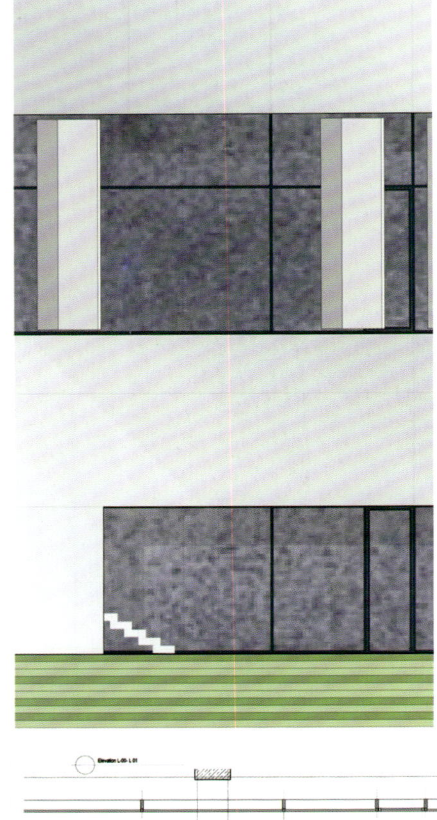

Canteen Building–Facade Details
餐厅——立面细节

建筑设计

主建筑的绿色内院体现了风景园的地形走势，并形成分级的感知空间，将办公室融入到公园景色之中。通过建筑底层的局部架空，使建筑艺术极其自然地过渡到风光美景中，在园中央构成一个可从不同角度观察的和谐整体。西面的主入口两侧是两个较小的入口建筑，主入口通过景观视觉主轴线与公园中央的食堂相连。各不相同的建筑品质沿着风景园的道路展现，并将自然与技术融合成一个节奏协调的整体。

Interior Design

The green internal courtyard of the main building continues the dynamism of the surrounding landscape terrain, extends to the stepped perceptual space and ingrates the office building into the landscape. The transparent building base uses various visual angles to create a mobile transition between the structure and the landscape empty space, and enters the center of the R&D Park.

室内设计

主楼绿色的内部庭院延续周围景观地形的动感，扩展到阶梯状的知觉空间，将办公室融入景观。透明的建筑基座以各种视角在结构和景观空地之间创造出流动的过渡，一直进入研发园中央位置。

Functional Layout

There are six buildings in total at the both sides of the large lake which are setting for offices and laboratories to form as a connective part of the site. It has added two small buildings that are prepared for the administrators, meetings and dinning facilities. The roads surrounding the site are only accessible for the pedestrians and vehicles are prohibited. Two smaller entrances are set at the two sides of the western main entrance, connecting with the dinning facilities by the axis.

功能布局

一座大湖两侧共有六座建筑物，内设办公室和实验室，形成研究园的连接部分。另外增加了两座为管理人员、会议和主要餐饮设施准备的小建筑。场地周围的道路不允许车辆通行，内部只允许行人进入。西入口两侧是两座小型入口建筑，通过轴线连接研发园中央的餐饮设施。

Nanopolis Suzhou

苏州纳米科技园

Features 项目亮点

The project is designed with the idea of “City of Nano”. The concept of “Nanotechnology Research and Development Park” is derived from traditional Chinese urban planning and modern science. The innovative technologies and environmental strategies will contribute to a more comfortable environment and reduce energy consumption.

项目以“纳米城邦”为设计理念。纳米技术研究和开发园的灵感概念来自传统中国城市规划和现代科学。创新技术和环境战略有利于设计一个更舒适的地方，同时减少其能源消费。

Keywords 关键词

Linear Planning
线性规划

Suzhou Gardens
苏州园林

Green Buffer Zone
绿色缓冲带

Location: Suzhou, Jiangsu, China
Developer: Suzhou Nanotech Co.,Ltd.
Architectural Design: HENN GmbH
Gross Floor Area: 400,000 m^2
Completion Date: 2013

项目地点：中国江苏省苏州市
开发商：苏州纳米科技发展有限公司
建筑设计：德国海茵建筑设计公司
总建筑面积：40 万 m^2
竣工时间：2013 年

Overview

Suzhou has established itself as the research and development area in the future. Except for the Biobay in the west, Nanopolis marks another key element of this strategy in this city.

项目概况

苏州确立了自己未来作为科研发展地区的目标。除了在城市西部的Biobay生物科技园，纳米科技园在城市标志着这一战略的另一个关键因素。

Design Idea

The project is designed with the idea of “City of Nano”. The concept of “Nanotechnology Research and Development Park” is derived from traditional Chinese urban planning and modern science. The innovative technologies and environmental strategies will contribute to a more comfortable environment and reducing energy consumption.

设计理念

项目以“纳米城邦”为设计理念。纳米技术研究和开发园的灵感概念来自传统中国城市规划和现代科学。创新技术和环境战略有利于设计一个更舒适的地方，同时减少其能源消费。

Site Plan 总平面图

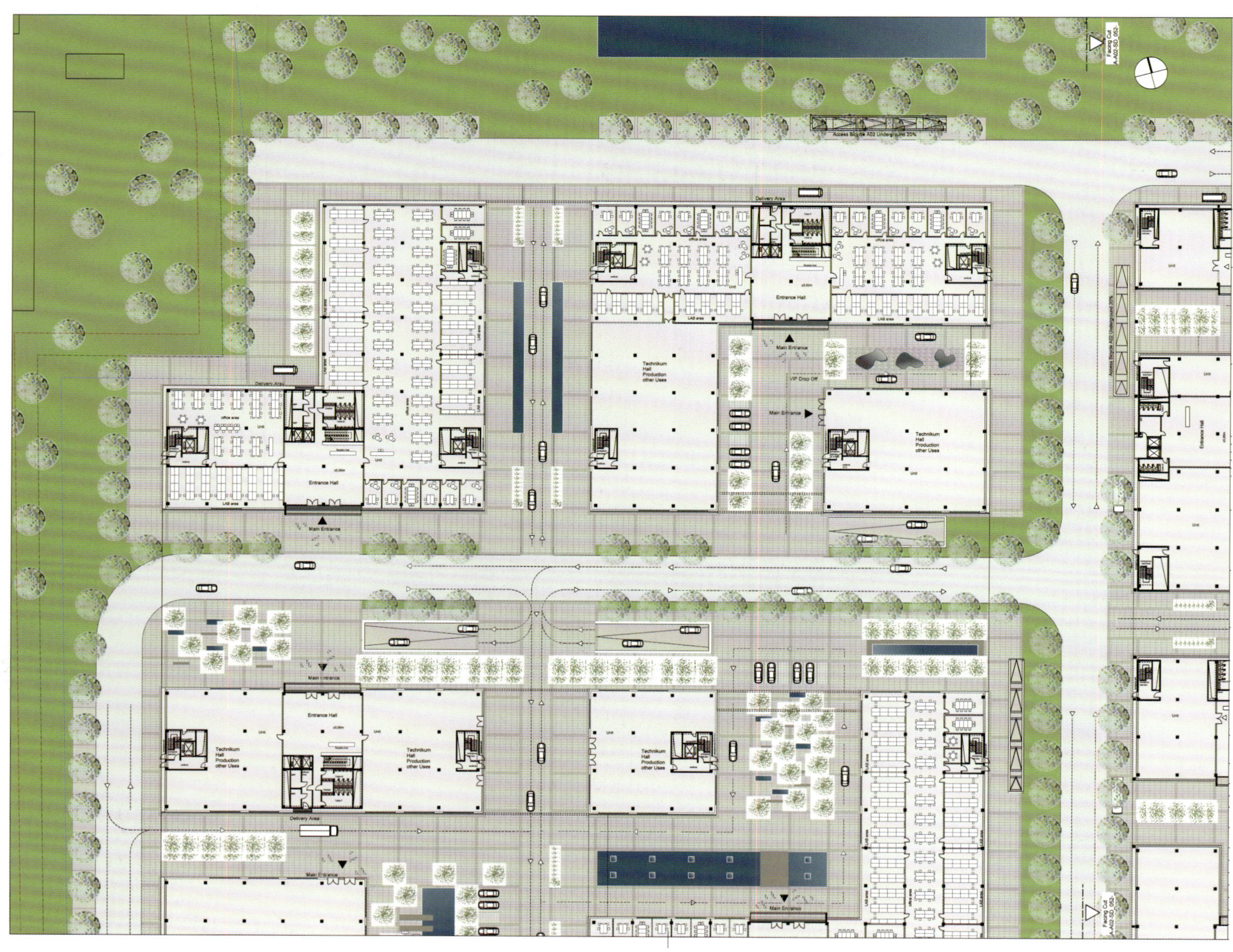

Site Plan Landscaping North 景观总平面图，北区

A/C Plot Topview A/C 地块总平面顶视图

Park Planning

Like many other ancient Chinese cities, the park is in a linear shape and with clear boundary and intricate inner connections. Many Chinese ancient cities have the concept of zoning, they divided the city into block, community, area, downtown and town in the order of small to large.

园区规划

如同中国许多古老的城市一般，项目园区形状呈线性，具有明确的边界，内部联系错综复杂。在中国的许多老城市中也具备这种分区理念，按小型到特大型的顺序，将城市划分为街区、小区、片区，到市区、市镇。

Functional Layout

The whole scheme contains a series of high-rise and mid-rise buildings, which surround a central square. The main functional buildings of this area are management office, exhibition center, conference center and some public spaces and temporary facilities. The central square is surrounded by a transitional green belt, and its form refers to the peculiar Chinese traditional garden model in Suzhou. A natural river passes through the site from east to west and moistens the ponds, ditches and waterscapes. The green buffer zone and the outer regions of the site connect through a series of shared roof landscapes, courtyards and corridors. The whole comprehensive park delimits through a dense R&D and production zone to separate its boundary with the surrounding areas.

功能布局

整个方案包括一系列环绕着一个中央广场的高层和中层建筑。该区域的建筑主要功能为管理办公、展览中心、会议中心，还有一些公共场所和临时设施。中心广场环绕着过渡性的“绿带”，其形式参考了苏州特有的中国传统园林造型。一条自然河流自东向西从场地穿过，滋润了这片地域的水塘、沟渠以及水景。这条绿色缓冲带与场地外环区域由一系列可共享的屋顶景观、庭院以及走廊串联起来。整个综合园区通过一条密集的研发与生产地带进行界定，将其边界与周边区域分隔开来。

Section A–A
A–A 剖面图

Section B–B
B–B 剖面图

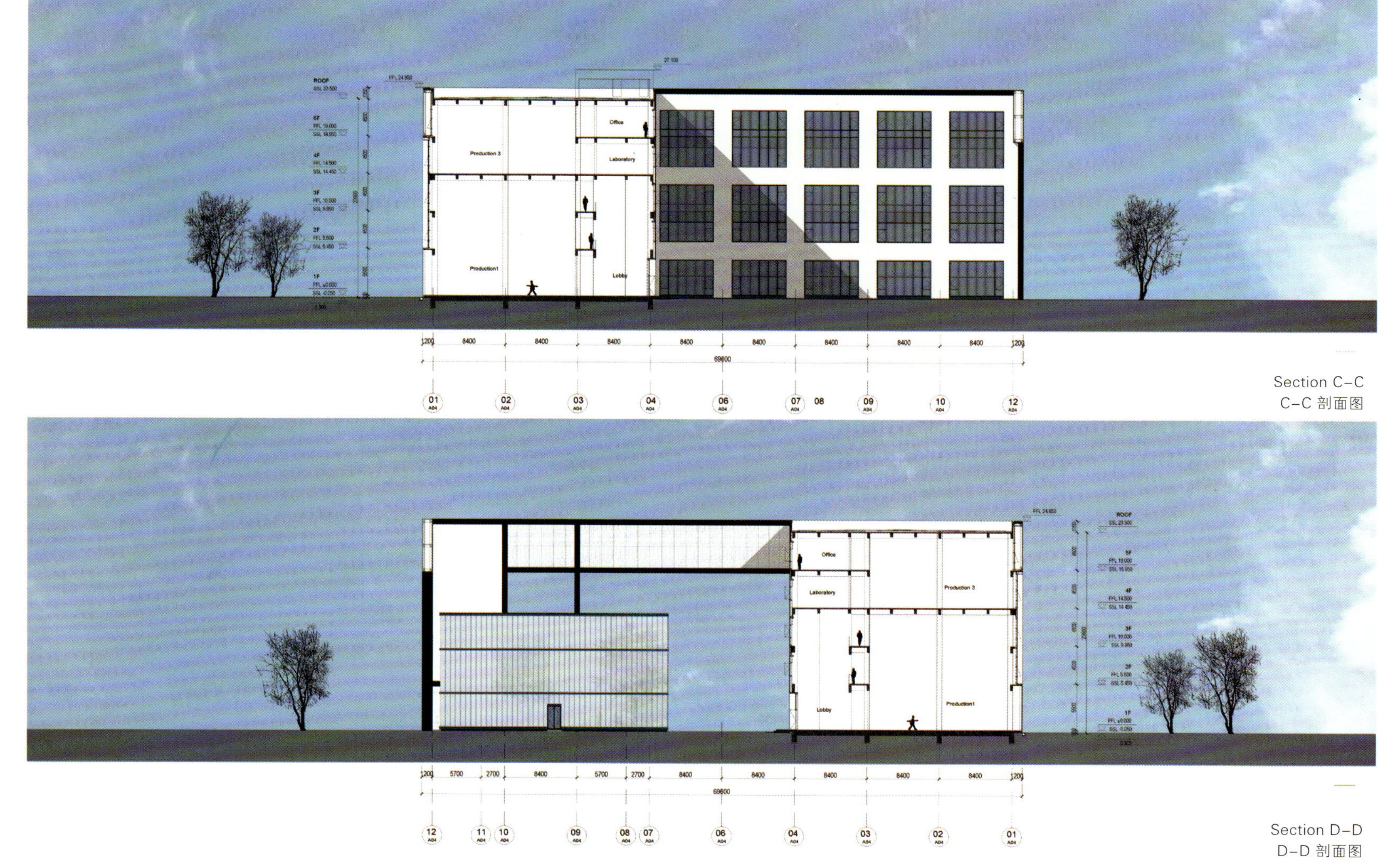

Section C–C
C–C 剖面图

Section D–D
D–D 剖面图

China Life R&D Center

中国人寿研发中心

Features 项目亮点

The main building of this R&D Center is in the shape of "Jade Seal", and with large body mass, which looks grand and magnificent.

研发中心主体建筑呈现“玉印”之形，建筑体量大，气势磅礴。

Keywords 关键词

Open Floor
楼层开放

Enclosed Garden
围合花园

Vertical Glass
垂直玻璃

Location: Haidian District, Beijing, China
Architectural Design: HENN GmbH
Cooperative Design: CCDI Group
Gross Floor Area: 397,000 m^2
Completion Date: 2013

项目地点：中国北京市海淀区
建筑设计：德国海茵建筑设计公司
合作设计：中建国际设计顾问有限公司
总建筑面积：397 000 m^2
竣工时间：2013 年

Overview

China Life R&D Center is located in the Environmental Science and Technology Demonstration Park of Zhongguancun, Daoxianghu, Haidian District, Beijing. The project is with a land area of 140,000 m^2, gross floor area of 397,000 m^2 and total investment of 6 billion yuan. This R&D Center is an agency directly under the headquarters of China Life and an important part of the "three centers in two places" IT implementation system of China Life, as well as a reconstruction project, which is mainly responsible for the IT R&D and test of off-site disaster recovery.

项目概况

中国人寿研发中心位于北京市海淀区稻香湖中关村环保科技示范园内。工程占地面积 14 万 m^2，总建筑面积 39.7 万 m^2，总投资 60 亿元。该研发中心是中国人寿总部直属机构，是中国人寿“两地三中心”IT 实施体系的重要组成部分，亦是重建项目，主要承担公司的 IT 研发、测试异地灾备等任务。

Design Idea

This R&D Center is on the green belt of Beijing edge. The architectural approach integrates the understanding of Chinese traditional buildings and people's future high demands of buildings on functions and techniques.

设计理念

研发中心位于北京城边的绿化带上。在建筑手法上糅合了对中国传统建筑的理解与将来人们对建筑物功能上、技术上的高要求。

Architectural Design

The main building of this R&D Center is in the shape of "Jade Seal", and with large body mass, which looks grand and magnificent. The central part of this project is a tranquil garden for communication and meditation. The exquisitely created courtyard adopts the classic Chinese residential style, and all the floors are open to the main building.

建筑设计

研发中心主体建筑呈现"玉印"之形，建筑体量大，气势磅礴。项目核心部分是一座安静的花园，是交流和沉思的地点。精心打造的庭院采用了经典的中国式住宅形式，所有的楼层都向主楼打开。

Functional Layout

The R&D Center is made up of a central office building, a R&D laboratory building and a training center with a dormitory. Like a Chinese seal, the central office building is square, which embodies the integrity and security of China Life in the field of life insurance. The center of this building is an enclosed garden. As a communication and leisure space, the garden becomes the center of the whole building .

功能布局

研发中心由一座中心办公楼、一座科研实验楼、一个带有宿舍的培训中心组成。就像中国的印章一样，中心办公楼呈方形，显示了中国人寿作为寿险行业所体现的诚信与保障。在这栋建筑物正中是一个围合的花园。花园作为交流、休闲空间成为整栋建筑物的中心。

Interior Design

The architectural idea of "garden within garden" not only inherits the Chinese traditional dwellings, but also creates a more spacious interior space. The vertical glass pieces on the two sides of the overhanging building facade add dynamism to the building. The jade colored glass visor can make a corresponding change according to the changes of the sunlight incident angles, so that to avoid covering the sight from the building interior to the surroundings of the inner courtyard or the buildings. In the lobby of the central office building, the lifts at two sides are all wrapped in red light boxes. The whole water surrounded site has received a better development under the architectural idea of sustainable development.

室内设计

"园中园"的建筑理念不但是中国传统民居的传承，而且创造了更加宽敞的室内空间。在两边出挑的建筑立面上的垂直的玻璃页片为建筑物增添了动感。翠三色的玻璃遮阳板可以根据阳光的入射角度的变化做出相应变动，从而避免遮挡到从建筑物内部到内庭或建筑物周边的视线。在中心办公楼的大堂里，两边的垂直电梯都被包裹在红色的灯箱里。整个被水环绕的基地在可持续性发展的建筑理念下得到了更好的开发。

Section 1–1
1–1 剖面图

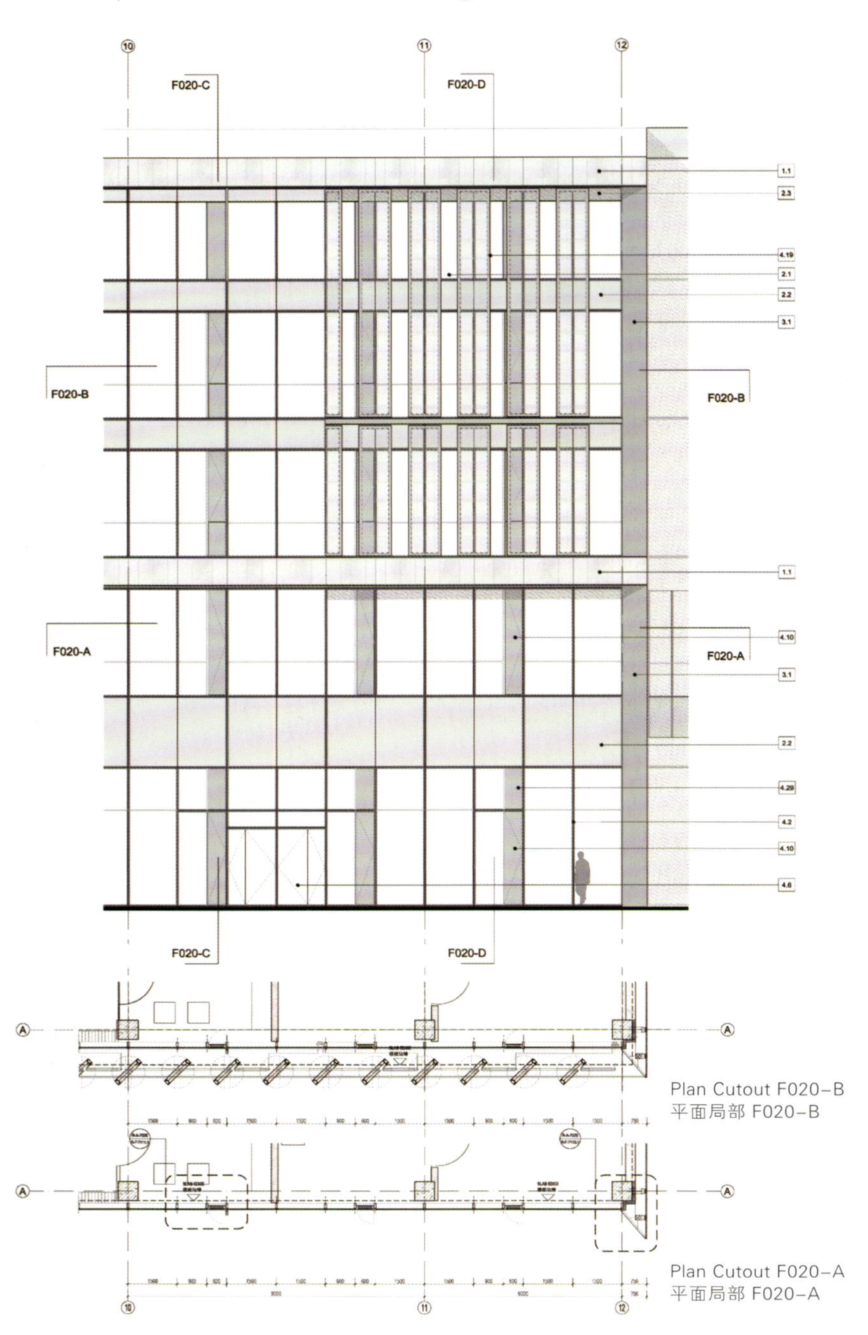

Plan Cutout F020-B
平面局部 F020-B

Plan Cutout F020-A
平面局部 F020-A

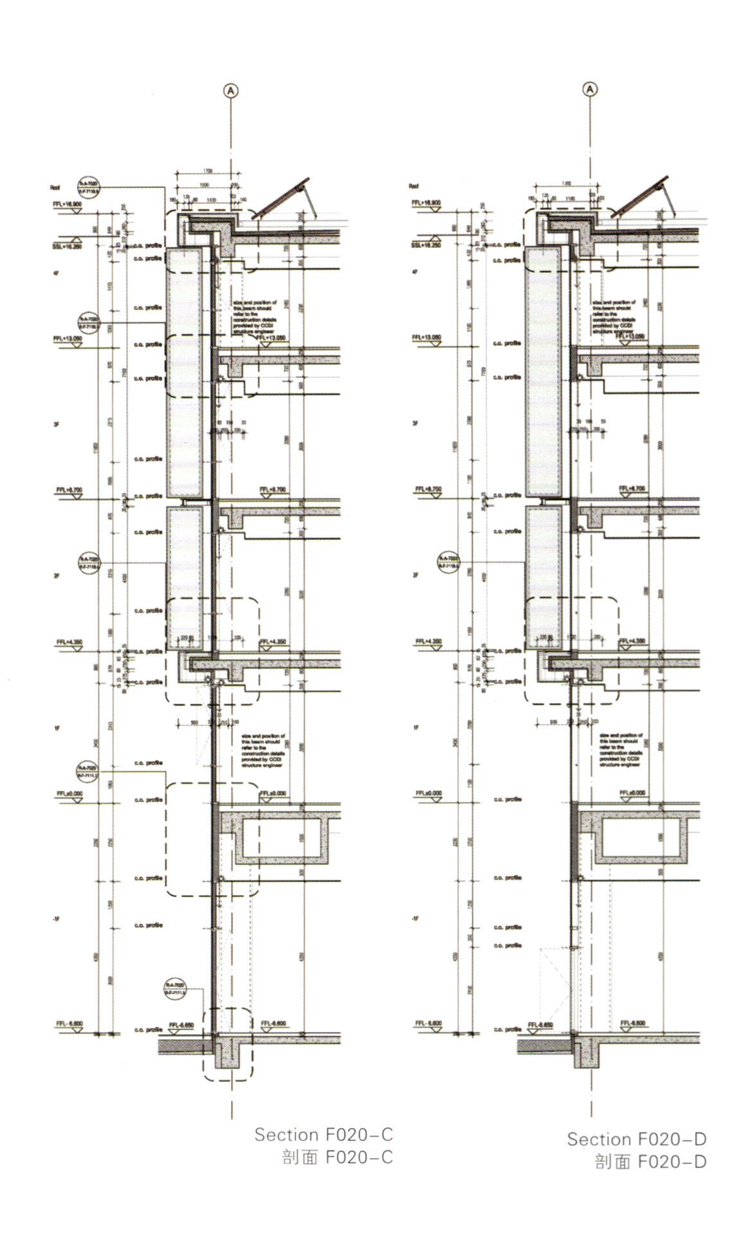

Section F020-C
剖面 F020-C

Section F020-D
剖面 F020-D

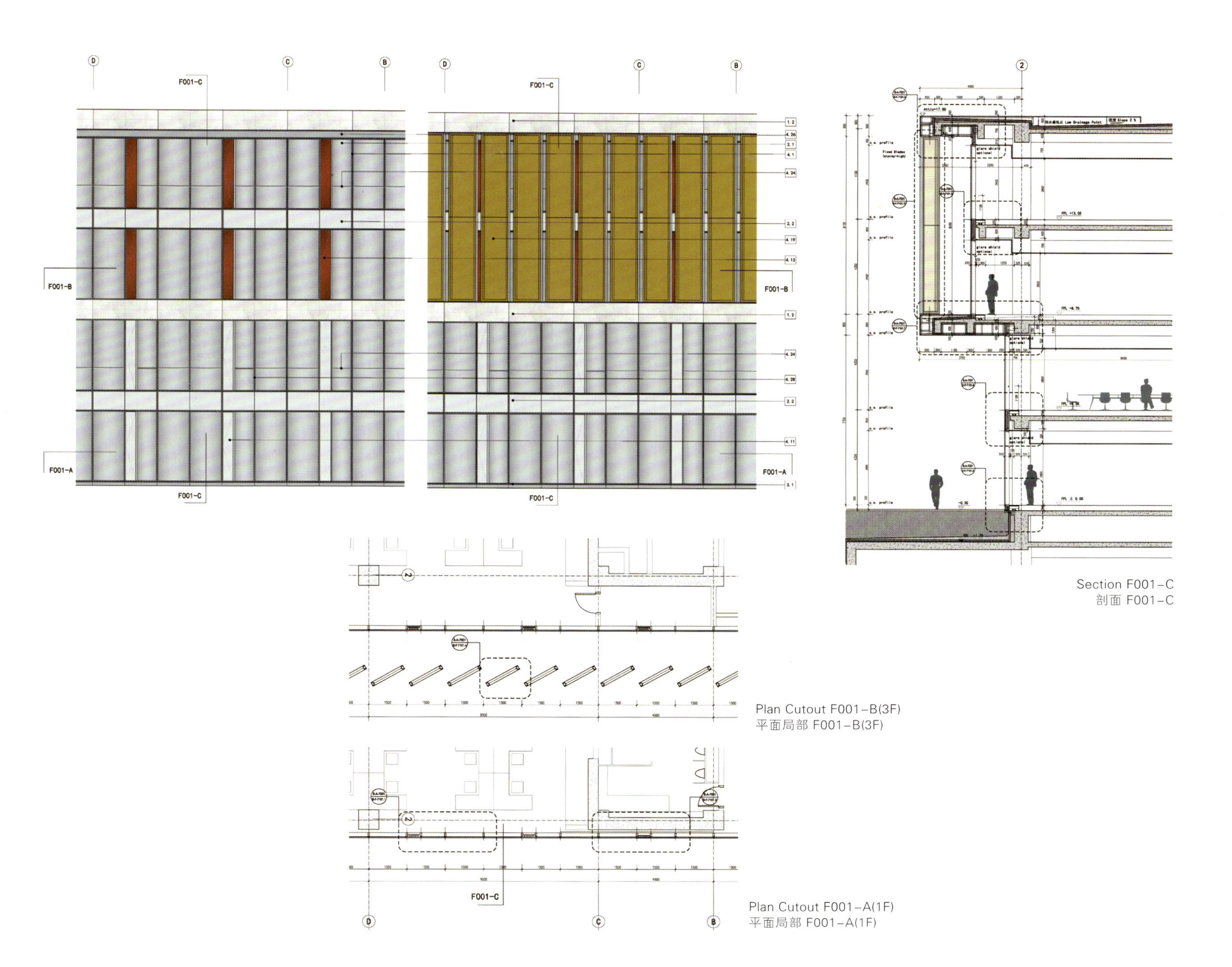

Section F001-C
剖面 F001-C

Plan Cutout F001-B(3F)
平面局部 F001-B(3F)

Plan Cutout F001-A(1F)
平面局部 F001-A(1F)

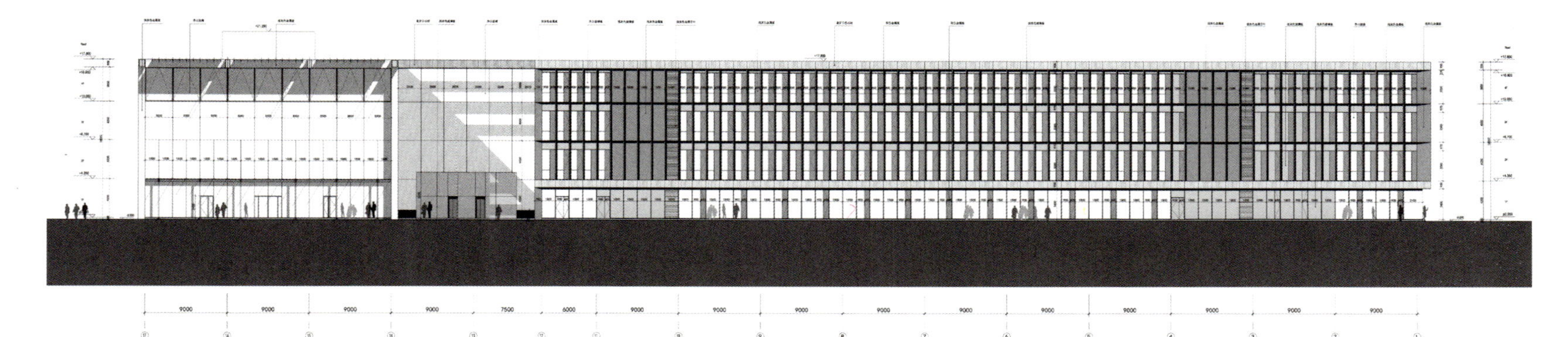

Section 2–2
2–2 剖面图

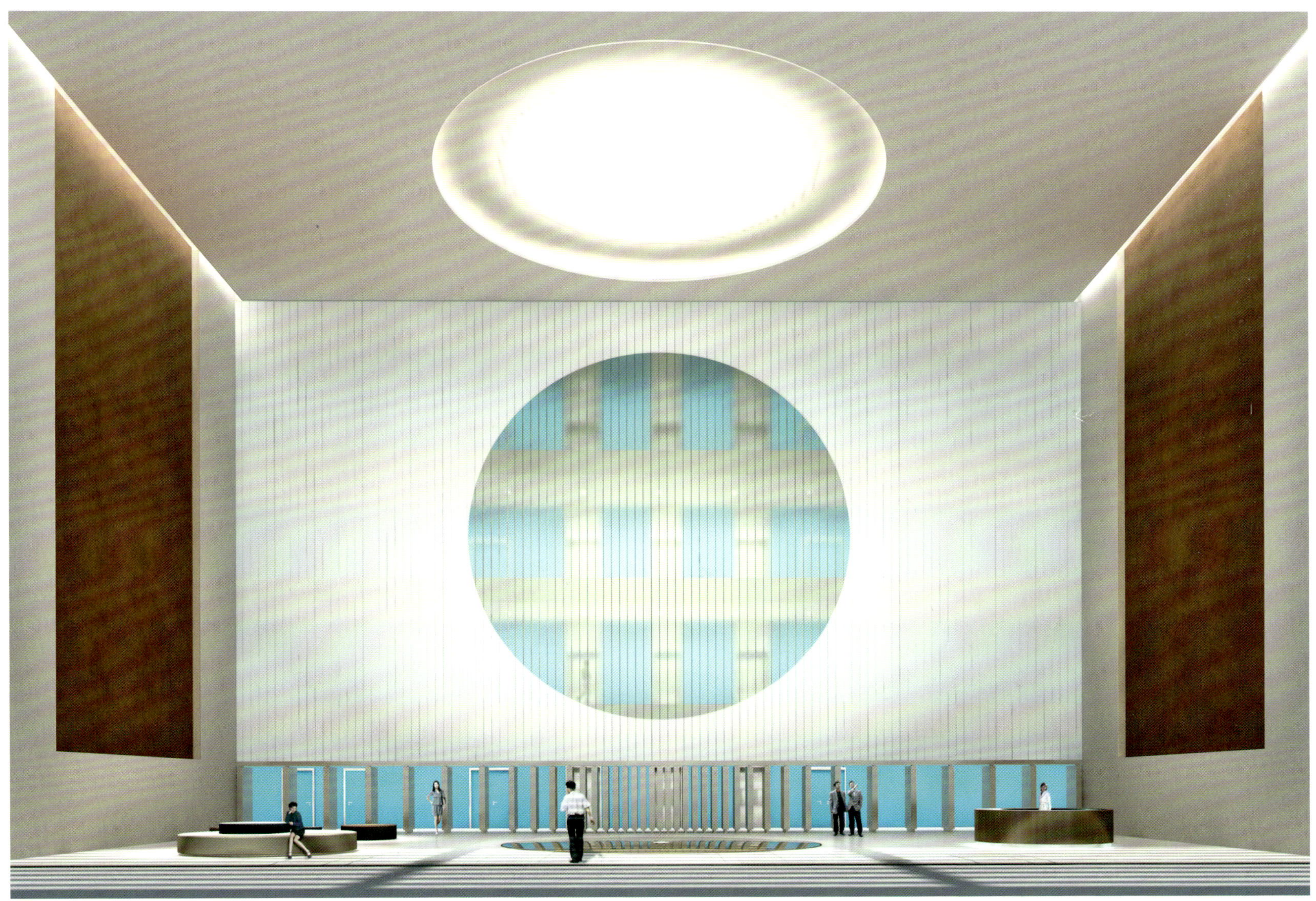

UFIDA Software Park No. 5 R&D Center

用友软件园 5 号研发中心

Features 项目亮点

Project's staff activity center achieved large-span space with arch steel frame, obtained better economy and also echoed with the arched service quarter in the south and east.

项目的员工活动中心用拱形钢架实现大跨空间，获得了较好的经济性，同时呼应了南侧和东侧的拱形服务用房。

Keywords 关键词

Large-span Space
大跨空间

Greening Water System
绿化水系

Calm and Quiet
从容宁静

Location: Haidian District, Beijing, China
Architecture Design: BCKJ Architects
Land Area: 60,496 m^2
Gross Floor Area: 33,000 m^2
Completion Date: 2007

项目地点：中国北京市海淀区
建筑设计：北京别处空间建筑设计事务所
用地面积：60 496 m^2
总建筑面积：33 000 m^2
竣工时间：2007 年

Overview

No. 5 R&D Center is located in the southwest corner of UFIDA Software Park, mainly including a set of staff quarters of 29,000 m^2, a 8,500 m^2 employee activity center (gym) and a 3,700 m^2 closed development office. As the whole park's energy and logistics center, the basement of employee activity center is equipped with nearly 180,000 m^2 energy center.

项目概况

5 号研发中心位于用友软件园的西南角，主要功能包括一组 2.9 万 m^2 的员工宿舍、一个 8 500 m^2 的员工活动中心（体育馆）和一座 3 700 m^2 的封闭研发办公楼。作为全园的能源中枢及后勤保障，员工活动中心地下室设有为近 18 万 m^2 建筑提供能源的能源中心。

Design Idea

Designers properly handle the relationship with the surrounding architectures and environment, making the project become a harmonious part of the whole park. R&D Center, from the design and performance, conveys the design philosophy of "ecological and natural, practical and simple, modern and national, economic and durable".

设计理念

设计师妥善处理与周边建筑及环境的关系，使之成为整个园区中和谐的组成部分。研发中心从设计和性能内涵上传达出“生态自然、实用简洁、现代民族、经济经久”的设计理念。

Planning Layout

According to the planning requirements of the overall software park, No. 5 R&D Center has three large parts: closed development office, staff apartment area and sports and leisure area composed of staff activity center, outdoor football field,

Site Plan 总平面图

basketball courts and tennis courts. Closed development office, in the northern end of the west, is adjacent to the greening water system in the north side with good views and can be easily linked to the No. 1 R&D Center in north shore through the bridge. E-shaped staff apartment, located in the center of the west land, fully enjoy the sunshine and ventilation of the north-south building, while the two introverted enclosed courtyards form calm and quiet residential atmosphere. The southwest side is back office and dormitory.

» 规划布局

根据整体软件园区的规划要求，5 号研发中心分为 3 个大的组成部分，使用性质分别为封闭研发办公区、员工公寓区以及由员工活动中心、室外足球场、篮球场、网球场组成的运动休闲区。西侧北端为封闭研发办公楼，该楼紧邻北侧的绿化水系，拥有好的景观，通过小桥能方便地与北岸的 1 号研发中心连系起来。员工公寓呈 E 字形布置在西段用地的中央，充分享用了南北向建筑的日照通风，同时围合出两个内向院落，形成从容宁静的居住气氛。其西南侧为后勤办公及宿舍。

Architectural Design

1. Adopting 10 badminton courts as the core, employee activity center comprehensively arranged the table tennis hall, billiards hall, gym and staff restaurants and other functional spaces, and added the training function after completion. Building achieved large-span space with arch steel frame, obtained better economy and also echoed with the arched service quarter in the south and east. Its arched steel columns provide strength for building, suggesting their motor function.

2. Closed development office and apartment are linked by a corridor, with convenient connection, now the former is used as a temporary president office. Architects think more about how to make the surrounding landscape integrate into architecture, while shared atrium and roof garden has become a place offering a variety of exchanges.

3. Multilayer staff apartment broke dull of determinant with a small angle of torsion so courtyard shows healthy tension, and echoes with the overall park plan. Original foundation is the old river way of which design made use handling height into localized underground bicycle storage, leaving ground for greening to create a good walking park environment. In facade, the balcony has small-scale dislocation in the surface of building and is endowed color changes to form their own rhythm.

4. Project preserves native trees within the site, so that the park gets mature green environment rapidly after the completion.

5. The main building used the grey and red brown cracked concrete block and the local steel shell nosing is plain and detailed.

6. In addition to energy-saving measures of airtight insulation, water reuse, etc., the architects designed the permeable ground and rainwater collection system to supplement landscape water. The whole western park's 180,000 m^2 energy provided by a large ground-source heat pump system under staff activity center which is the largest ground source heat pump station in North China when completed.

7. The entire project has barrier-free facilities and provides personalized service.

建筑设计

1. 员工活动中心以 10 个羽毛球场地为核心，综合布置了乒乓球馆、台球馆、健身房及职工餐厅等功能用房，竣工后根据需要又增加了培训的功能。建筑用拱形钢架实现大跨空间，获得了较好的经济性，同时呼应了南侧和东侧的拱形服务用房。其落地的拱形钢柱脚，为建筑提供了力度，暗示其运动功能。

2. 封闭研发楼与公寓有连廊相接，联系方便，目前暂时用作集团总裁办公用房。设计中更多地考虑了如何让周边景色纳入建筑中来，同时让共享的中庭和屋顶花园成为提供多种交流的场所。

3. 多层员工公寓用小角度的扭转打破了行列式的呆板，使院落呈现出良性的张力，并且与整体的园区规划相呼应。原地基为旧河道，设计中利用地基处理高度做了局部地下自行车库，给院落绿化让出地面，创造出好的步行园区环境。立面处理上，阳台在建筑表层小尺度错动，并赋予色彩的变化，形成自身的韵律节奏。

4. 项目尽量保留了用地内的原生树木，使建成后的园区迅速获得了较为成熟的绿化环境。

5. 建筑主体使用了灰色和红褐色露裂混凝土砌块，局部钢构收口，朴素而有细节。

6. 除保温密闭、中水回用等节能措施外，本工程设计了渗水地面及雨水收集系统，用收集的雨水补充景观用水。全园西部 18 万 m^2 的能耗由员工活动中心地下的大型地源热泵系统提供，竣工时是华北地区最大的地源热泵站。

7. 本工程全程设有无障碍设施，提供人性化的服务。

12

China Automotive Technology & Research Center (CATARC)

中国汽车技术研究中心

Features 项目亮点

The planning process combines overhead, receding, high, middle and low architectural forms with various approaches to build a well proportioned skyline with scattered layering and to form a moving routes with changeable landscapes.

在规划过程中采用架空、退台、高中低建筑形态结合等多种手法构筑高低层次错落、疏密有致的天际线，以形成移步换景的活动路线。

Keywords 关键词

Changeable Landscape
移步换景

Scattered Layering
高低错落

Well Proportioned
疏密有致

Location: Dongli District, Tianjin, China
Architectural Design: KaziaLi Design Collaborative Inc.
Land area:298,326.5 m²
Gross Floor Area: 205,352 m²
Design Time: 2009

项目地点：中国天津市东丽区
建筑设计：凯佳李建筑设计事务所
用地面积：298 326.50 m²
总建筑面积：205 352 m²
设计时间：2009 年

Overview

The CATARC new base is located at the intersection of East Xianfeng Road and Sanjing Road, Dongli District, Tianjin. The plot is an industrial land with an available construction area of 298,326.5 m² and a plot ratio of no less than 0.6.

项目概况

中国汽车技术研究中心新院区建设项目位于天津市东丽区先锋东路与三经路交口。该处地块为工业用地，可建设用地面积为 298 326.5 m²，容积率不小于 0.6。

Design Idea

As a major project of Tianjin city, the research center aims to be the condensing core and driving force of the development of Dongli District. In addition to having a good practical function, it also requires extraordinary charm and vitality. The design needs to reflect the mutual promotion of function, richness of space and interactivity of landscape to highlight the scientific and technological progress.

设计理念

作为天津市的重点项目，研究中心旨在成为东丽区发展的凝聚核心和推动力。除了要具备良好的实用功能外，还要求具有非同一般的魅力和活力。项目在设计上要体现出功能互促性、空间丰富性、景观互动性，以彰显科技进步。

Architectural Design

The style of this architectural design realizes the consistent form that guides in the planning, as well as emphasizes conciseness, brightness and extreme ambiance of times. The architectural design strives for the integrity of its own as well as within the groups. It uses relaxation relationships of the concise architectural forms to create a rich and smooth outer space and emphasizes the rhythm of the buildings. The interior space of the groups is ingenious and changeable. The buildings along

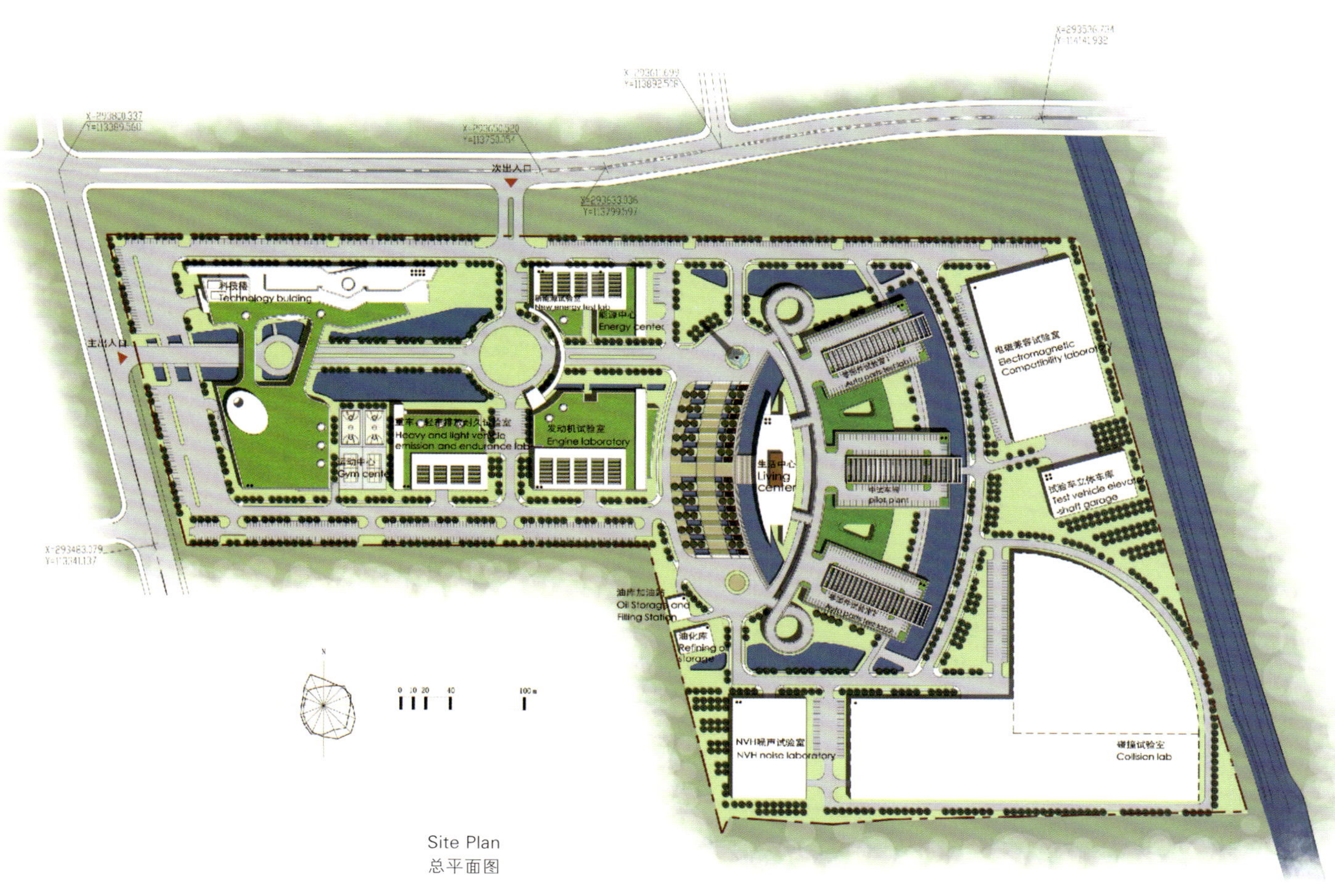

Site Plan
总平面图

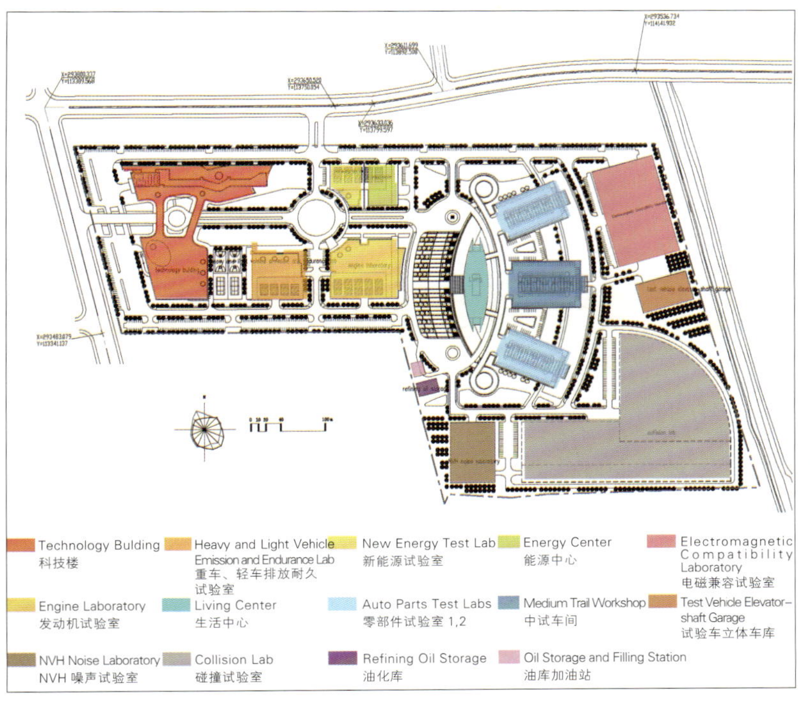

Function Analysis 功能分析图

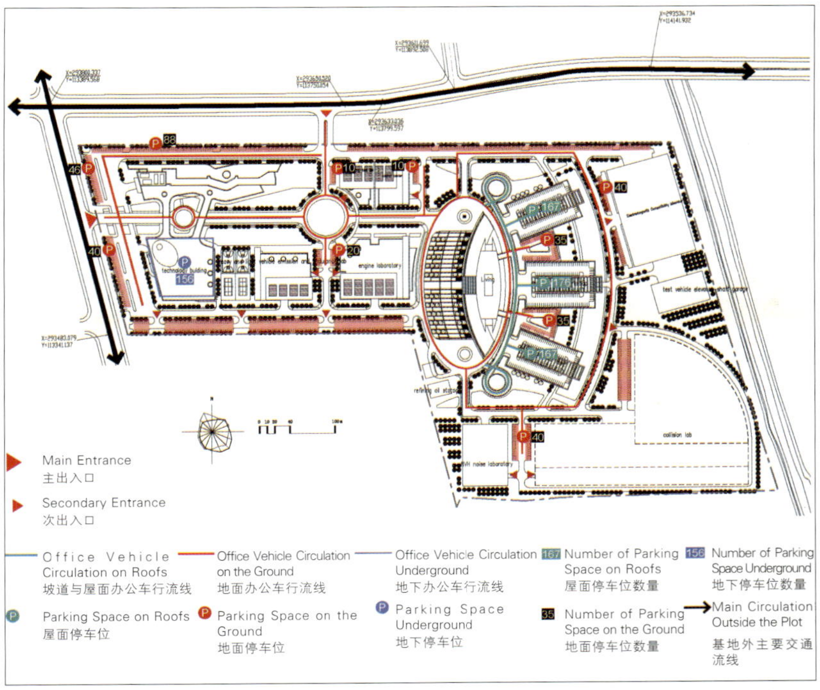

Main Entrance
主出入口

Secondary Entrance
次出入口

Office Vehicle Circulation on Roofs
坡道与屋面办公车行流线

Office Vehicle Circulation on the Ground
地面办公车行流线

Office Vehicle Circulation Underground
地下办公车行流线

Number of Parking Space on Roofs
屋面停车位数量

Number of Parking Space Underground
地下停车位数量

Parking Space on Roofs
屋面停车位

Parking Space on the Ground
地面停车位

Parking Space Underground
地下停车位

Number of Parking Space on the Ground
地面停车位数量

Main Circulation Outside the Plot
基地外主要交通流线

Office Vehicle Circulation & Parking Space 办公车行流线与停车位

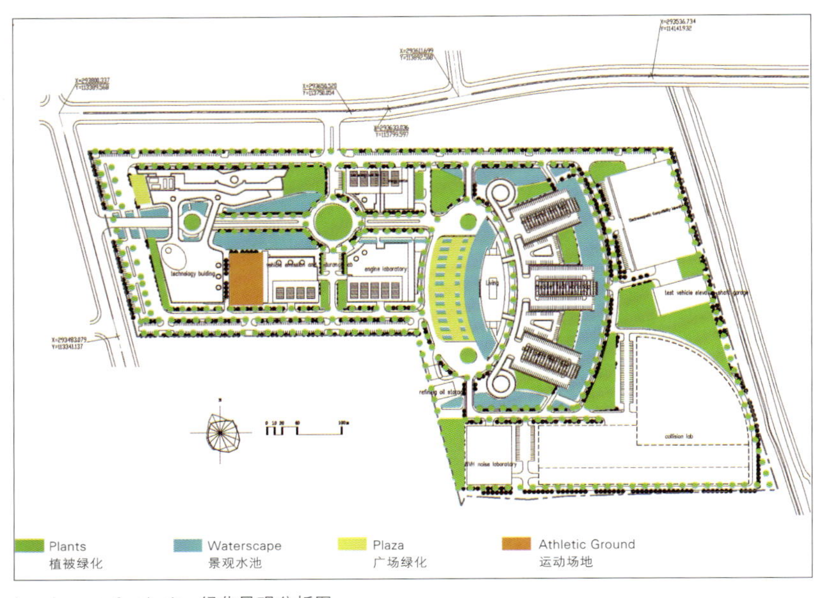

Landscape Analysis 绿化景观分析图

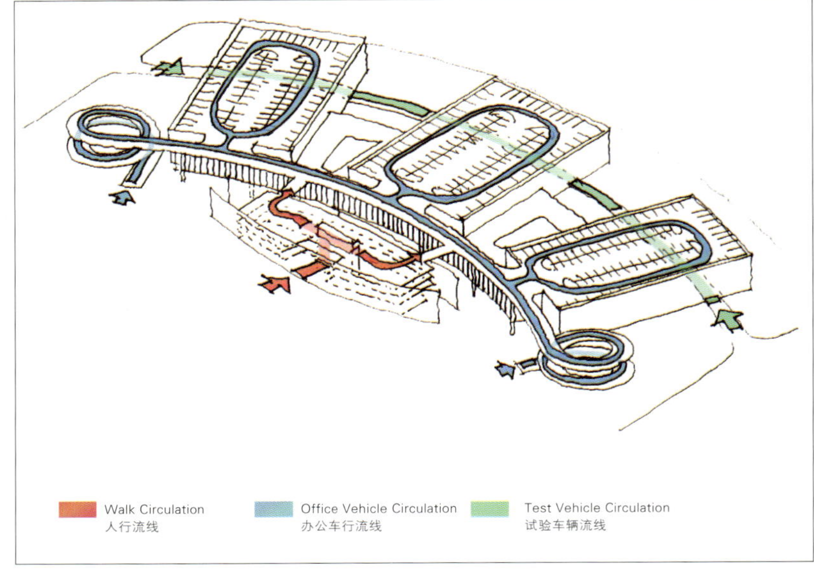

Roof Parking Analysis 屋面停车分析

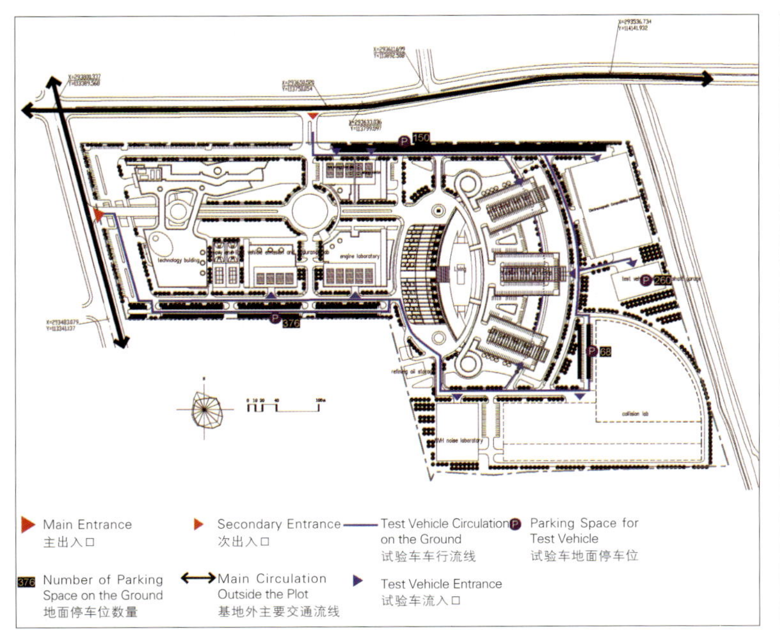

Test Vehicle Circulation & Parking Space 实验车行流线与停车位

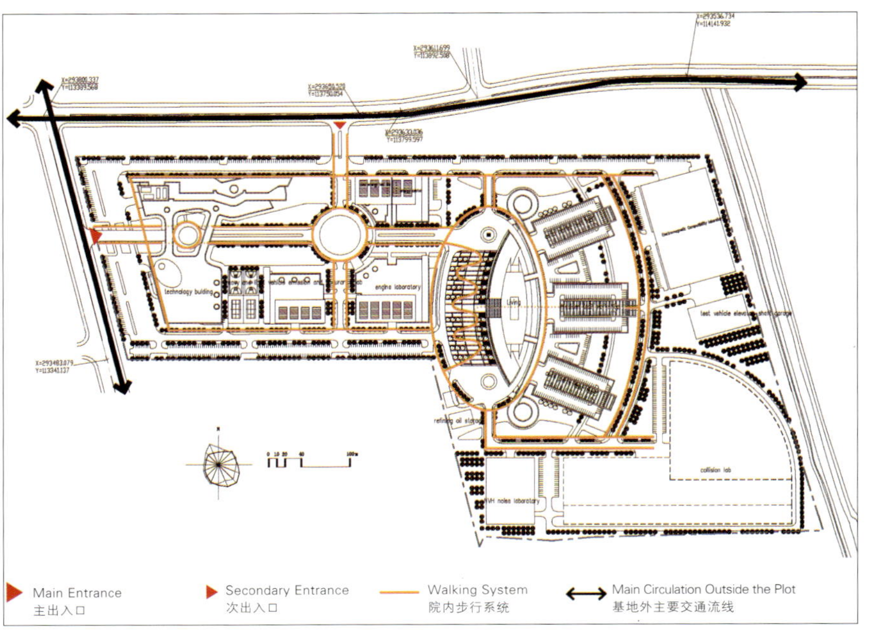

Walking System 院内步行系统

the street needs to show the continuity of the street and to achieve harmony in terms of height, color, material identification and so on. The street should have strong layering and rich changes.

The building materials are diversified on the basis of integrity. The facade design and the architectural tone are mainly modern and concise. Glass wall, metal and stone are considered as the main materials. And the design also pays attention to the comparisons of the material on hard and soft, virtual and actual.

The planning process combines overhead, receding, high, middle and low architectural forms with various approaches to build a well proportioned skyline with scattered layering and to form a moving routes with changeable landscapes.

建筑设计

建筑设计的风格满足规划中所引导的形态一致，同时强调简洁、明快、极具时代气息，建筑设计力求自身的完整和群落之间的整体性，以简洁的建筑形体张弛关系创造丰富流畅的外部空间，强调建筑节奏。组合内部空间巧妙多变。沿街建筑要体现街道界面的连续性，在高度、色彩、材质标志等方面取得统一协调，街面层次感强、富有变化。

建筑材料在统一的前提下多样化，立面设计和建筑色调以现代简洁为主，材质考虑玻璃幕墙、金属、石材等，并注意材料本身的软硬对比、虚实关系对比。

在规划过程中采用架空，退台，高、中低建筑形态结合等多种手法构筑高低层次错落、疏密有致的天际线，以形成移步换景的活动路线。

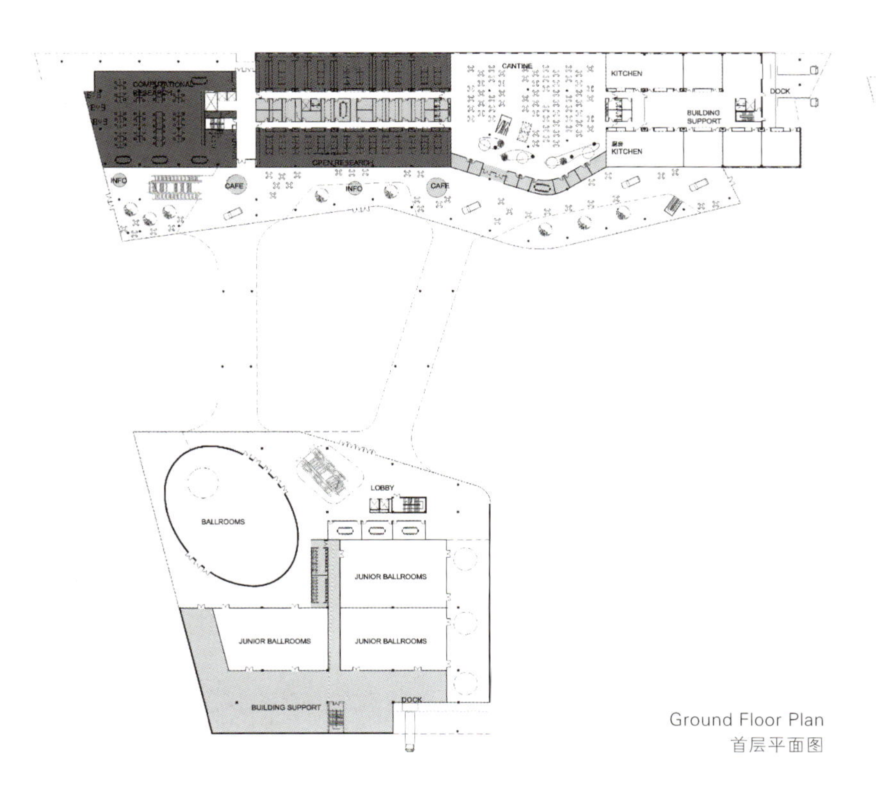

Ground Floor Plan
首层平面图

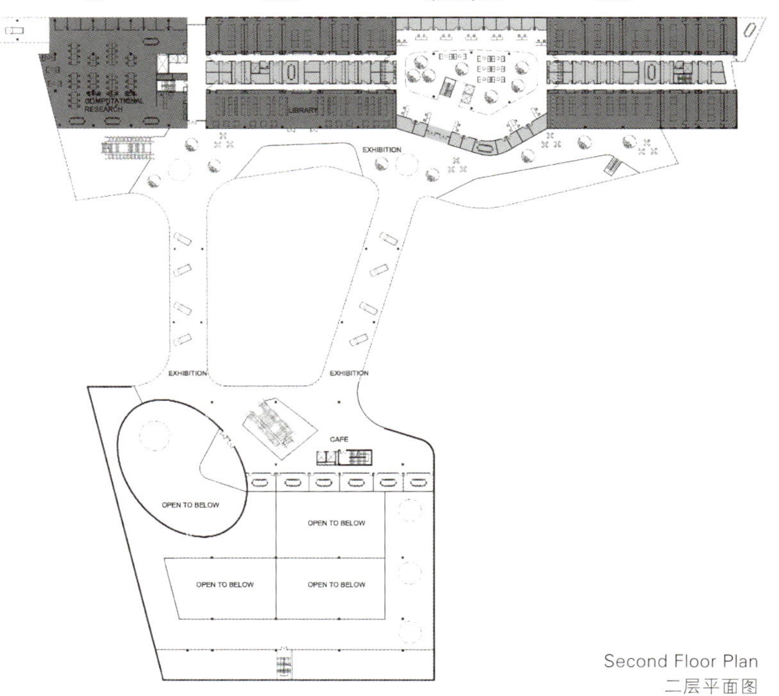

Second Floor Plan
二层平面图

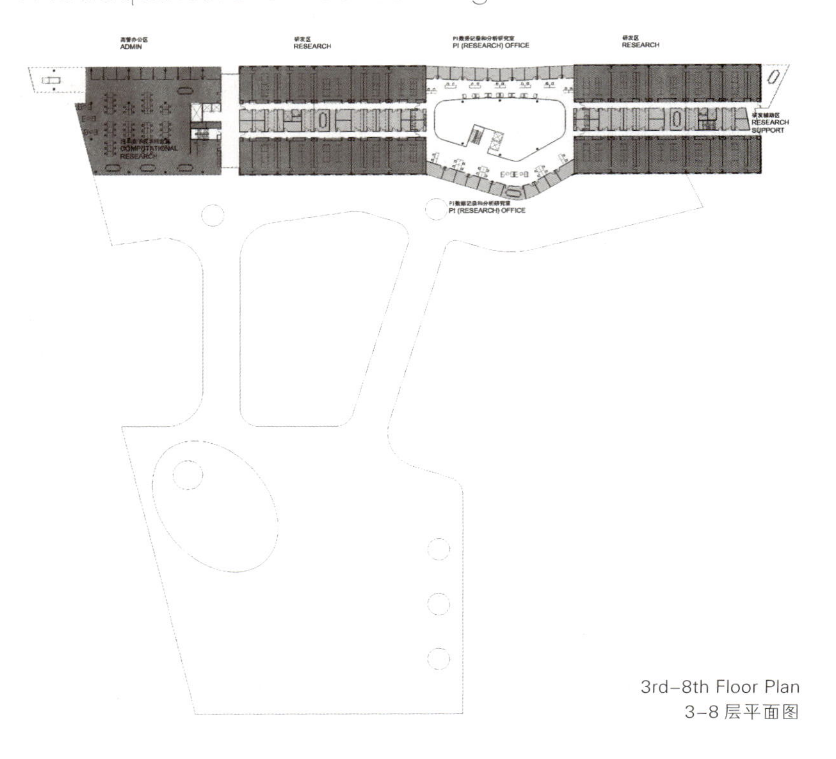

3rd–8th Floor Plan
3–8 层平面图

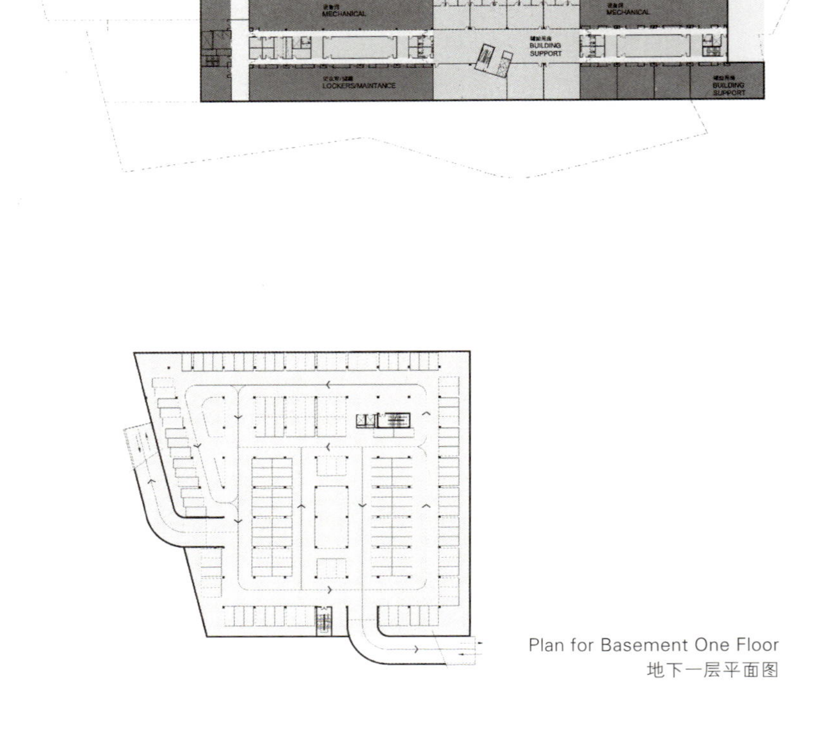

Plan for Basement One Floor
地下一层平面图

Spacial Layout

The architects strive to create an intensive and rich landscape space in the park and to make it shared within the park so that to improve the quality of the office space:

1. Using the enclosure and guidance of the buildings, planning a centralized square space to let the workers in the office easily get in touch with the beautiful exterior environment.

2. Using waterscape which is referred as the unified landscape element to connect each landscape node, organically combining each space together to improve the continuity of the space.

3. Arranging symbolic landscape element through important square node to enhance and enrich the guidance, layering and interest of the space.

空间布局

设计师着力打造集中、丰富的园区内景观空间，使其共享于园区，提升办公空间的品质：

1. 利用建筑的围合性与导向性，规划出集中的广场空间，使各建筑内的办公人员能够更容易地接触到优美的室外环境。

2. 用水景这一统一的景观元素将各景观节点串联起来，使各空间有机地结合在一起，提升空间的连贯性。

3. 通过重要的广场节点设置标志性景观元素，使空间的引导性增强，更富于层次感与情趣性。

CATARC

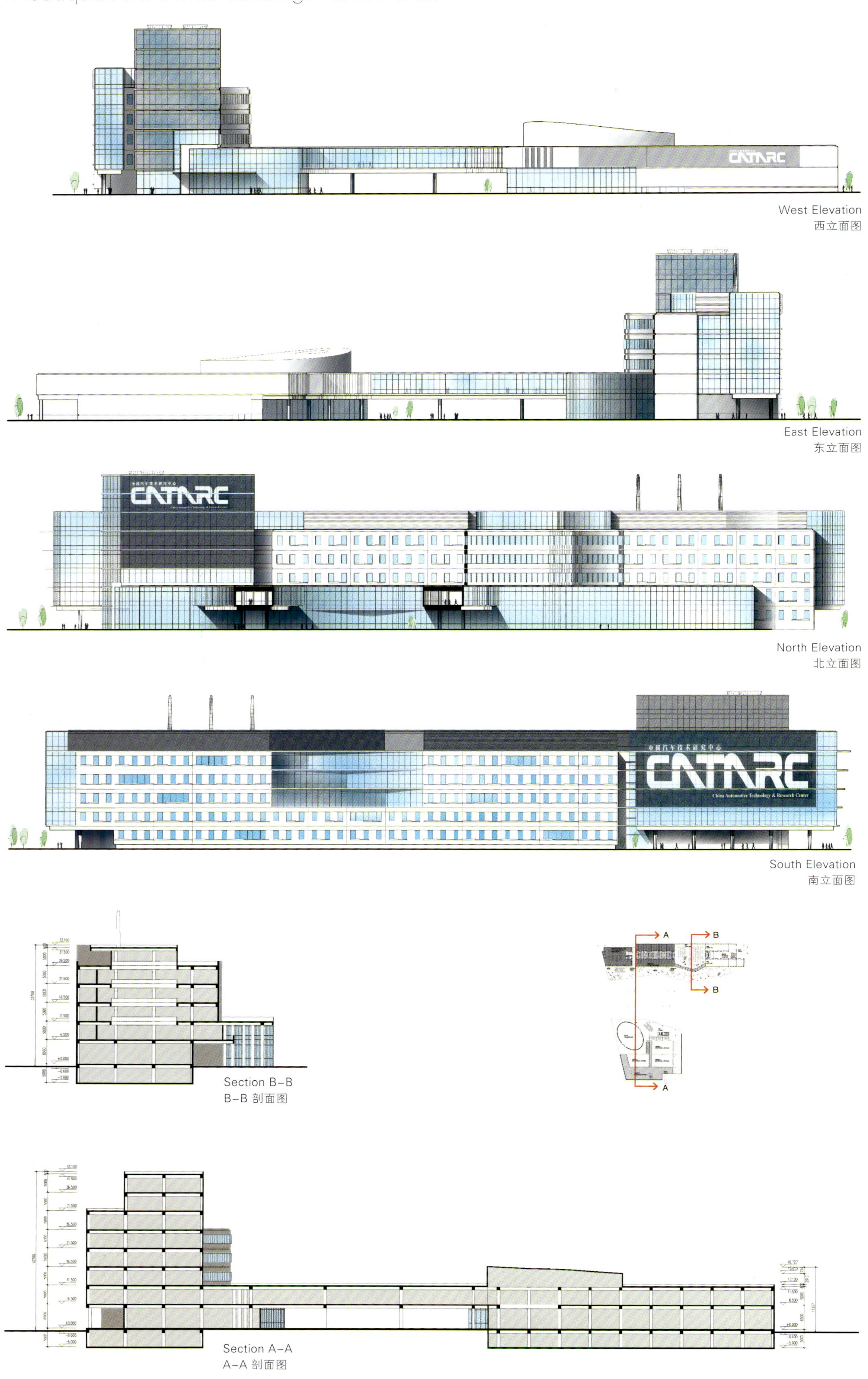

West Elevation
西立面图

East Elevation
东立面图

North Elevation
北立面图

South Elevation
南立面图

Section B–B
B–B 剖面图

Section A–A
A–A 剖面图

CATARC
CATARC
CATARC
China Automotive Technology & Research Center

Southwest China Industrial Headquarters Base, Guiyang

贵阳市西南工业总部基地

Features 项目亮点

With emphasis on stylish and dignified architectural style, the base is focusing on the headquarters buildings, forming a three-dimensional eco landscape system, and creating a newly defined working space for the enterprises.

项目强调个性、尊贵的建筑风格，总部基地以全景总部大楼为建筑主体，以园林、空间、人文构筑立体生态景观，为企业打造全新定义的工作空间。

Keywords 关键词

Headquarters Economy
总部经济

Identified and Dignified
个性尊贵

Working Space
工作空间

Location: Guiyang, Guizhou, China
Investor: HK IRE Real Estate Investment Group Co., Ltd.
Planning and Design: W&R Group
Land Area: 374,000 m^2
Total Floor Area: 285,675 m^2
Design Date: 2010
Status: Under Construction

项目地点：中国贵州省贵阳市
投资机构：香港阳晨集团
规划设计：美国 W&R 国际设计集团
用地面积：374 000 m^2
总建筑面积：285 675 m^2
设计时间：2010 年
项目状态：在建

Overview

Located in Wudang District of Guiyang City, Southwest China Industrial Headquarters Base occupies a total land area of 374,000 m^2. It got a total investment of about 1.8 billion yuan and the construction will take about five years. It is a key industrial project in Wudang District of Guiyang City, and it will be a headquarters business park with advantaged economic environment, innovative working space and iconic appearance.

项目概况

西南工业总部基地位于贵阳市乌当区，总占地面积约 374 000 m^2，总投资约 18 亿元，建设周期约 5 年。这是贵阳市乌当区重点发展的工业园区地产项目，项目定位为西南乃至中国最具有优秀总部经济环境、创新的工作空间和标志性外观的总部经济园区。

Location Analysis

Located in Wudang District of Guiyang City, on the south of Torch Avenue and on the west of Shuidong Road, the project is about 8 km from downtown Guiyang, 8 km from Guiyang Longdongbao International Airport, and 12 km from Guiyang Railway Station. It is close to Guiyang Royal Hotspring and Regal Poly Guiyang Hotel. And the newly built avenue as well as the Ring Expressway will connect the headquarters base well with its surroundings.

区位分析

本项目位于贵阳市乌当区火炬大道南侧，水东路西侧；距贵阳市中心约 8 km、贵阳国际机场约 8 km、贵阳火车站约 12 km，紧邻贵阳御温泉和保利国际温泉酒店。新添大道和绕城高速为总部基地与外界的联通提供了便利的交通条件。

JUNLAI HOTEL
JUNLAI HOTEL

Landesgirokasse

Design Idea

With emphasis on stylish and dignified architectural style, the base is focusing on the headquarters buildings, forming a three-dimensional eco landscape system, and creating a newly defined working space for the enterprises. The headquarters buildings within the site are used for the low-density courtyard offices, the public activity spaces, and the supporting facilities, creating a mixed-use headquarters business park.

设计理念

项目强调个性、尊贵的建筑风格，总部基地以全景总部大楼为建筑主体，以园林、空间、人文构筑立体生态景观，为企业打造全新定义的工作空间。基地内各栋总部楼的功能涵盖低密度庭院办公、企业级活性空间、全方位综合配套三大功能，开创“多功能总部商务模式”新格局。

Architectural Design

To highlight its identity as a modern international headquarters base, the buildings are designed in modern elegant style, and at the same time, they are inspired by the elements and styles of global headquarters buildings to meet different enterprises' requirements. The headquarters buildings mainly employ double glazed glass curtain walls, stones and red bricks, which are elegant, functional and economical. And the iconic high-rise buildings are designed with large-area glass curtain wall which is interspersed with U-shaped glass and printed glass to form colorful facade effect, shaping contrast with the solid walls covered by dry hang stones or light-colored coatings.

建筑设计

出于体现现代国际级工业总部基地的形象要求，项目的建筑单体设计采用了简洁的现代建筑风格，汇聚世界总部建筑元素与风格，满足世界各类企业总部的文化与功能要求。总部楼群主要采用的建筑材料为保温节能的中空双层玻璃显框幕墙、石材、红砖，兼顾美观、实用和经济性。而标志性高层单体建筑则采用较大面积的玻璃幕墙，局部使用 U 形玻璃以及印刷玻璃，形成丰富的立面效果，同时与干挂石材或浅色高级外墙涂料为表面的较实的墙面构成虚实的对比。

Sci-tech Parks
科技产业园

Frontier Base
前沿基地

Sci-tech Innovation
科创一体

Industrial Cluster
集群产业

Scale Effect
规模效应

Tian'an Digital City, Dongguan

东莞天安数码城

Features 项目亮点

Guangzhou–Shenzhen Road in the west is "separated" from the project while the beautiful scenery of Dongguan Canal forms part of the view of the project.

项目对西侧广深公路采用"阻隔"方法，对北侧东莞运河采用合理借景方法。

Keywords 关键词

Borrowing Landscape Reasonably
合理借景

Obscure Bent Axle
隐性曲轴

Protophyte
原生植物

Location: Dongguan, Guangdong, China
Landscape Design: W&R Group
Floor Area: 12,000 m^2(Phase Ⅰ)
Landscape Area: 45,000 m^2
Design Time: 2011

项目地点：中国广东省东莞市
景观设计：水石国际
建筑面积：12 000 m^2（一期）
景观面积：45 000 m^2
设计时间：2011 年

Overview

The project is located in Dongguan city and providing research and development, office and marketing exhibition space, effective information logistics facilities as well as the facilities for relaxation, communication and residence for private enterprise which possess concentrated intelligence and innovation ability. The first phase is built in an area of 48,000 m^2 of which the buildings occupy 12,000 m^2 and green space 4,300 m^2.

项目概况

数码新城项目位于东莞市，园区以为智力密集及创新特点的民营企业提供研发、办公、营销展示的空间和完善高效的信息物流辅助设施，以及为完善以上功能所必需的舒适合理的休息、交流、生活空间的设施。一期规划总用地为 48 000 m^2。其中建筑占地面积约为 12 000 m^2，配合市政绿化面积为 4 300 m^2。

Theme Orientation

It is an urban high and new technology industry park with balanced ecosystem. It aims to become a fresh, concise, energetic and comprehensive park, suitable for both business and residence.

主题定位

城市型生态高新科技产业园。基于总体定位，拟重点创造一个清新简洁、充满活力、宜商宜居的多功能综合园区。

Landscape Layout

Considering the characteristics and functions of the main buildings in Block A, the designers of Tian'an Digital City, Dongguan used an obscure bent axle to piece out the entrance square, welcoming square, water feature and sunken squares. Besides, there are also "temporary green spaces" and "dream gardens" as centralized green land.

天安数码城

Site Plan 总平面图

景观布局

结合东莞天安数码城A区各个主建筑的特性和功能，在景观总体布局上用一条隐性的曲轴串连起4个主要结点广场空间，即入口广场、迎宾广场、中心水景广场、下沉花园广场。同时塑造“临时绿地”和“梦想庭院”作为集中绿化块面。

Landscape Design

The designers strive for balancing the effect and cost to acquire best cost performance. Guangzhou-Shenzhen Road in the west has “separated” from the project while the beautiful scenery of Dongguan Canal forms part of the view of the project. The paving materials and decorations of the project unify it while also giving it some variety; for the design of landscape features and other landscape elements, interests, function and style are top priorities.

景观设计

力争景观塑造在效果、成本的良好平衡前提下，获得最高的性价比。一是处理好和地块周边的关系，对西侧广深公路采用“阻隔”方法，对北侧东莞运河采用合理借景方法。二是重点关注铺装材料和小品设计，即在铺装材料方面在统一中求一定的变化和丰富性；在景观小品以及其他景观元素设计中，突出趣味性、功能化、风格化。

» Development by Stages

The main landscape belt is arranged by the functional requirement of architectures. The original protophyte and temporary green spaces in different phases are well preserved and so that to promote the value of the project's environment.

» 分期开发

根据建筑功能需求布置主景观带，充分利用地块内现有的原生植物，并且处理好分期建设进程中的临时绿化用地，利用景观环境提升项目价值。

东莞天安数码城
DONGGUAN TIANAN CYBER PARK

只为成功寻找方法，不为失败罗列借口
LED

壳牌石油
同时也
的运输物流

每一个时代都因
创新
这是一群
产业变革的力量
改变时代
的人。
超越自己。
EEC

Park Office District

国际单位

Features 项目亮点

Lots of glass has been adopted to make the space more transparent, and designers create intangible interaction between outdoor landscape and interior office to let the view out of the window become a part of the interior.

玻璃的大量使用增加了空间的通透性，为了使“窗外的风景成为室内的一部分”，设计师在设计时使室外绿化景观与办公区产生了无形的互动。

Keywords 关键词

Modern and Concise
现代简约

Sense of Art
艺术气息

Transparent Space
空间通透性

Location: Guangzhou, Guangdong, China
Architectural Design: TECON Limited Liability Company
Landscape Design: Atelier cnS
Land Area: 178,000 m^2 (20,000 m^2 for Phase Ⅰ, 56,000 m^2 for Phase Ⅱ)

项目地点：中国广东省广州市
建筑设计：广州德空建筑设计有限公司
景观设计：广州市竖梁社建筑设计有限公司
占地面积：17.8 万 m^2（一期占地 2 万 m^2，二期占地 5.6 万 m^2）

Overview

Located on the extended area of north of Baiyun New Town, Park Office District is 15 minutes' drive away from Guangzhou Baiyun International Airport, 10 minutes away from Guangzhou Railway Station and Guangdong Provincial Bus Station, besides, lots of roads are across here such as Guangzhou-Qingyuan Expressway, Beijing-Zhuhai Expressway, North Second Ring Road, Southern Expressway, Airport Expressway, No. 105 and No. 106 National Roads.

项目概况

国际单位位于白云新城北部延伸区，距白云国际机场约 15 分钟车程，距离广州火车站、广东省汽车站约 10 分钟车程。广清高速、京珠高速、北二环高速、华南快速干线、机场高速、105 国道和 106 国道纵横交错。

Design Concept

Following the original art concept of Times Property, this project has provided strong sense of art to the public, with hundreds of well-known corporations in the categories of innovation, science & technology and arts, the sales center is modern and concise and full of fashion that is echoing with the fashion orientation of Park Office District.

设计理念

项目秉承了时代地产一直以来的艺术理念，国际单位整体艺术感强烈，目前已经有数百家创新、科技、艺术类的知名企业进驻，其招商中心现代简约而富有时尚艺术气息的设计感，可谓与国际单位的时尚创意定位极其契合。

中国广州服务外包示范园

國際單位
CULTURE
COMPLEX
中国广州服务外包示范园
创意村120–1500m²
写字楼火热招商中
3674 8888

記
城市
IMPRINT PARK
GUANGZHOU

Park Office
國際單位
PARK OFFICE DISTRICT

COSTA COFFEE
完美手工咖啡，
来到您的身边
國際單位

B2

B5
1949

COSTA
COFFEE

Interior Design

Offices on the second floor of sales center have adopted lots of glass to make the space more transparent, and designers create intangible interaction between outdoor landscape and interior office to let the view out of the window become a part of the interior. Well-proportioned office layout with every single corner being fully used, balancing functions with sense of fashion which have all met the design concept of Times Property.

室内设计

国际单位招商中心二楼办公区域，玻璃的大量使用增加了空间的通透性，为了使“窗外的风景成为室内的一部分”，设计师在设计时使室外绿化景观与办公区产生了无形的互动。错落有致的办公设置，合理地利用每一处空间，在艺术感十足中兼具了功能化，这一点十分符合时代地产的设计理念。

Functional Layout

Park Office District occupies the gross land area of 178,000 m^2.

Phase Ⅰ is for science and technology center with the land area of 20,000 m^2, which has involved 135 innovation, science and technology enterprises.

Phase Ⅱ, with the land area of 56,000 m^2, is designed as Guangzhou city imprint park including peasant-worker museum, industry transformation and upgrading exhibition corridor, original streets of urban village, live-action area, undergraduate innovation park, industrial design center, Park Office District Arts Center and so on.

Phase Ⅲ, Ⅳ and Ⅴ are aimed to be the complementary and supporting facilities, as well as upgrading the quality of phase Ⅱ, the planned culture creative leisure park, originality streets, idea hotel and creative industry headquarter will be the new landmark of creative cultural industry of Guangzhou even the whole country.

功能布局

国际单位创意园总占地 17.8 万 m^2。

一期定位为科技创意园，占地 2 万 m^2，已吸引 135 家科技、创意类企业进驻。

二期为广州城市印记公园，占地 5.6 万 m^2，包括农民工博物馆、产业转型升级展示长廊、城中村原貌街道、生活实景区、大学生创业园、工业设计中心、国际单位艺术中心等。

三、四、五期的主要功能定位为文化创意休闲公园、创意街市、创意酒店、创意产业总部，是二期功能的完善、配套的加强以及品质的进一步升级，将成为广州市乃至全国创意文化产业新的标杆。

农民工博物馆

UANGZHOU

广州市城市印记公园手模印记
广州市城市印记园修建的农民手模印记一共向全省各地征集了

Wuhan Raycom

武汉融科·智谷

Features 项目亮点

By adopting the traditional residential mode of Chinese quadrangle courtyards, the independent business buildings reveal the eastern residential aesthetics in modern ways and reach artistic living experience under the high plot ratio.

独栋商务花园采用中国传统四合院居住模式，用现代手段演绎东方居住美学，在高容积率下实现艺术性的生活体验。

Keywords 关键词

Eastern Aesthetics
东方美学

Plot Planning
岛区规划

Chinese Quadrangle Courtyard
四合院模式

Location: Hongshan District, Wuhan, Hubei, China
Developer: Raycom (Wuhan) Real Estate Development Co., Ltd.
Architectural Design: Coast Palisade Consulting Group Ltd.
Land Area: 275,000 m^2
Design Time: 2013

项目地点：中国湖北省武汉市洪山区
开发商：融科智地武汉公司
设计单位：加拿大 C.P.C. 建筑设计顾问有限公司
用地面积：275 000 m^2
设计时间：2013 年

Overview

Located in the Liqiao Fishery, Hongshan District, Wuhan, this project is at the south-west of the cross of Third Ring Road and Wenhua Avenue with great transportation while all the planned roads have been built. Third Ring Road interchange is on its north-east that connected Tunkou and Donghu development zones at the west and east ends respectively, and it is easy access to the cross through Luoshi South Road, while Third Ring Road, Luoshi South Road and Wenhua Avenue are the city expressways with great road conditions to connect suburb and urban area.

项目概况

融科•智谷项目位于武汉市洪山区李桥渔场，三环线与文化路的交叉口西南片。项目周边道路已按规划实施建成，路网较为完善。项目东北部为三环线立交，通过三环线向西可至沌口开发区，向东可至东湖开发区；北部通过珞狮南路可直达街道口，地块通达性较好。三环线、珞狮南路、文化大道为城市快速路，主要承担远城区与中心城区的交通联系和过境性交通，路况条件良好。

Design Concept

The project is designed to be a new industry park that applies to the development tendency of the innovative and scientific research industry. After the study on the new office mode, city development, exterior environment of the project, as well as sustainable development, architects have raised up four concepts for the architectural design, which are cloud wisdom (new office mode), city property (city development), “park—office” (exterior environment of the project) and “ecology—office” (sustainable development) respectively.

设计理念

项目旨在为创意科研产业提供符合行业发展趋势的新型产业园区。通过对于新型办公模式，城市发展，园区外部环境，可持续发展四方面的研究，提出了本园区规划建筑设计的四个概念。它们分别为云智慧（新型办公模式）、都市属性（城市发展）、“公园—办公”（园区外部环境）及“生态—办公”（可持续发展）。

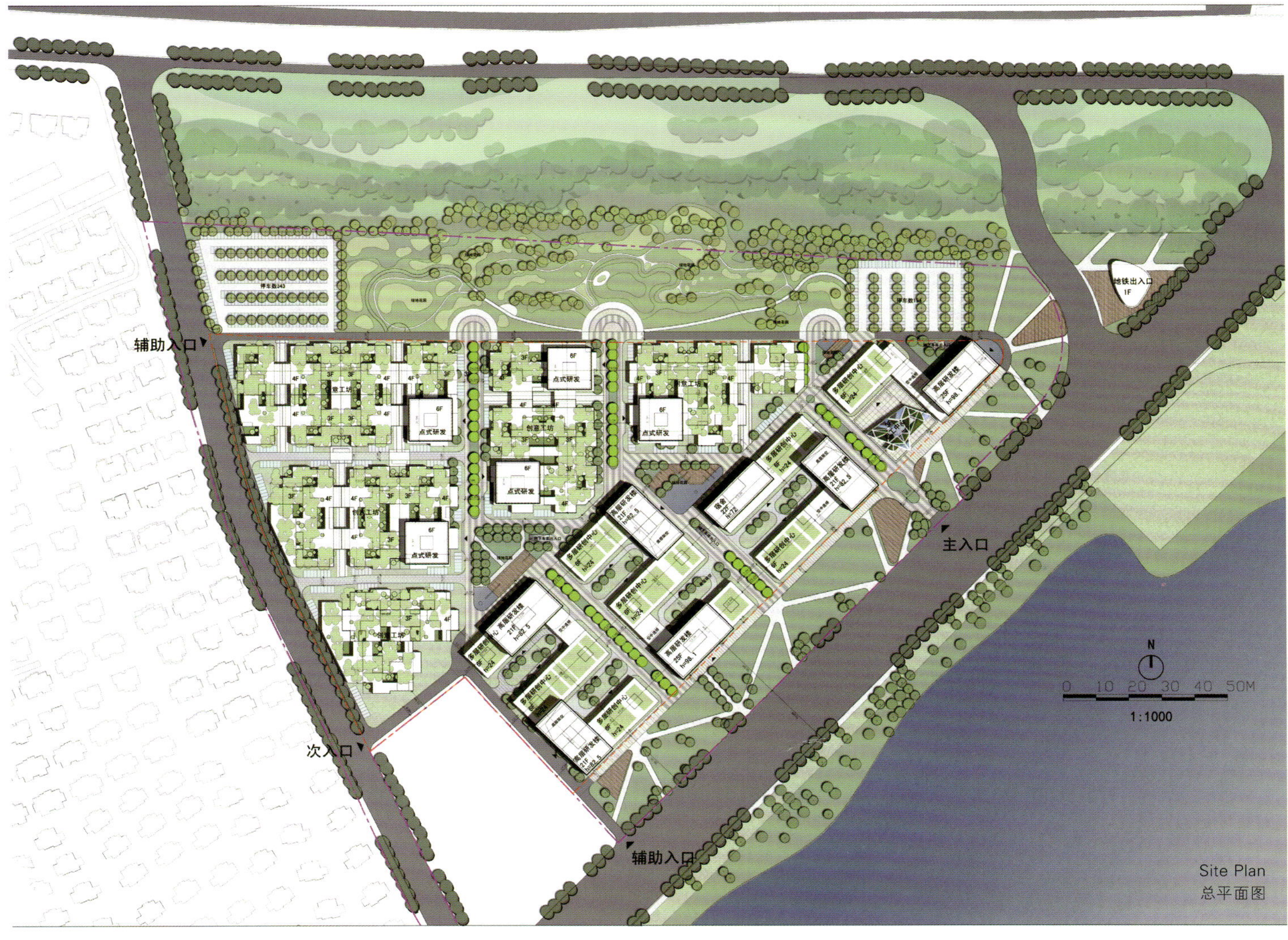

Site Plan
总平面图

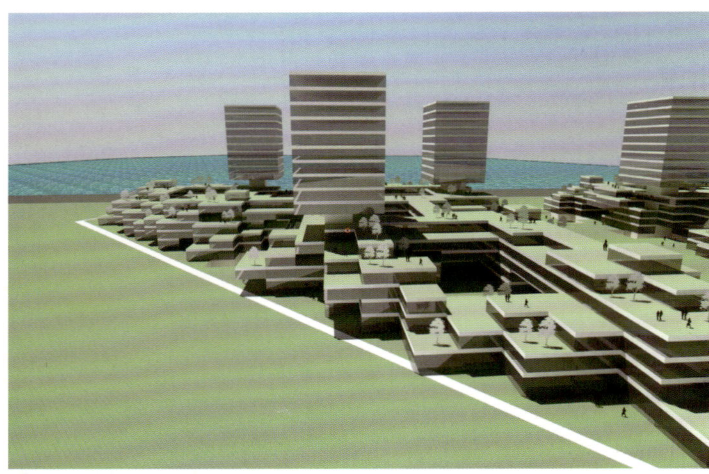

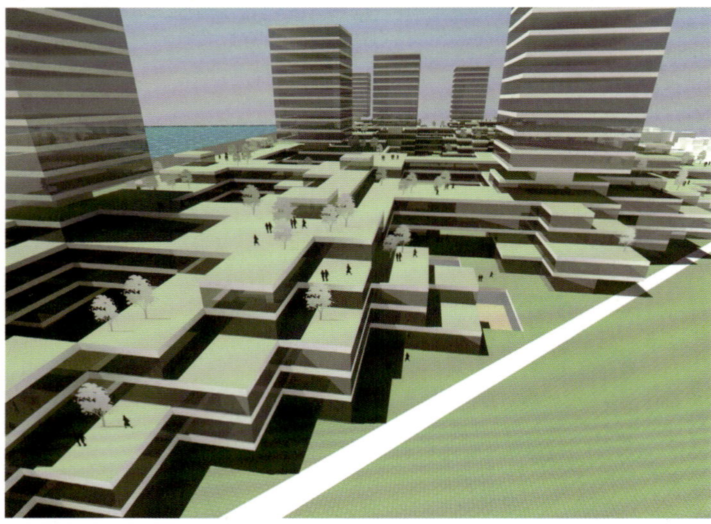

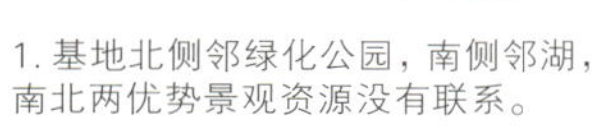

1. 基地北侧邻绿化公园，南侧邻湖，南北两优势景观资源没有联系。

2. 将北侧公园景观与南侧湖景引入基地内部，将景观资源进行深度整合，构建更完整的景观系统。

3. 南北景观整合联动形成“超级公园”，其景观要素将基地切分为三个岛区，每个岛区与景观实现最大化的接触与互动。

4. 将每个岛区规划赋予主题板块与组团，将办公室建设于“超级公园”之中，绿岛之上，规划结构上，实现“公园式办公”。

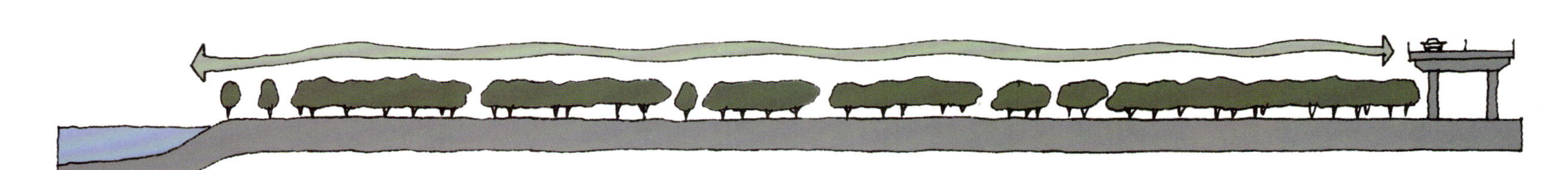

1. 未建设的基地与公园、湖景的关系。

2. 低层与多层建筑通过引入架空空间、下沉庭院、屋顶花园等空间要素与周边景观资源交融互动形成整体“地景”。

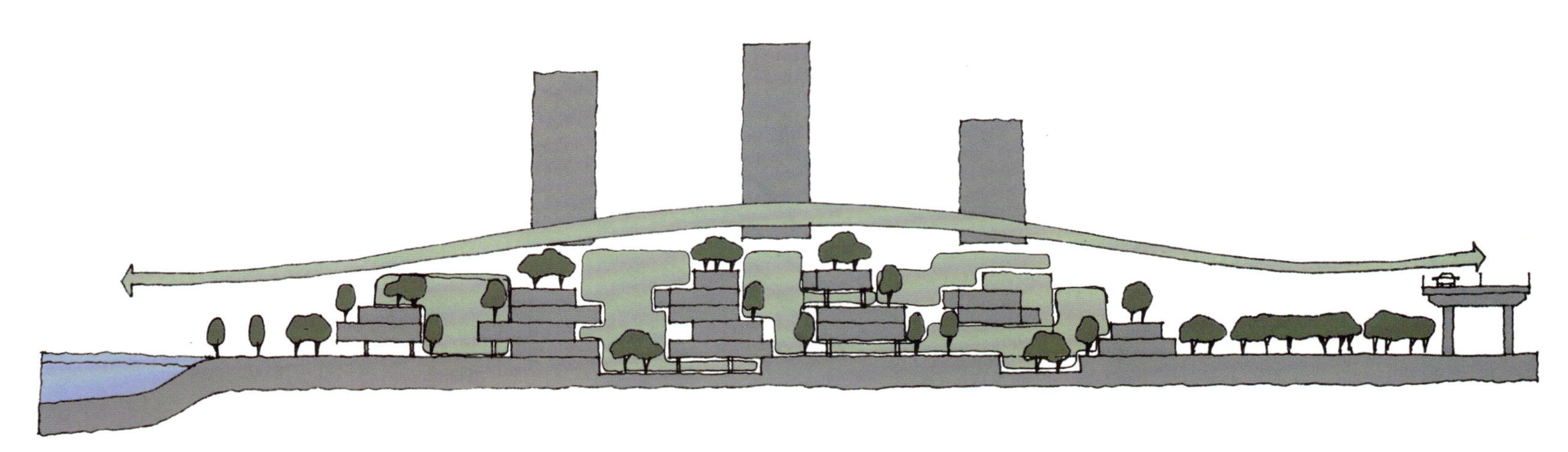

3. 高层凌空于“地景”之上，对城市展现“城市雕塑”形象特征，对用户尽享高区湖景资源。

Landscape System
立体绿化系统

三环高架车流观湖视线不被高层排布完全遮挡，为城市预留透景视廊

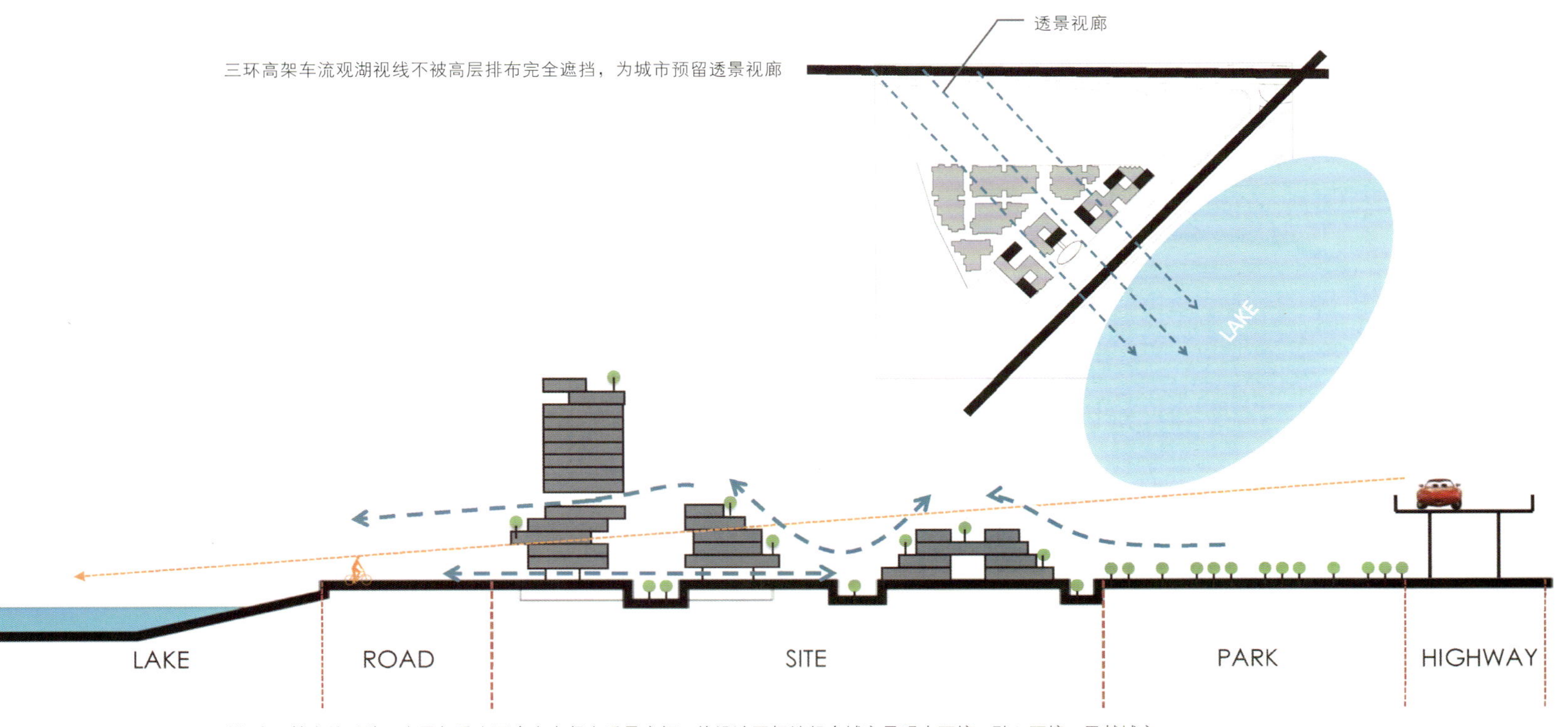

通过架空、镂空等手法，水平与垂直两个方向保有透景空间，使设计更好地契合城市景观大环境，融入环境，贡献城市。

Perspective Strategy
城市透景策略

Planning Structure Analysis
规划结构分析

Traffic Drawing
交通分析图

Architectural Design

By adopting the traditional residential mode of Chinese quadrangle courtyards, the independent business buildings reveal the eastern residential aesthetics in modern ways and reach artistic living experience under the high plot ratio. By making the units built layers upon layers, the land and sky resources are being fully used. Architects place the buildings backwards to leave more spaces for the sunshine and wind and to create pleasant living environment with various paths reaching to the buildings. The backwards and enclosed buildings have cut off the noise outside so that each unit can enjoy their own quiet courtyard, which has representing the ideas of harmony between nature and man and pursuit of nature. Thematic plots and groups are planned for the project by building the office buildings in the "super park" and over the green island, to achieve garden offices in planning structure.

建筑设计

独栋商务花园采用中国传统四合院居住模式，用现代手段演绎东方居住美学，在高容积率下实现艺术性的生活体验。户与户向上叠加，充分占有了土地资源和天空资源。建筑层叠的退让，给阳光和风留出通道，创造良好的住宅环境，每户独有的回家的路径。利用建筑的围合退叠，隔绝外部的喧嚣，让每户均能拥有独占天空的静谧院子，符合天人合一、追崇自然的思想。岛区规划赋予主题板块与组团，将办公室建设于"超级公园"之中，绿岛之上，规划结构上实现"公园式办公"。

Landscape View Analysis
景观视线分析图

Functional Layout

Three thematic plots are planned to be built for this project, shaping a letter of Y to form independent plots, which are Culture Corner, Animation World and Knowledge Garden respectively.

Culture Corner: news, translation and international cultural communication, educational training and consulting, national publishing industry.

Animation World: internet and software engineering.

Knowledge Garden: digital technology and industrial design, engineering design and construction, trade and modern service industry.

Each plot is designed as different themes, and by introducing axises and ring roads into or between each plot to create complicated and interacted public space system.

功能布局

园区规划构建三个产业主题板块，以Y字形主轴为界，形成各自相对独立的岛区，分别为文化天地、动漫世界、创智家园。

文化天地：新闻、翻译与国际文化交流、教育培训及咨询、民族出版产业。

动漫世界：互联网、软件工程。

创智家园：数字科技及工业设计、工程设计与施工、贸易与现代服务业。

每个岛区赋以不同的板块主题氛围，岛内及岛间引入轴线或环路，综合构建复合化、交互式的公共空间系统。

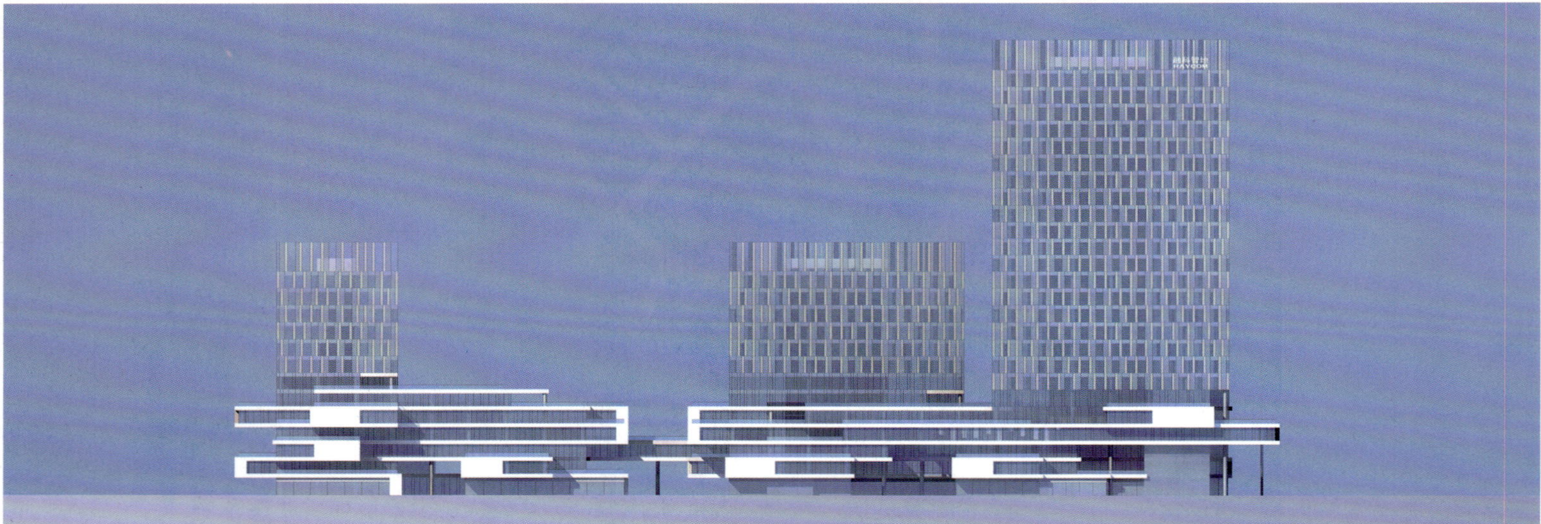

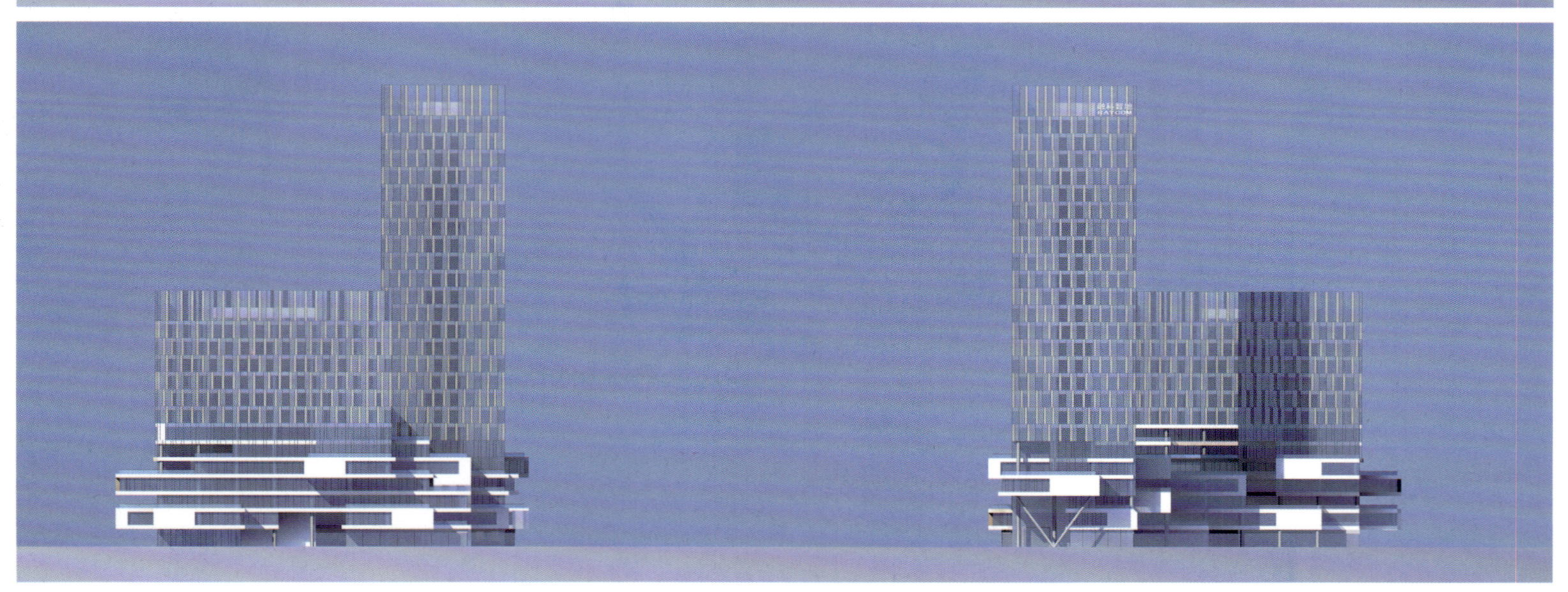

lenovo
Microsoft
RAYCOM
融科智谷
Farrow
O.P.C GROUP
RAVINTOLA
fennia

Microsoft

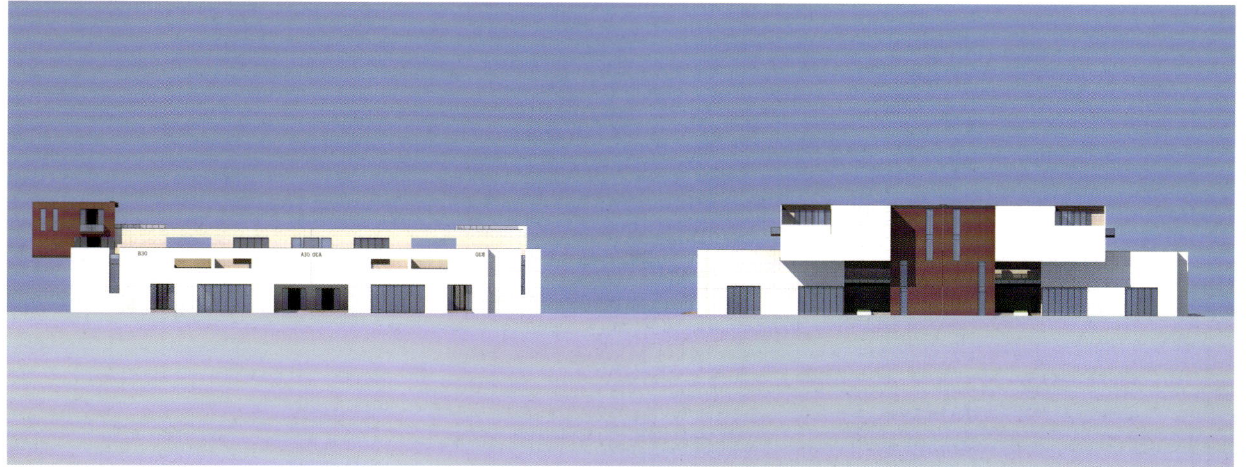

B07

SAMSUNG
RAYCOM
lenovo
融科智谷

B08
A01

IBM

B30

C21
C30
A21
A30

Tianchen Science and Technology Park, Tianjin

天津天辰科技园

Features 项目亮点

The whole buildings are light and bright. Light blue glass are cooperated well with light grey stone and silver metal dividing lines to create decorative effect with a strong sense of modernization.

整栋建筑物色彩清淡而明快，浅蓝色玻璃和浅灰色石材巧妙地搭配，加上亮银色的金属分割条，形成一种现代感极强的装饰效果。

Keywords 关键词

Glass Curtain Wall
玻璃幕墙

Interlocked
形体咬合

Sense of Modern Beauty
现代美感

Location: Beichen District, Tianjin, China
Developer: China Tianchen Engineering Corporation
Architectural Design: Coast Palisade Consulting Group Ltd.
Land Area: 37,684.6 m^2
Gross Floor Area: 190,000 m^2
Design Time: 2008

项目地点：中国天津市北辰区
开发商：中国天辰化学工程公司
设计单位：加拿大 C.P.C. 建筑设计顾问有限公司
占地面积：37 684.6 m^2
总建筑面积：190 000 m^2
设计时间：2008 年

Overview

Located in the plot of Beichen District of Tianjin which is between Wuqing New Town and central urban area, with advantaged location and environment, this project is placed lying on the North Canal and with low- and middle-storey buildings around, closing to lots of schools, hospitals and communities, also connecting with Beijing urban area through Beijing-Tianjin Road. It is the entrance to Tianjin City and has a gross land area of 37,684.6 m^2.

项目概况

基地所在地为天津北辰区，该区位于“一轴”的武清新城及中心城区之间，周边分布有各级大中学校、医院及住区，背靠北运河，紧邻中心城区同北京相连的交通要道京津路，区位和环境优势极为突出。基地周边建筑以中低层为主，地处入津的重要通道。基地总面积为 37 684.6 m^2。

Design Idea

Architects are trying to make it as a landmark building of Beichen District that on the way from Beijing to Tianjin. Regarded as the new office building of China Tianchen Engineering Corporation, it is supposed to be modern and concise, and also an image for a modern enterprise with great goals, full of dynamic and entrepreneurial spirit. In the meanwhile, it should also contain the restrained and low-key content, just like other remembered buildings that can stand up the tests of weather and time.

设计理念

设计师试图将其诠释为北辰区由北京进入天津中心城区的地标性建筑，成为北辰区的一个象征。作为中国天辰化学工程公司的新办公楼，它应当是现代的、简洁的，象征公司作为一个现代化企业所具备的高瞻远瞩，充满活力和进取精神的形象。同时，它又应当有适当的内敛和低调，像无数历经岁月仍生机蓬勃，为人们所记忆的建筑，经得起时间的考验。

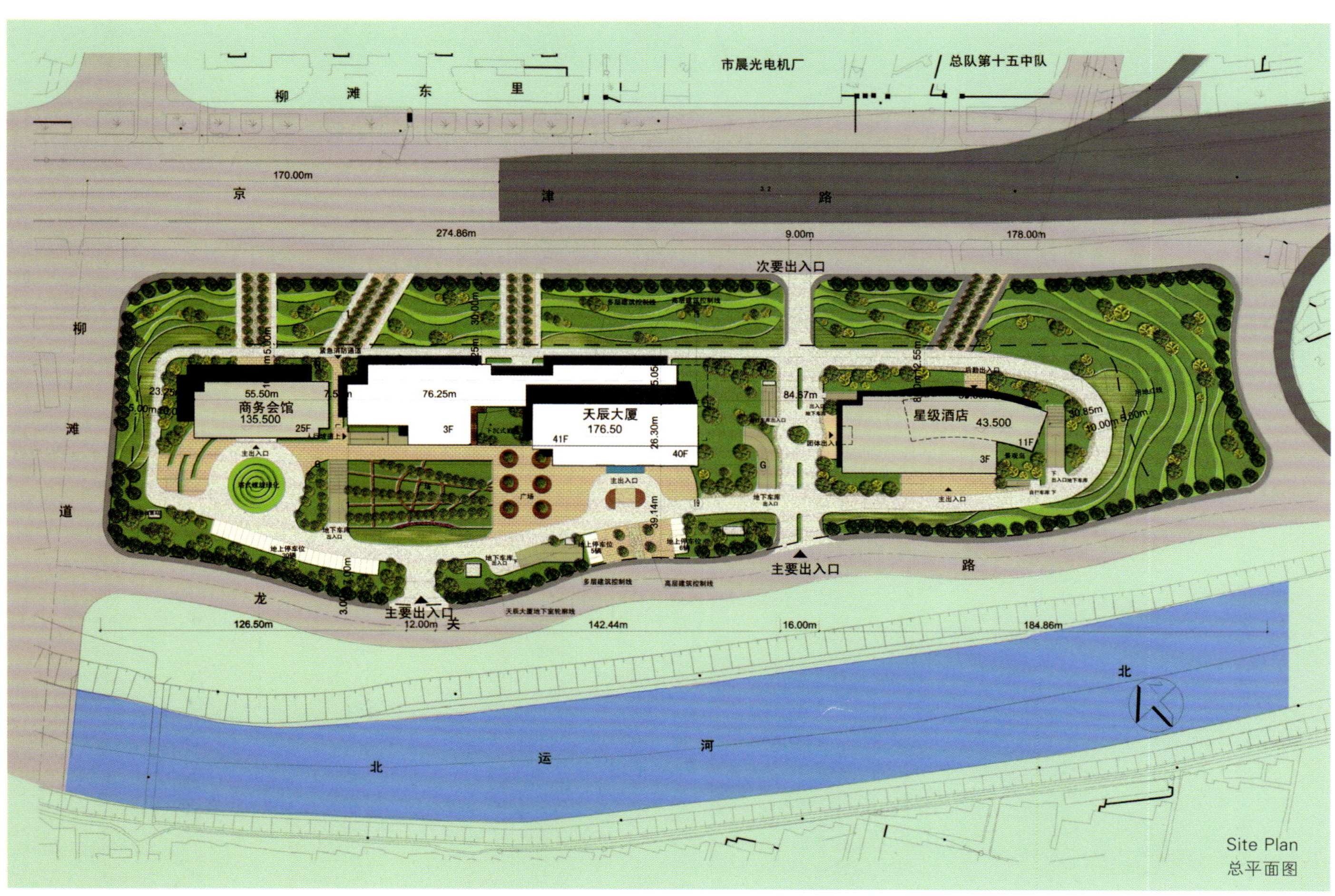

Site Plan
总平面图

South Elevation
南立面图

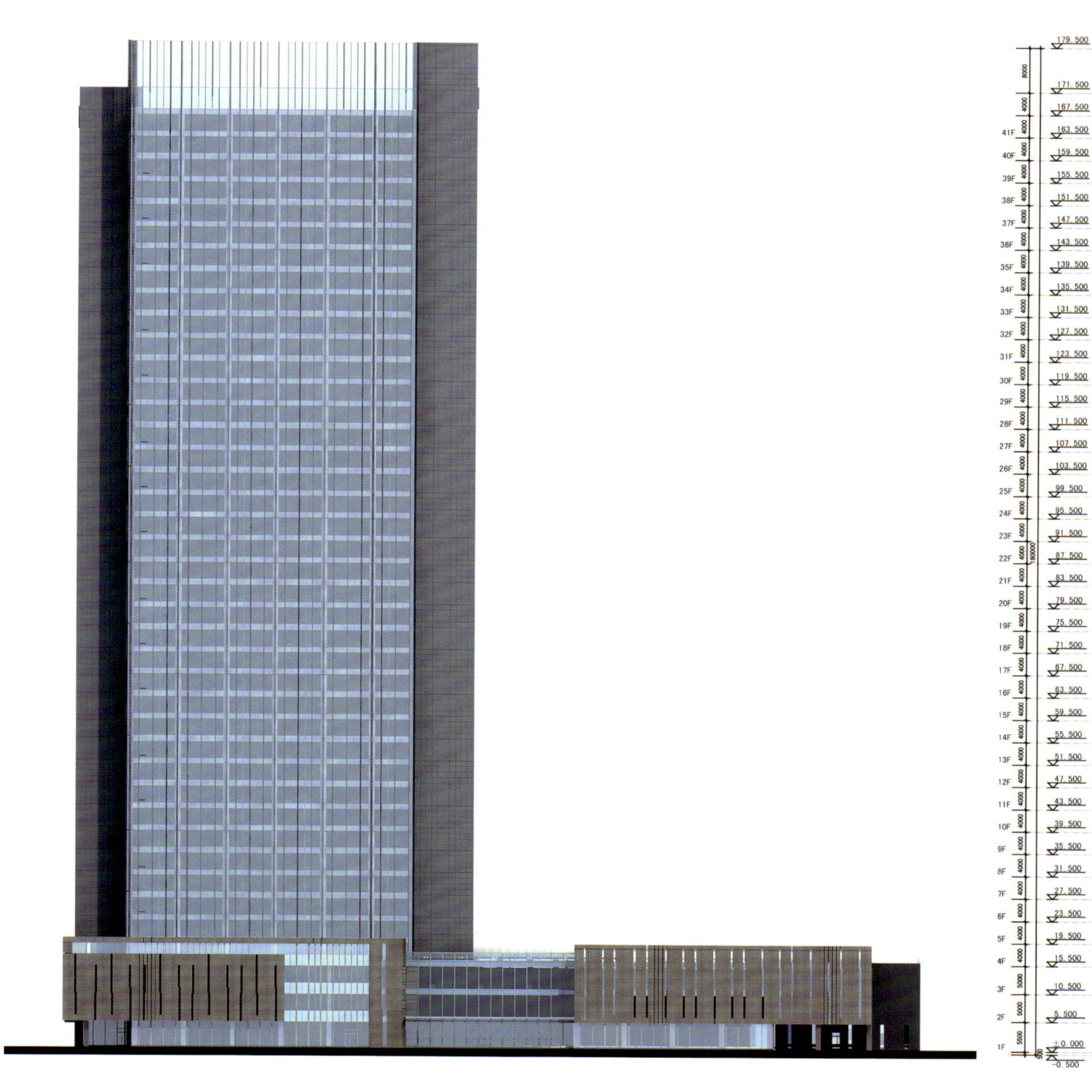

North Elevation
北立面图

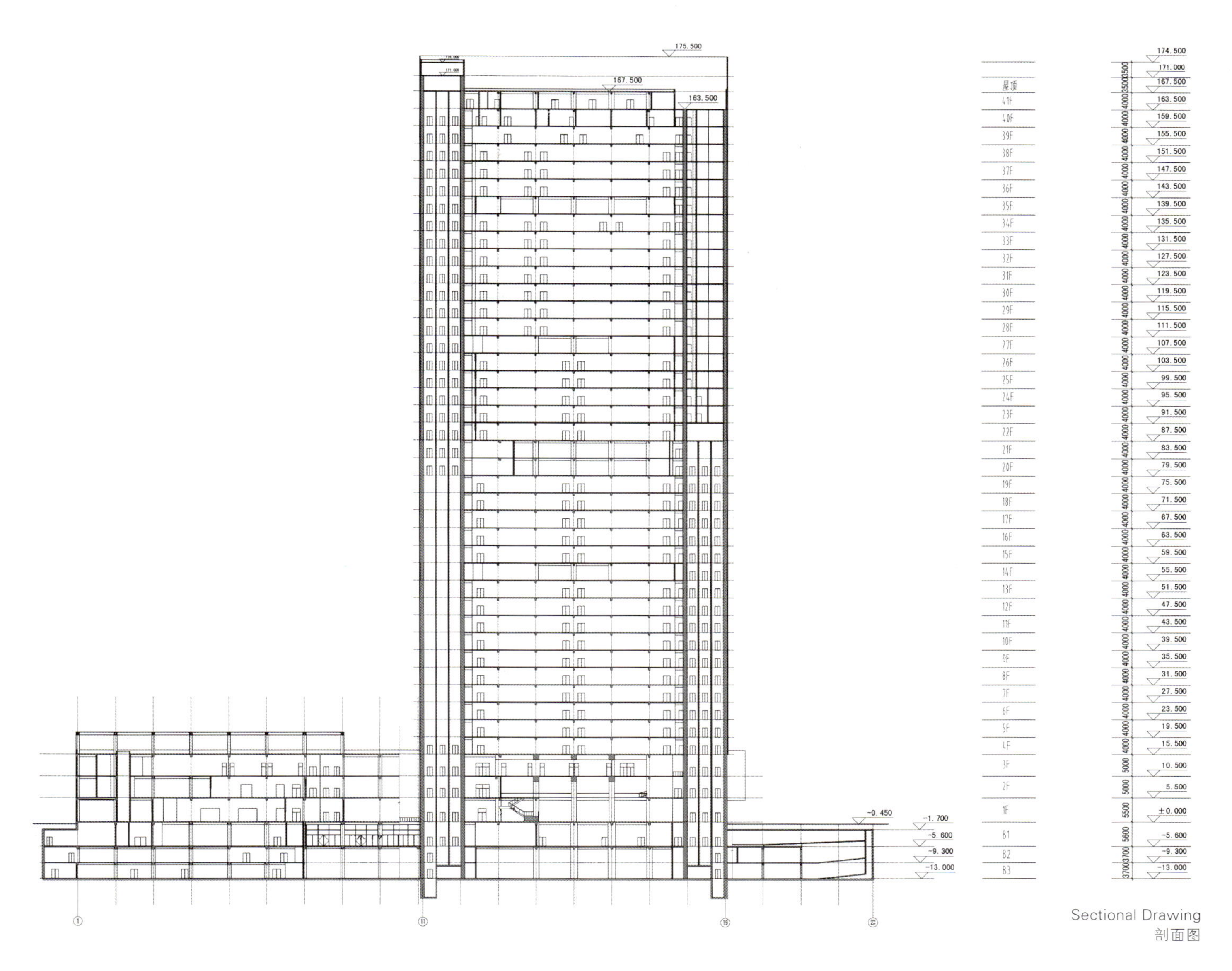

Sectional Drawing
剖面图

Architectural Design

Modern methods are used for the main buildings, while these two buildings in simple shapes are interlocked with each other and the exterior surfaces have expressed their plane function, stones used for the core tubes and glass curtain wall adopted for the offices have represented the light and delicate characteristics of modern office buildings. The top floor is designed almost same as the standard floors, but choosing different materials and more detailed components. The annex uses massive stones as main materials in modern methods that make the whole buildings light and bright, in the meanwhile, light blue glass are cooperated well with light grey stone and silver mental dividing lines to create decorative effect with a strong sense of modernization.

Selecting glass as the building skin, it is crystal and clear, coupling with delicate mental dividing lines and canopy to express the sense of modernization. The represented surreal beauty and modern atmosphere have fully met the enterprise spirit of Tianchen.

Adopting grey stones as exterior decorative materials of the annex to make it massive but also with quality, architects choose large glass for the first shop floor, and flexible windows for second and third floor, while use heavy materials for the base of main buildings to form sharp contrasts with light upper storeys.

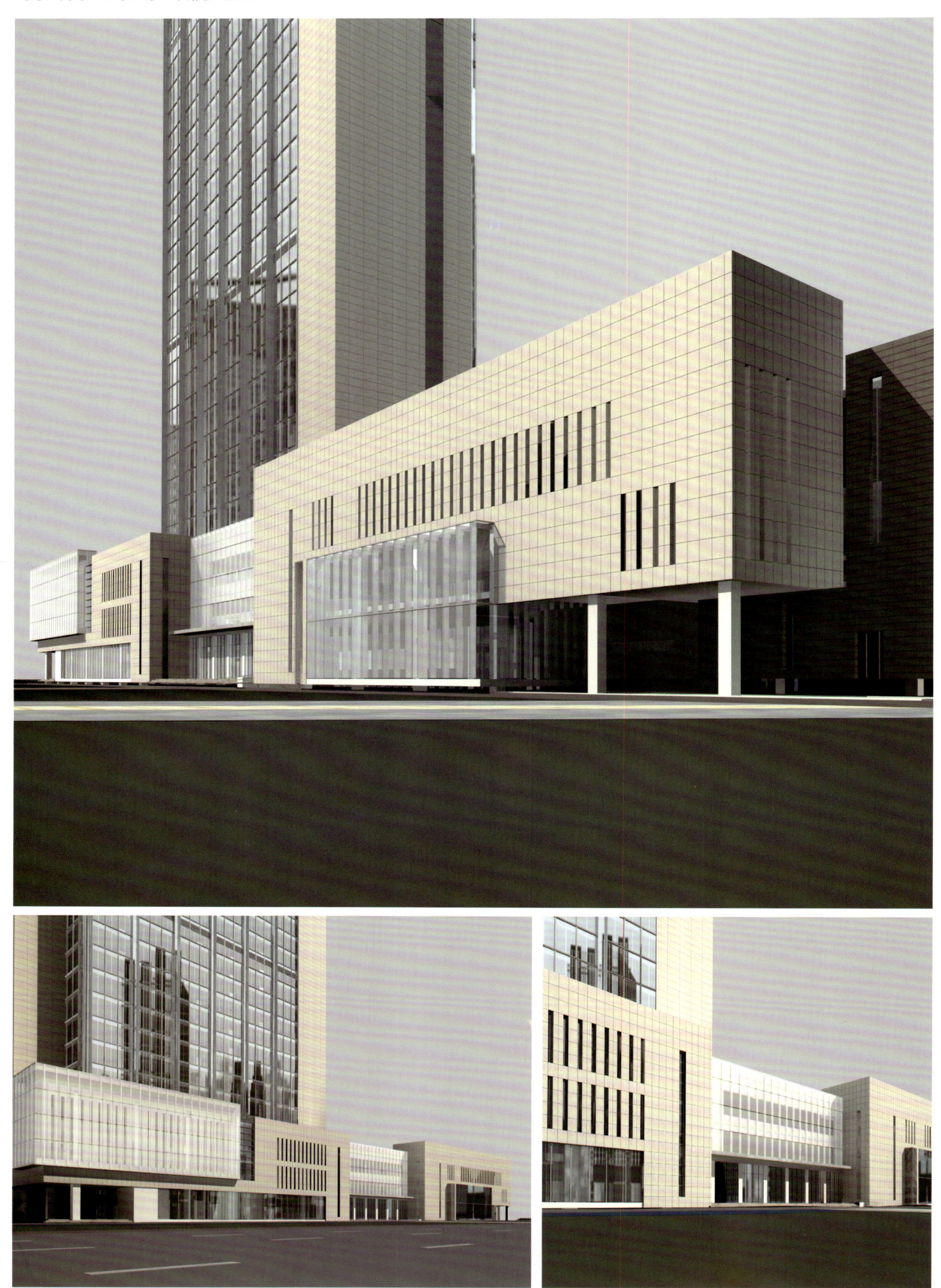

» 建筑设计

主楼采用极其现代的处理手法，形体上的两个极简单的形体互相咬合，外表皮充分反映了平面功能，核心筒部分采用石材，办公部分为明框玻璃幕墙，体现了现代办公楼的轻盈、精致。顶部处理与标准层基本一致，仅通过材质的变化和细部构件的处理来达到设计效果，裙房侧以较厚重的石材为主要材料，同样采用极现代的手法，整栋建筑物色彩清淡而明快，浅蓝色玻璃和浅灰色石材巧妙地搭配，加上壳银色的金属分割条，形成一种现代感极强的装饰效果。

玻璃作为建筑表皮，晶莹通透，细致的金属分割条和雨篷极具现代感。整个建筑表达的超现实美感和现代气息符合天辰的企业精神。

裙房采用灰色石材做外饰材料，厚重又富质感。低层为商铺，以大玻璃面为主，二、三层以灵活的开窗作为母题。厚重的材质作为主楼的基座，与轻巧的主楼形成对比。

» Functional Layout

Two office buildings are connected by the annex to share its functions. On the vertical design, the slope of the land extended to the annex's roof has made full use of architectural space as well as roof space, in the meanwhile, the green roof has also made its contribution to energy-saving of the buildings.

In order to make full use of plane, each building is adopted rectangle plane, while core tubes are placed separately to meet different requirements and to be manageable.

For the height of annex, it regards meeting requirement of functions as main element, cooperating facades with the vertical changes to make the facades full of its own characteristics.

North-south corridors are set inside the buildings to keep north and south in touch, and also make the facades more transparent.

Three-storey basement is designed as an integrated garage but with individual management for each storey.

» 功能布局

两个办公楼通过裙房连接，功能上考虑可为两个楼公用。在竖向设计上，通过斜坡从地面转换至裙房屋面，充分地利用了建筑空间和屋面空间。同时，绿色屋面也为建筑的节能创造了良好的条件。

建筑单体为平面利用率最高的长方形平面，考虑到不同功能的要求，核心筒分开设置以便于分别管理。

裙房的层高设计以满足功能要求为主要因素，外立面与竖向变化相结合，使得立面形态富有特点。

在建筑中部设置了穿越式的通道，既解决了基地南北向的交通联系，又使建筑外立面更有通透感。

地下室为三层，停车库为一体式设计，但可分别管理。

Vernacular City—Xinyu Vogue Photoelectricity Factory, Jiangxi

乡土城市——江西新余沃格光电厂区

Features 项目亮点

The project is trying to clear up the boundary between city and country, to break the conceptual barrier of industry and nature and to connect the mechanical industrial manufacture and enjoyable nature together.

项目试图消解城市和乡村的界限，试图打破工业和自然之间的概念壁垒，试图将枯燥机械的工业生产与享受自然的环境拼接在一起。

Keywords 关键词

Rural Flavor
田园气息

Visit Corridor
参观走廊

Vernacular Flavor
乡土气息

Location: Xinyu, Jiangxi, China
Architectural Design: aYa Arch Studio
Architects: Xu Yixing, Xue Shanshan, Chen Zhebin, Huang Lingxuan, Shen Jun
Land Area: 29,400 m^2
Design Time: May, 2010
Completion Date: May, 2012

项目地点：中国江西省新余市
设计单位：aYa 阿尼那建造生活
设计师：许义兴 薛珊珊 陈浙彬 黄龄萱 申俊
占地面积：29 400 m^2
设计时间：2010 年 5 月
竣工时间：2012 年 5 月

» Overview

This project is for a factory that mainly producing hi-tech glass products and its main functions are for production and processing, dormitory, office, restaurant, entertainment and so on. It used to be the farmland, Xinyu is becoming a city which is industrialized and urbanized rapidly in a context of processing industry moving from southeast China to southwest, the green land in the country has become industrial district that is full of factories, while the industrial district and the rest of country are connected strangely to form the landscape with special humanity desertification. In this case, architects decide to choose a balanced method to reach harmony between these two landscapes.

» 项目概况

项目是一个以生产高科技玻璃产品为主的工厂，主要功能是生产加工、住宿、办公、饮食、娱乐等，厂区用地原为乡村耕地，在东南沿海加工业向西南迁移的大背景下，江西新余成为一个承接这些加工业的城市，迅速的工业化和城市化，使乡村的田园绿野巨变为厂房林立的工业区，工业区和尚未改变的乡村陌生地拼接在一起，形成一个特殊的人文荒漠化的景观。在这个现实面前，设计师们选择用一个中间的策略，试图让两种景观融洽地共处。

» Design Idea

Vernacular "city"—as a stage of urbanization, industrial district is transforming from country and has cut off the inheritance of regional cultural traditions. It is the responsibility for the environment design to repair the damage and bring the environmental quality of country with the harmony between human and nature into the factory. In the meanwhile, as the main users, namely the workers are mostly from country, the rural flavor in the factory will remind them their hometowns and release the mental depression and tension that caused by mechanical work and boring interpersonal relationships to achieve physical and psychological health.

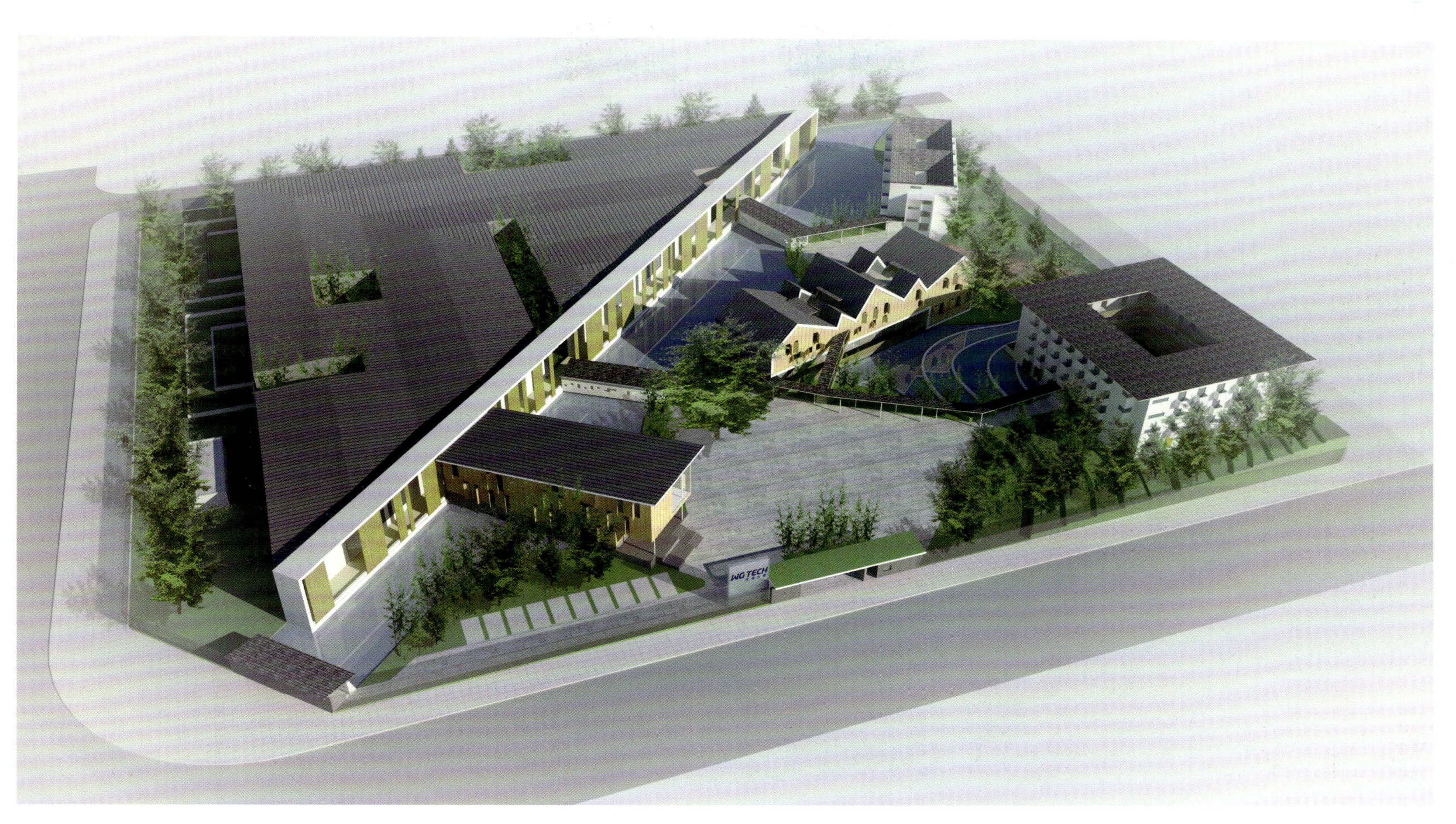

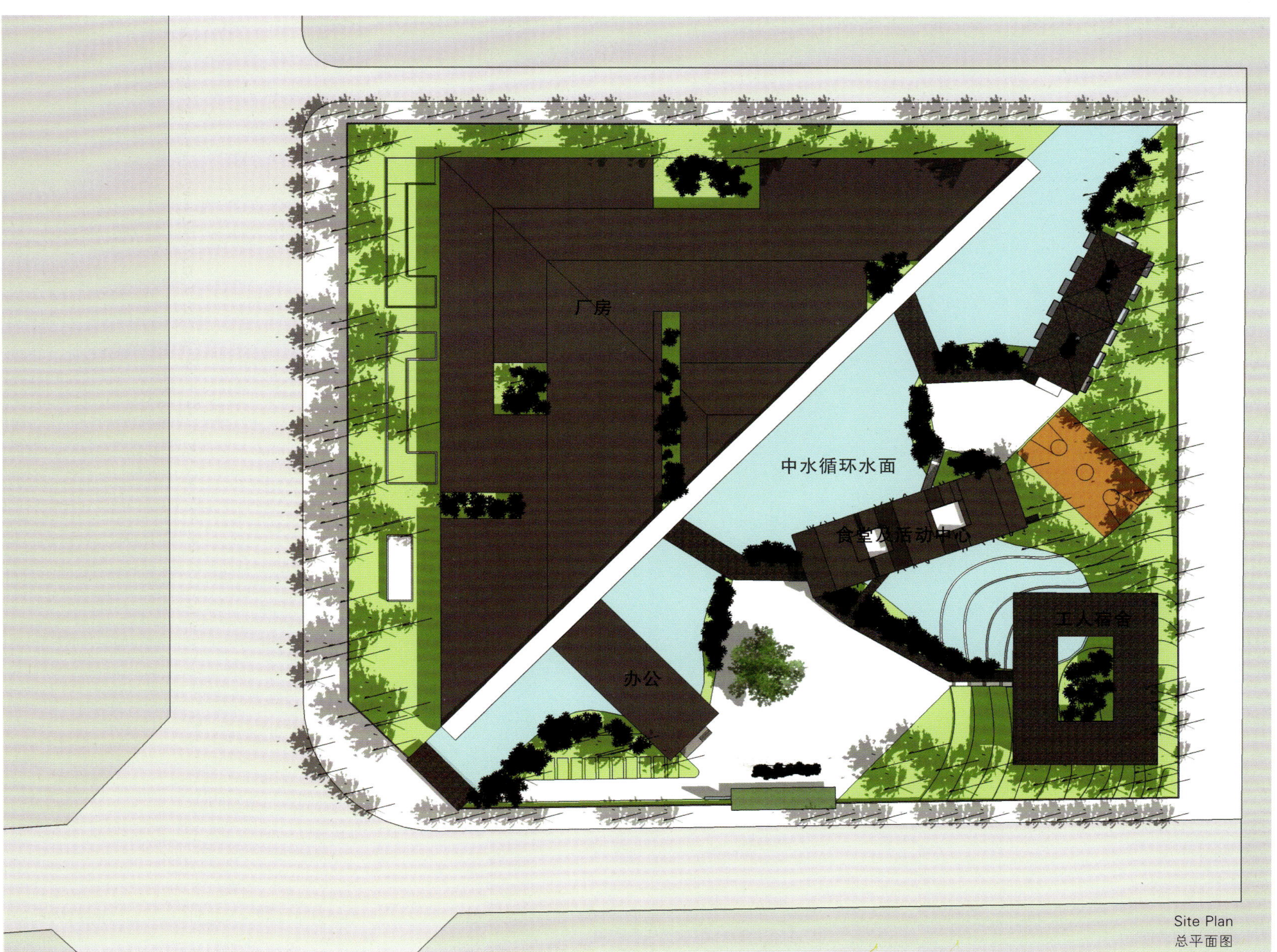

Site Plan
总平面图

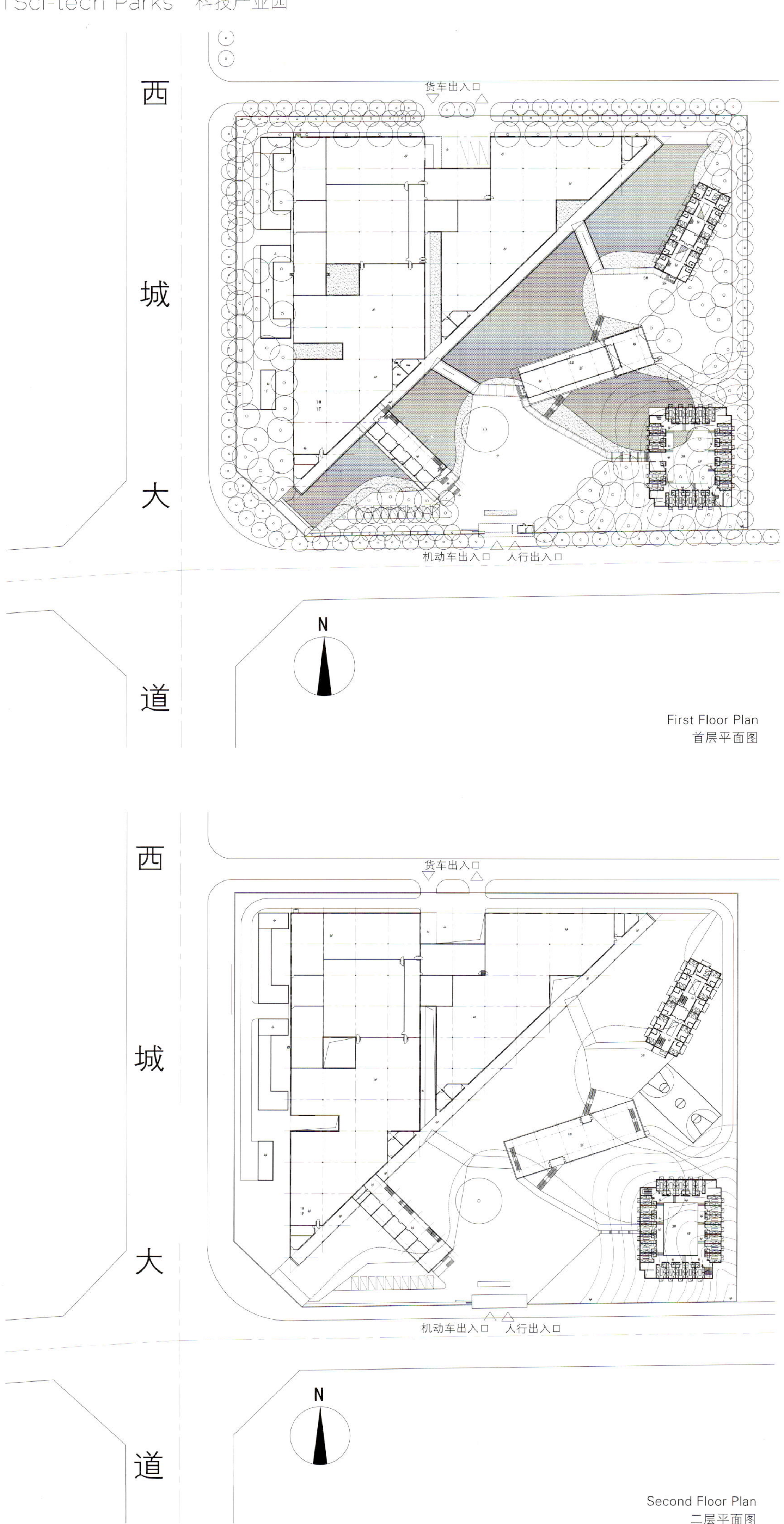

First Floor Plan
首层平面图

Second Floor Plan
二层平面图

设计理念

乡土的“城市”——工业区作为城市化的一个阶段，从乡村空间巨变而来，从而割裂了地域文脉和传统文化的传承。有责任的环境建设，应该修复这种断层，把乡土的环境质量因人们与自然融合为一体的本能带回厂区内。同时也因为主要的使用者——打工者大多来自乡村，工厂环境的乡土气息能够为他们带来故乡的抚慰，平衡机械的劳动和简单的人际关系所形成的心理抑郁和紧张，增进身心健康。

Functional Layout

A diagonal has divided the project into two parts—“city” and “country”. The primary triangle workshop is set in the corner with the long edge facing to another part, while this long edge is designed to be a long corridor with a height of over 10 m as a medium of transportation and landscape of two parts. A pond close to the corridor shows the reflection of both banks, the bridge that over the pond is the passageway connecting two different landscape, it forms another kind of landscape while workers walking on the corridor and bridge. Offices are connected with primary workshop and across the pond, and both sides of the offices can enjoy the landscape of the pond. The placements of canteen, activity room and administrative staff dormitory are echoing with landscape of pond and corridor. High-rise standard workshop in the southeast corner has placed its auxiliary functions to the north and west to keep out the western sun and cold air and then lead eastern and southern sunshine and air into the factory. For the space between buildings, architects create landscape that is full of rural flavor to make the “country” zone to be a place that workers can relax and recall the memory of their hometowns. While for the space in the workshop, architects embed a landscape corridor that likes the patio of southern China as a visit path, as well as for the workers to enjoy natural landscape when working.

功能布局

项目的场地沿对角线分成两部分——“城市”与“乡村”，主要的大厂房呈三角形偏于一隅，长边面对另一半场地，这个长边被赋予一条长长的10多米高的外廊，作为厂房的交通和两部分景观沟通的媒介。水池依附外廊获得两岸两个方向的倒影，跨越水池的桥即是两种环境切换的过程通道，工人在外廊和桥上的行走成为一种景观。办公室按功能要求与大厂房连接并横跨水池，两侧都可以借水池风景。食堂和活动室以及管理人员宿舍分别对应水池和长廊的景观布置。东南角高起的标准厂房，将附属功能独立到建筑的北侧和西侧，作为遮挡西晒和冷空气的屏障，将东向和南向的阳光空气导入。在建筑之间的区域，营造田园气息的园林景观，让“乡村”区域成为工人可以放松休息忆念故土的地方。在厂房中间，嵌入一条带有江南天井意向的参观走廊，是外来参观的路线，也是工人在工作中可以看见自然的空间。

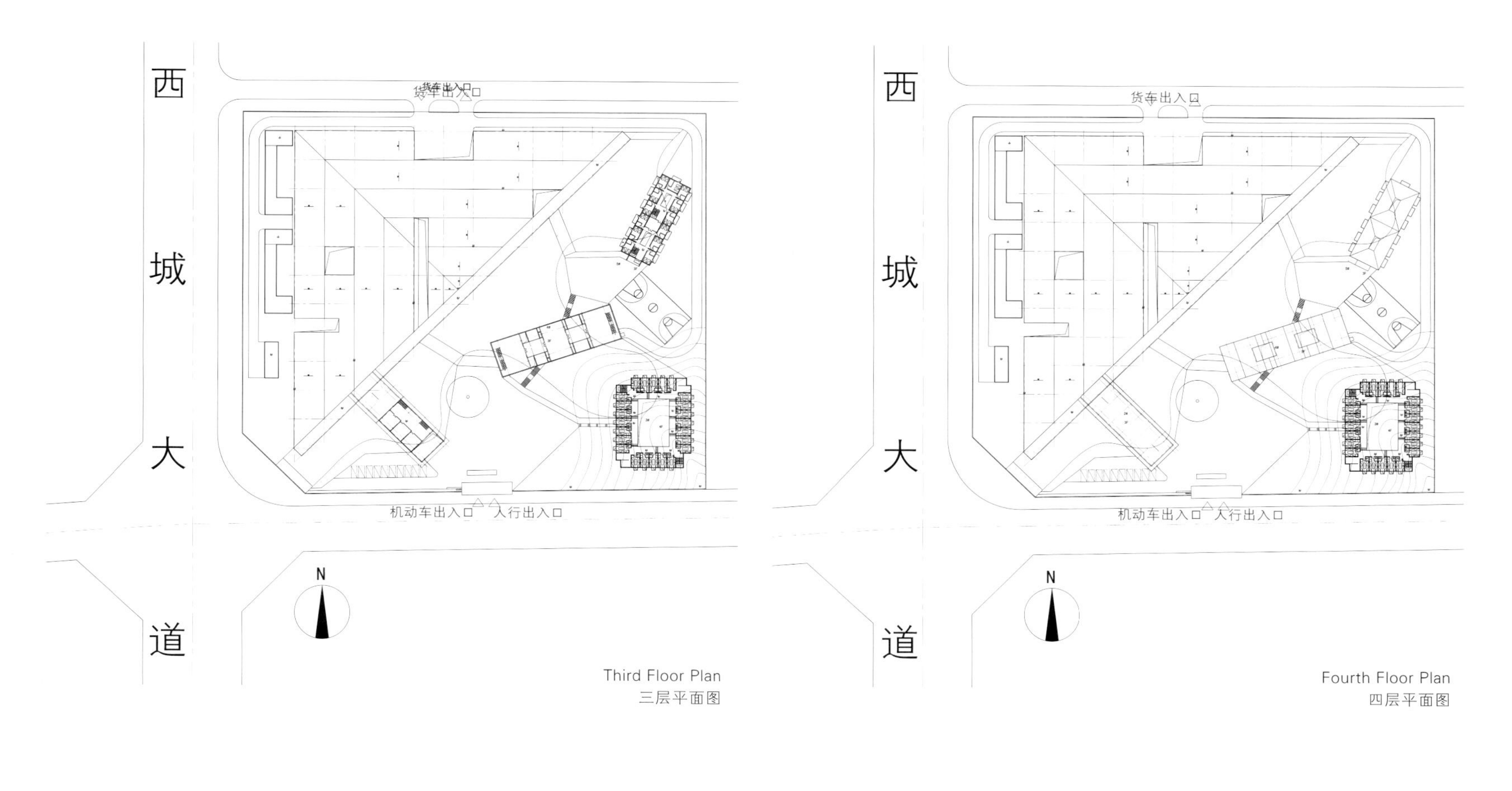

Third Floor Plan
三层平面图

Fourth Floor Plan
四层平面图

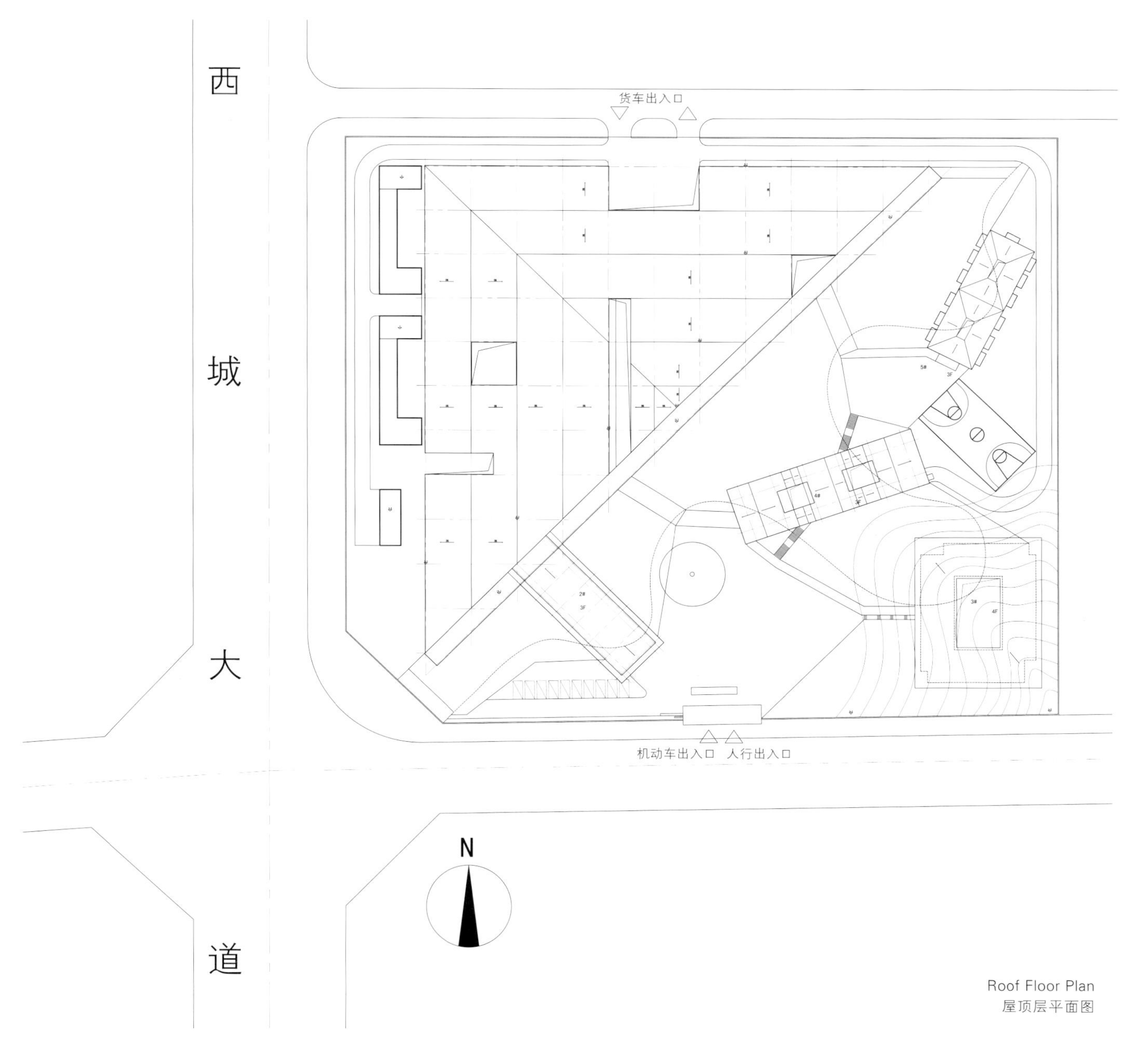

Roof Floor Plan
屋顶层平面图

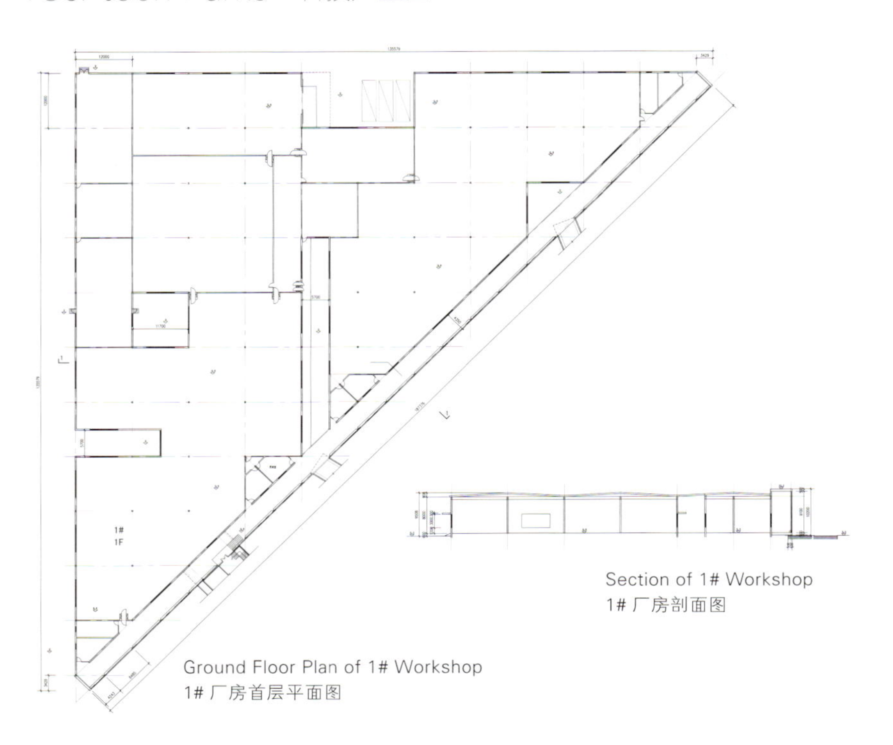

Section of 1# Workshop
1# 厂房剖面图

Ground Floor Plan of 1# Workshop
1# 厂房首层平面图

Mezzanine Plan of 1# Workshop
1# 厂房夹层平面图

Southeast Elevation of Workshop
厂房东南立面图

Ground Floor Plan of 2# Office Building
2# 办公室 首层平面图

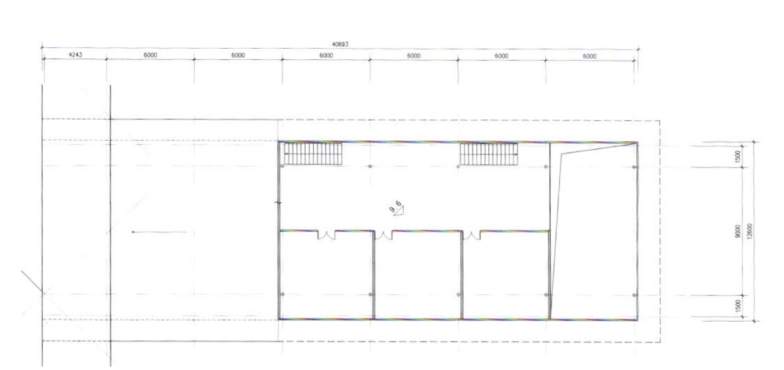

3rd Floor Plan of 2# Office Building
2# 办公室 三层平面图

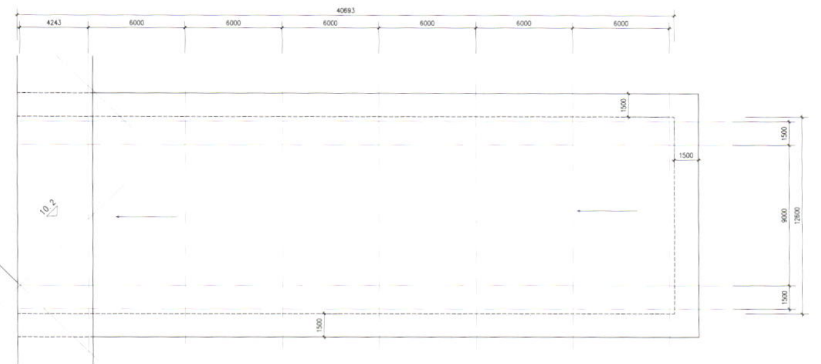

Roof Plan of 2# Office Building
2# 办公室 屋顶层平面图

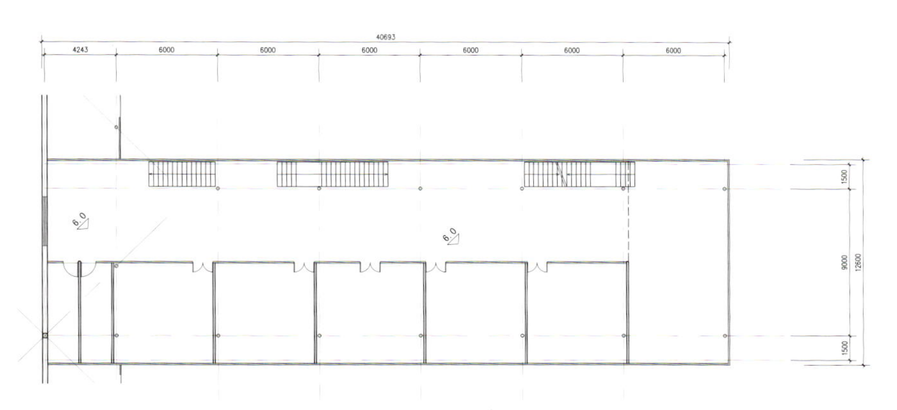

2nd Floor Plan of 2# Office Building
2# 办公室 二层平面图

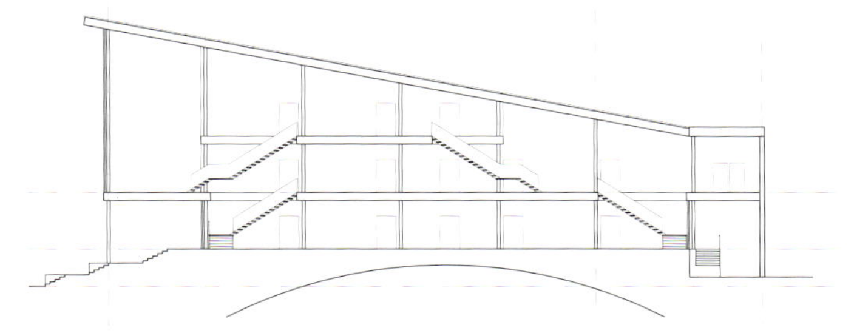

Section 1–1 of 2# Office Building
2# 办公室 1–1 剖面图

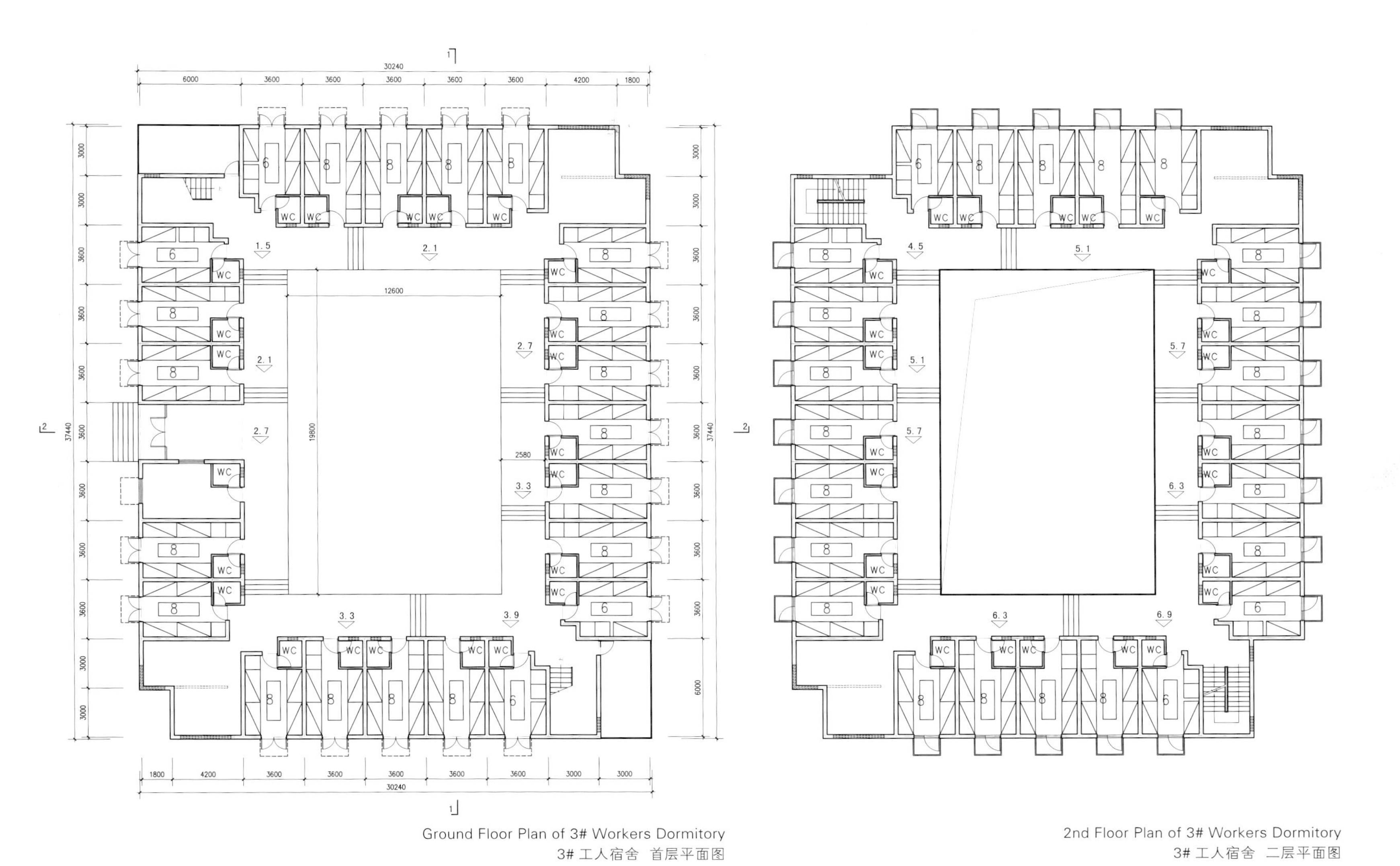

Ground Floor Plan of 3# Workers Dormitory
3# 工人宿舍　首层平面图

2nd Floor Plan of 3# Workers Dormitory
3# 工人宿舍　二层平面图

3rd Floor Plan of 3# Workers Dormitory
3# 工人宿舍　三层平面图

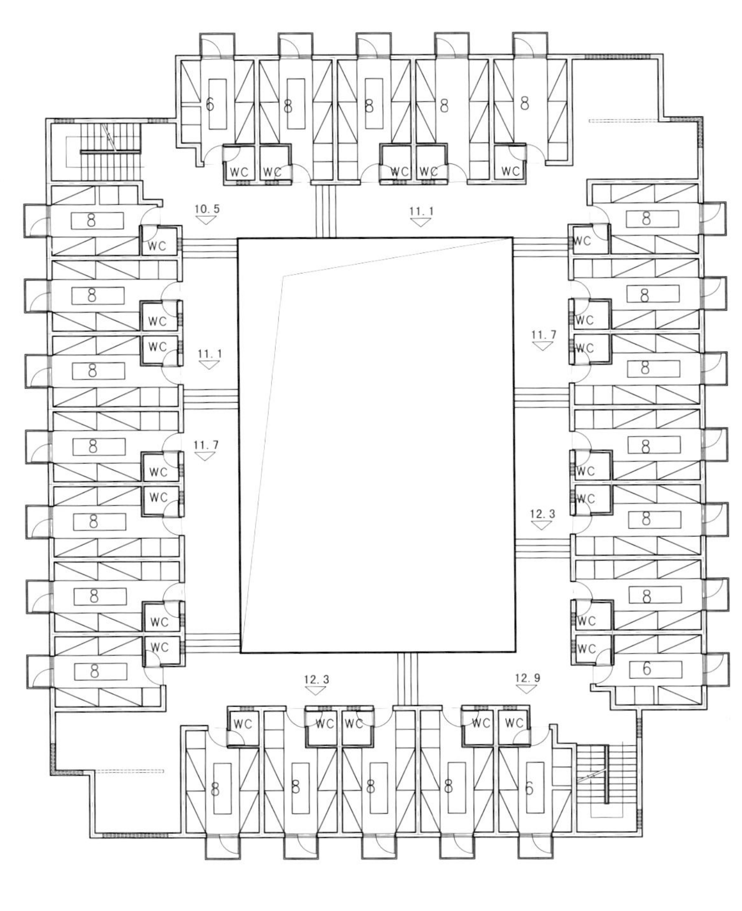

4th Floor Plan of 3# Workers Dormitory
3# 工人宿舍　四层平面图

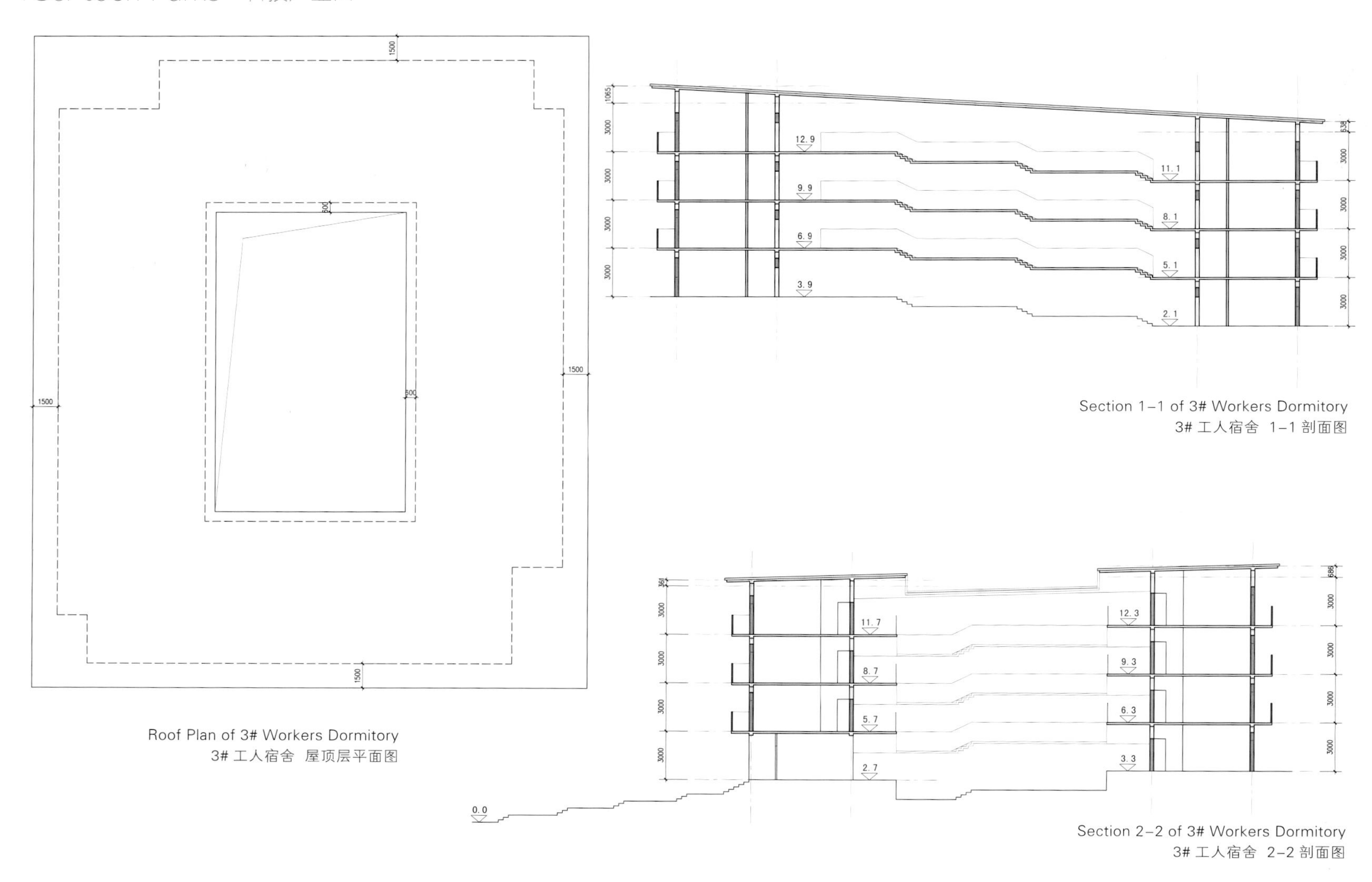

Roof Plan of 3# Workers Dormitory
3# 工人宿舍 屋顶层平面图

Section 1–1 of 3# Workers Dormitory
3# 工人宿舍 1–1 剖面图

Section 2–2 of 3# Workers Dormitory
3# 工人宿舍 2–2 剖面图

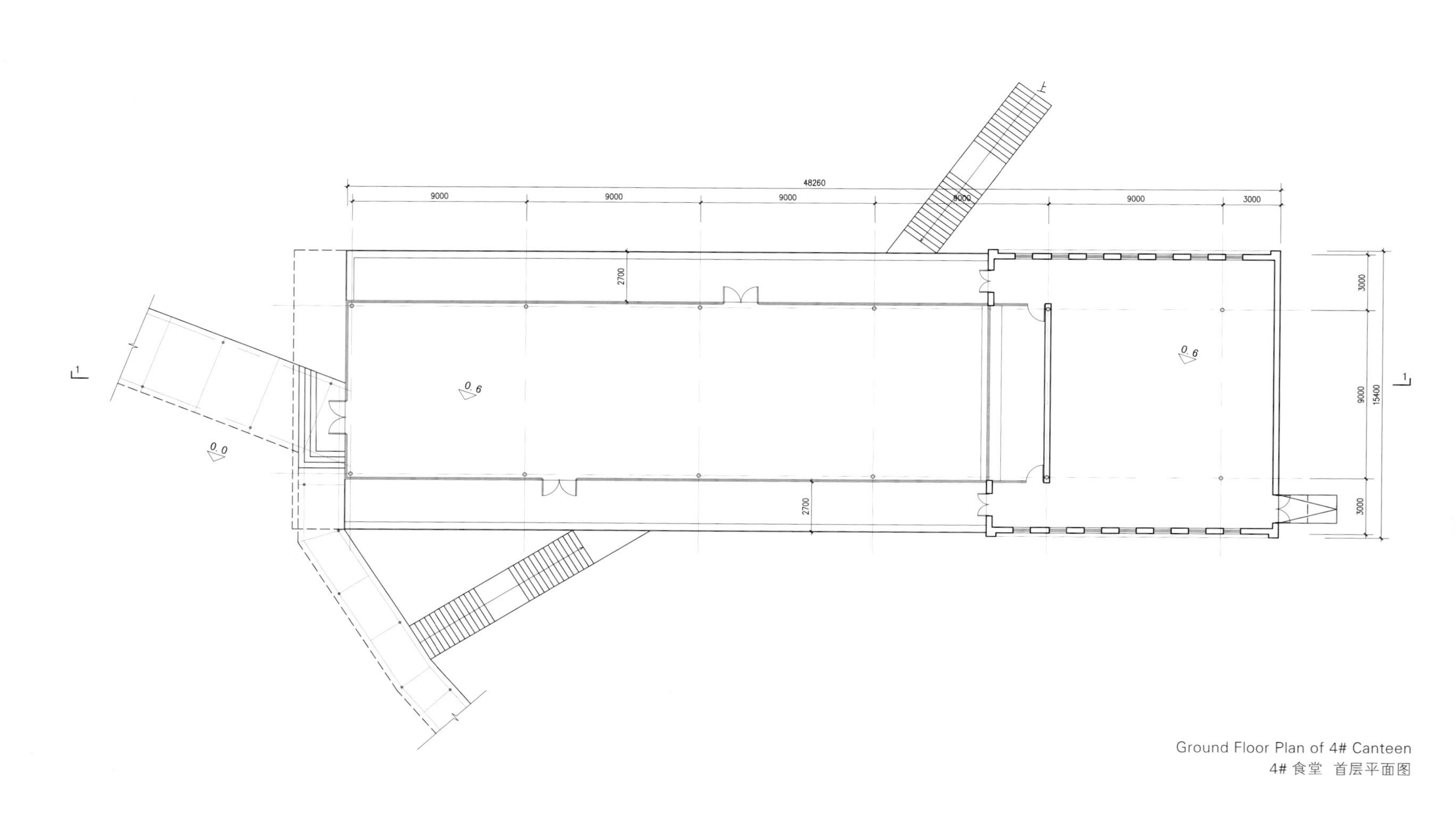

Ground Floor Plan of 4# Canteen
4# 食堂 首层平面图

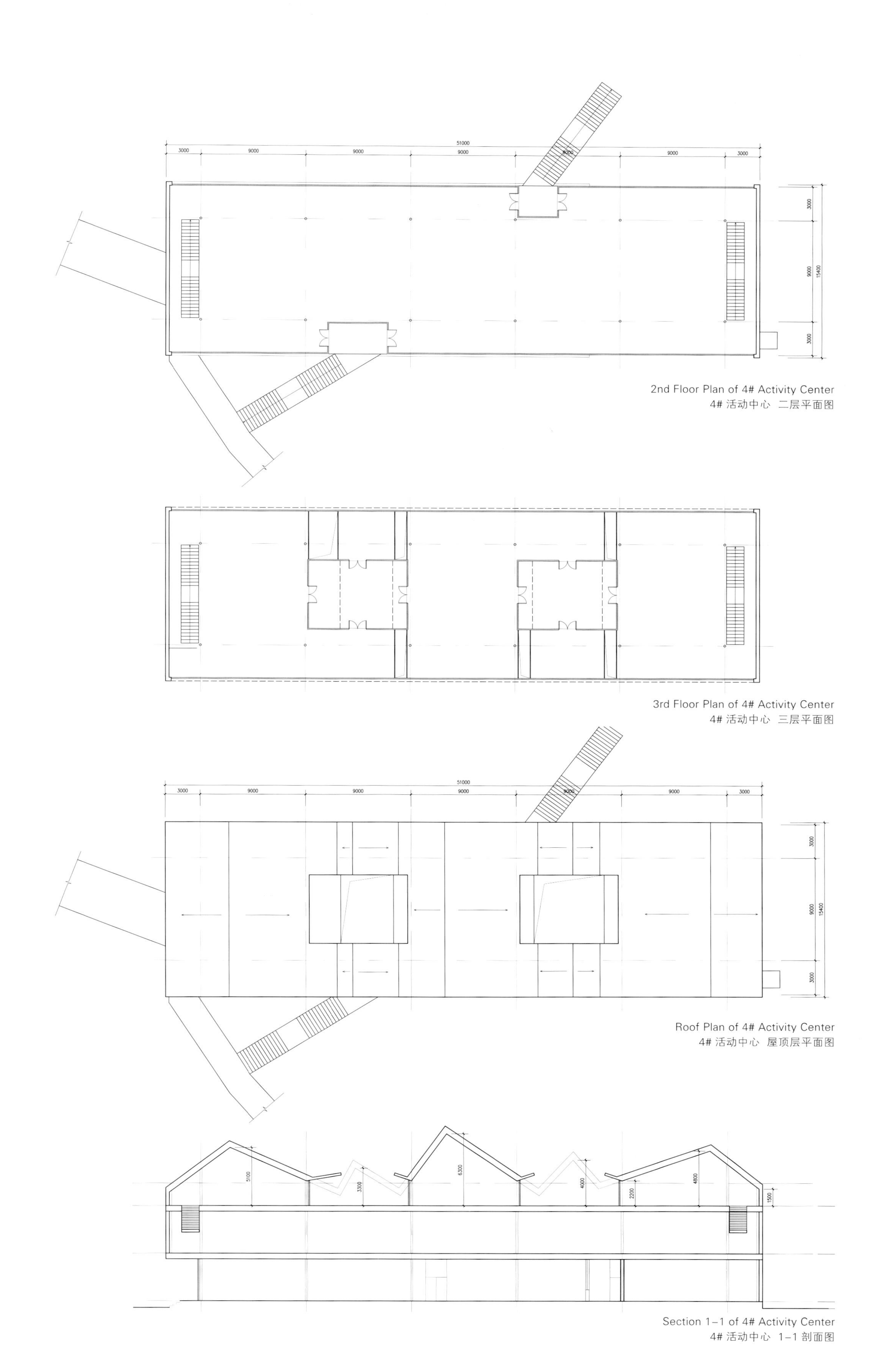

2nd Floor Plan of 4# Activity Center
4# 活动中心 二层平面图

3rd Floor Plan of 4# Activity Center
4# 活动中心 三层平面图

Roof Plan of 4# Activity Center
4# 活动中心 屋顶层平面图

Section 1–1 of 4# Activity Center
4# 活动中心 1–1 剖面图

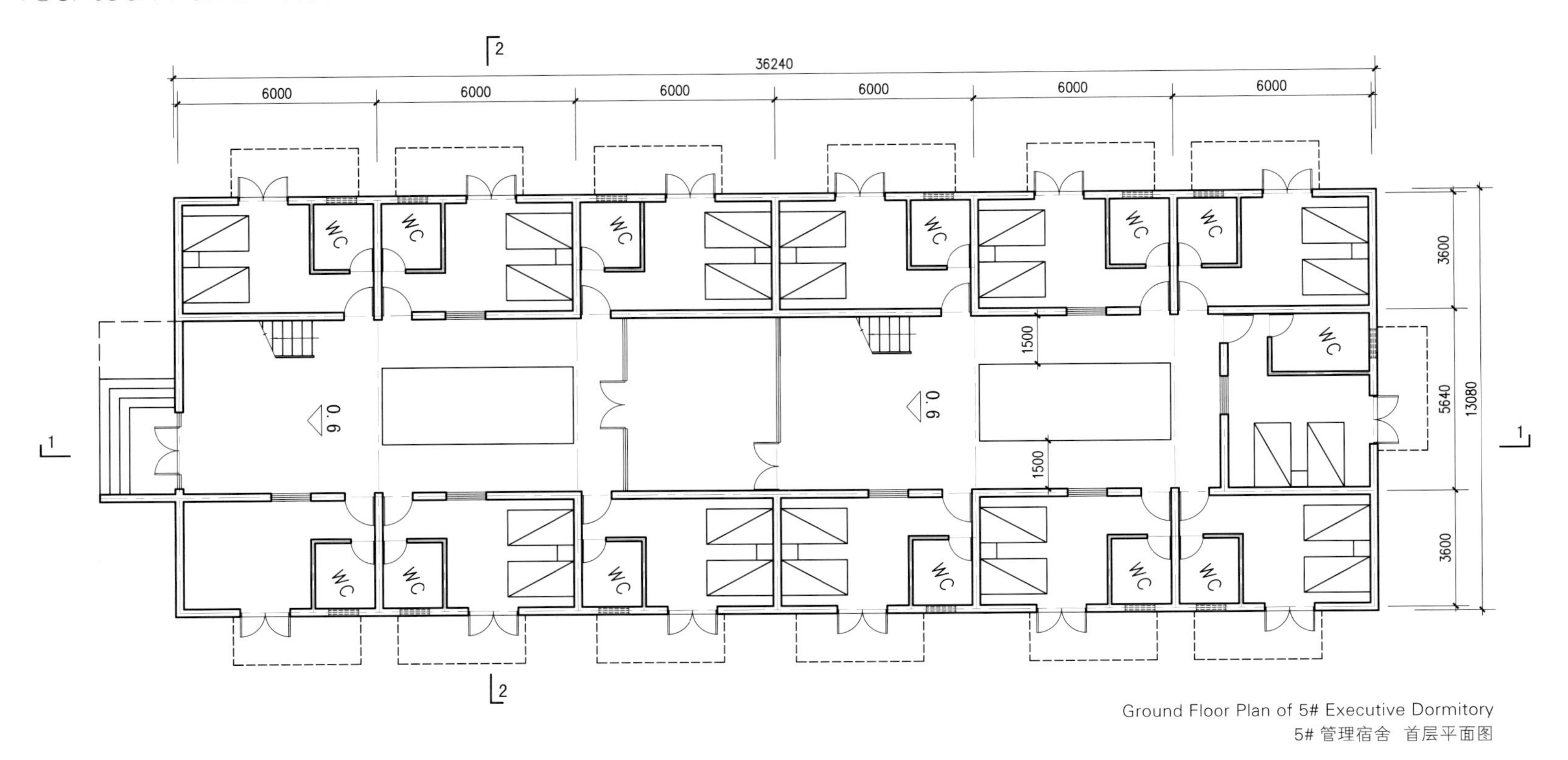

Ground Floor Plan of 5# Executive Dormitory
5# 管理宿舍 首层平面图

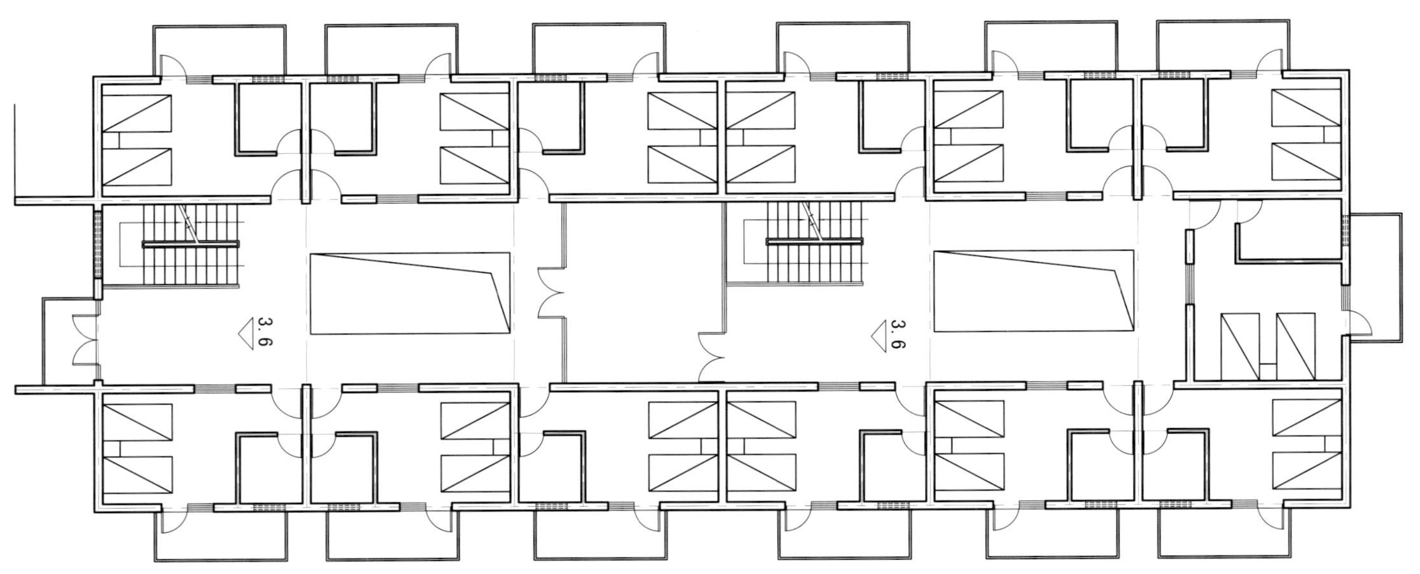

2nd Floor Plan of 5# Executive Dormitory
5# 管理宿舍 二层平面图

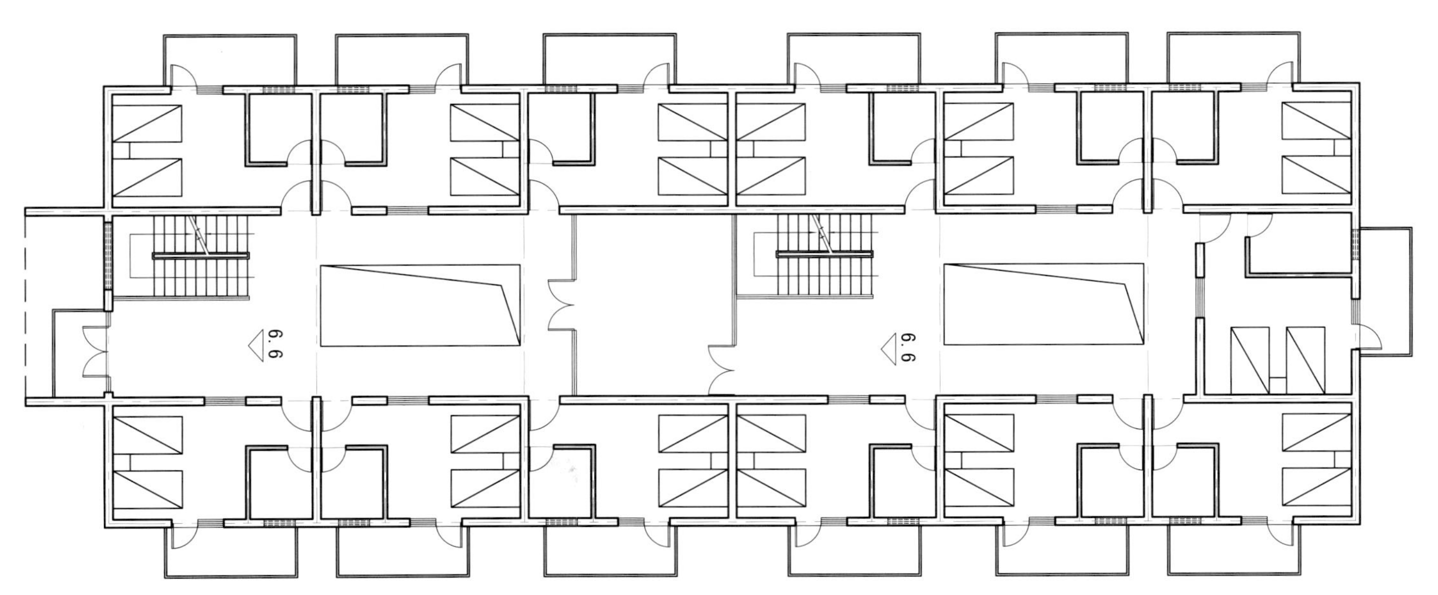

3rd Floor Plan of 5# Executive Dormitory
5# 管理宿舍 三层平面图

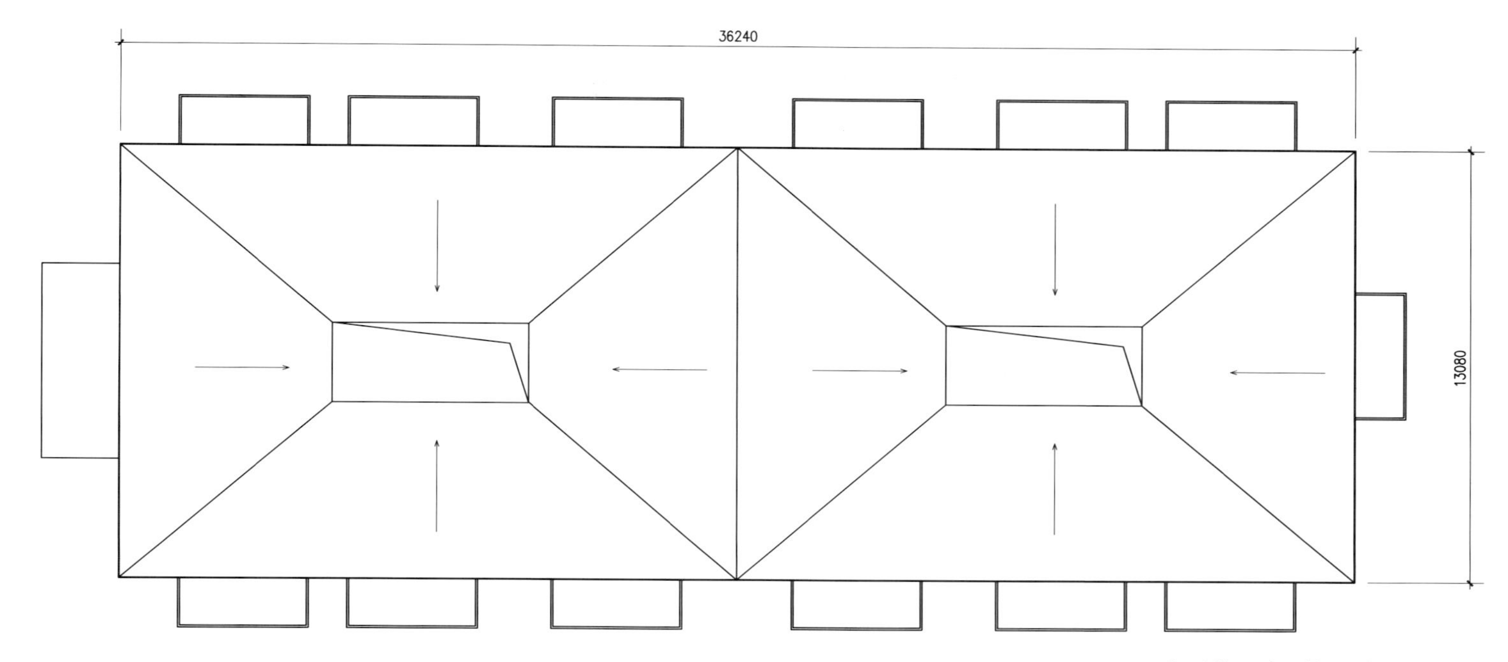

Roof Plan of 5# Executive Dormitory
5# 管理宿舍 屋顶层平面图

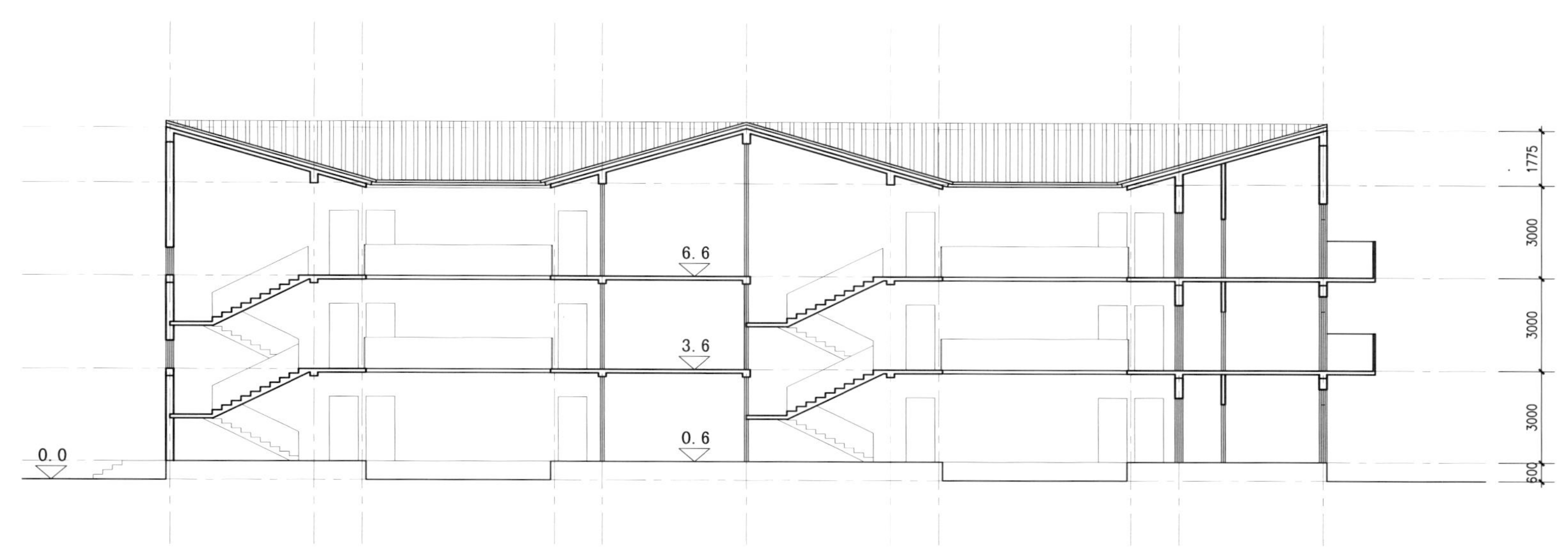

Section 1–1 of 5# Executive Dormitory
5# 管理宿舍 1–1 剖面图

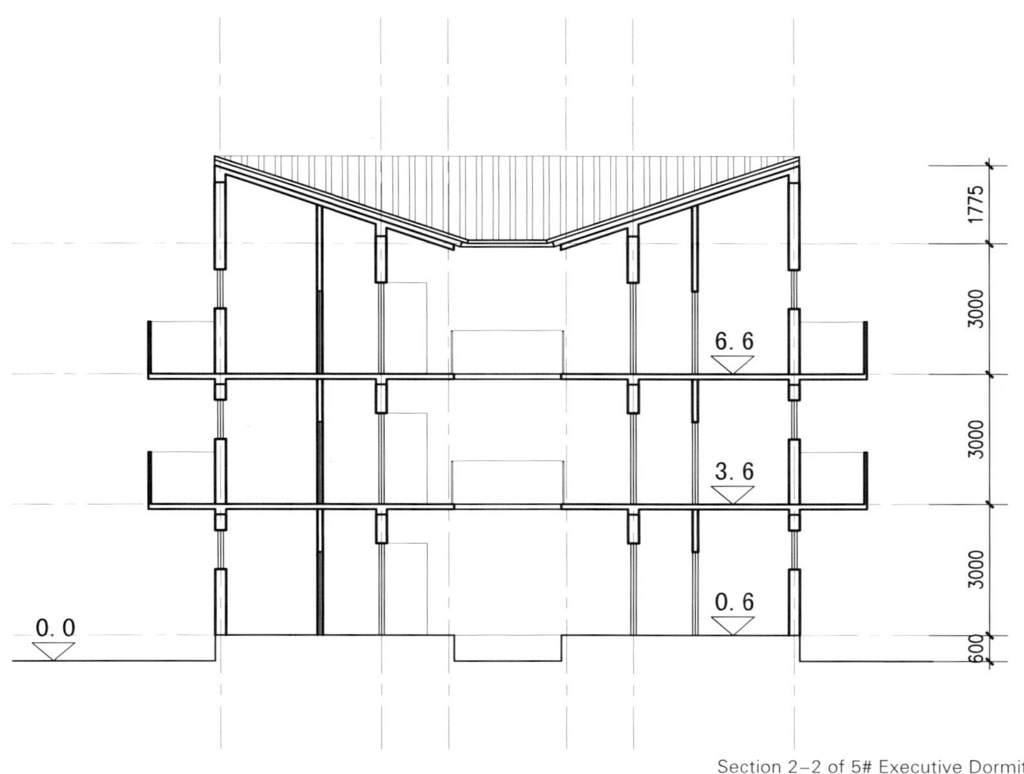

Section 2–2 of 5# Executive Dormitory
5# 管理宿舍 2–2 剖面图

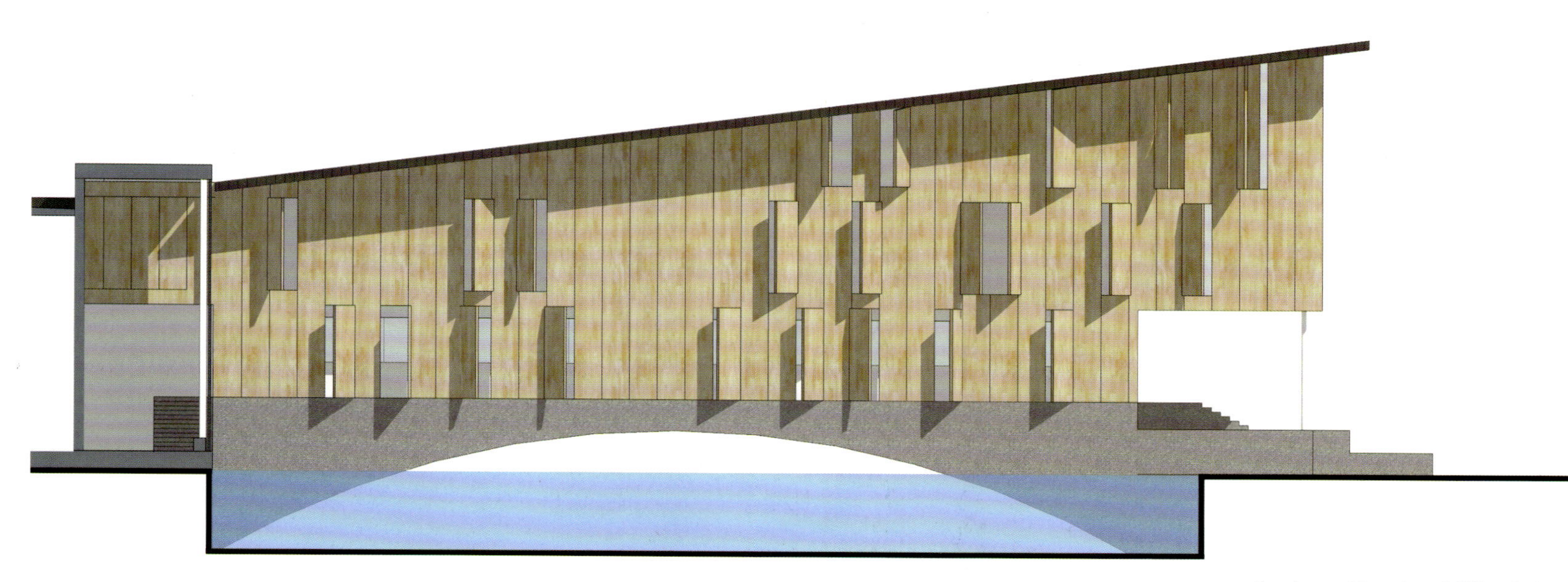

Southwest Elevation of Office Building
办公楼　西南立面图

West Elevation of Staff Dormitory
员工宿舍　西立面图

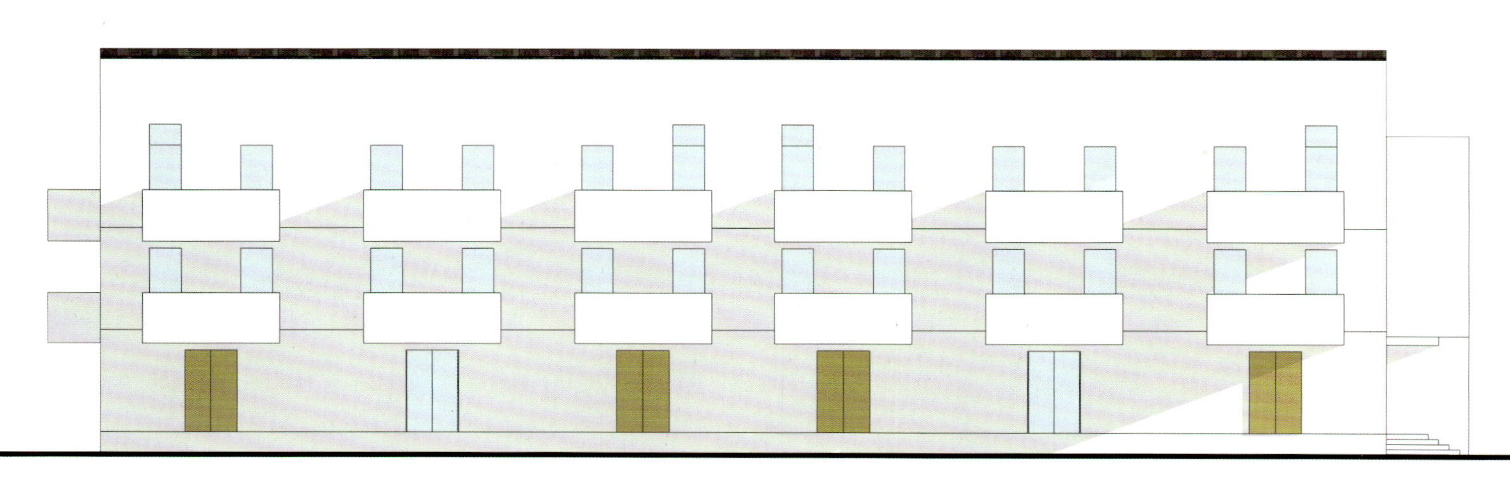

Northwest Elevation of Executive Dormitory
管理宿舍　西北立面图

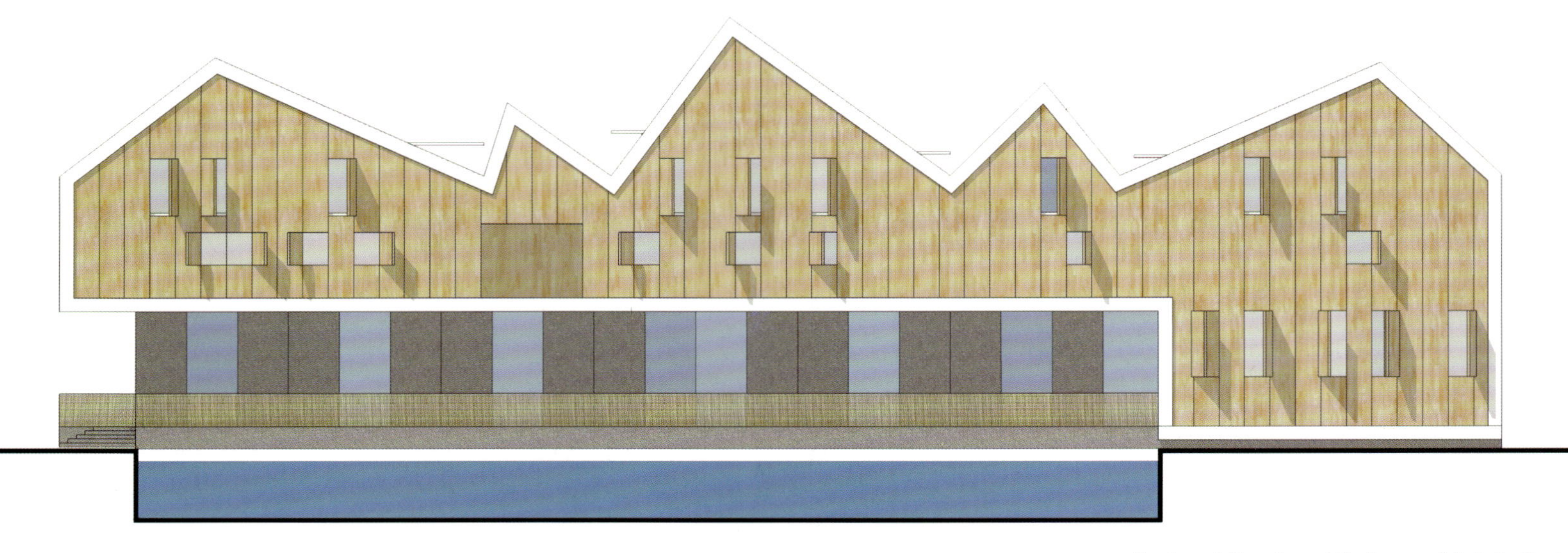

Southeast Elevation of Canteen and Activity Center
食堂与活动中心 东南立面图

Peak International Outsourcing Industrial Park, Huaqiao, Kunshan

颠峰（昆山花桥）国际外包产业园

Features 项目亮点

The aim of this planning is to create a modern industrial park with strong symbolism and neat and orderly regional boundaries; to properly lead in and collocate the landscape resources of Jimingtang River and to provide a maximum shared landscape for the office and living areas.

本规划旨在创造一个区域边界严整有序、地段标志性强的现代化产业园区；合理地引入和配置鸡鸣塘河景观资源，为办公和生活区提供最大化的景观共享。

Keywords 关键词

River Landscape
河流景观

Neat & Orderly
严整有序

Strong Symbolism
地标性强

Location: Suzhou, Jiangsu, China
Architectural Design: 9town-studio
Floor Area: 115,206 m^2
Design Time: November, 2010
Status: Under Construction

项目地点：中国江苏省苏州市
建筑设计：九城都市建筑设计有限公司
建筑面积：115 206 m^2
设计时间：2010 年 11 月
项目状态：在建

Overview

Peak International Outsourcing Industrial Park, is located in Huaqiao, Kunshan, at the junction of Jiangsu and Shanghai and near the Wusong River and the Jimingtang River. This plot is with rich waterfront resources.The site is just at the first node of Shanghai-Nanjing economic corridor—the main channel of Shanghai's external economy, and is only 25 km to Hongqiao Airport and 80 km to Pudong Airport. The business types of this industrial park will include knowledge outsourcing, financial BPO, IT infrastructure service and so on.

项目概况

颠峰（花桥）国际外包产业园基地地处苏沪交界处的昆山市花桥镇，临近吴淞江和鸡鸣塘河，地块滨水资源丰富。基地正好处于沪宁经济走廊这一上海对外经济主通道的第一节点，距虹桥机场仅 25 km，浦东机场 80 km。产业园业态将涵盖知识外包业务、金融 BPO 业务以及 IT 基础设施服务等业务。

Functional Layout

The site is divided into four basic functional areas: high-density office area, low-density office area, central office area and living area. The aim of this planning is to create a modern industrial park with strong symbolism and neat and orderly regional boundaries; to properly lead in and collocate the landscape resources of Jimingtang River and to provide a maximum shared landscape for the office and living areas. By integrating the living area with the office area, the design manages to introduce the urban public space into the industrial park, and then creates a sci-tech industrial park with distinct regional and industrial features.

功能布局

基地分为四个基本的功能区：高密度办公区、低密度办公区、核心办公区和生活服务区。本规划旨在创造一个区域边界严整有序、地段标志性强的现代化产业园区；合理地引入和配置鸡鸣塘河景观资源，为办公和生活区提供最大化的景观共享；促进生活区与办公区的适度混合、将城市公共空间引入产业园区；从而打造一个具有显著地域特点与产业特点的科技产业园区。

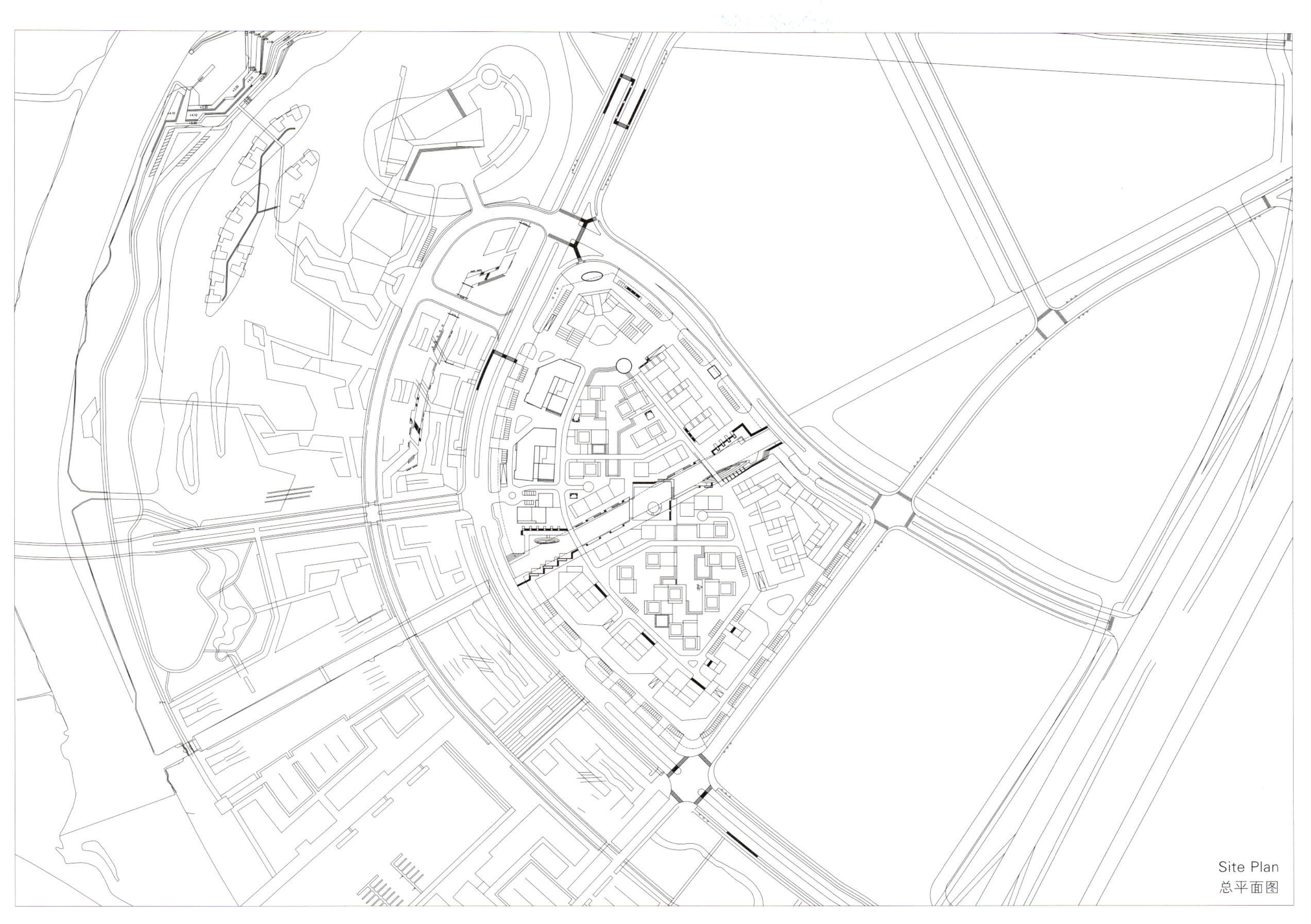

Site Plan
总平面图

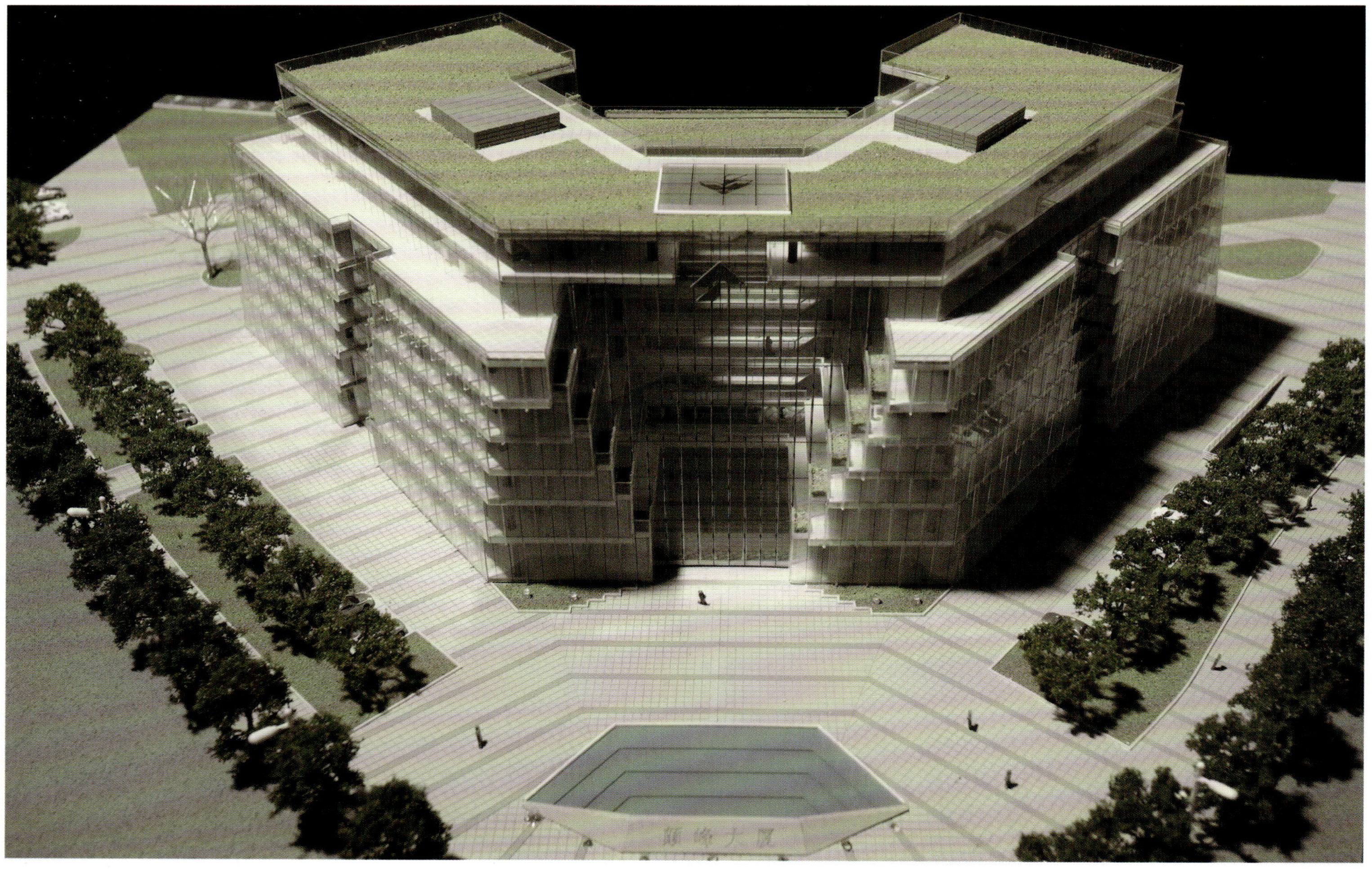

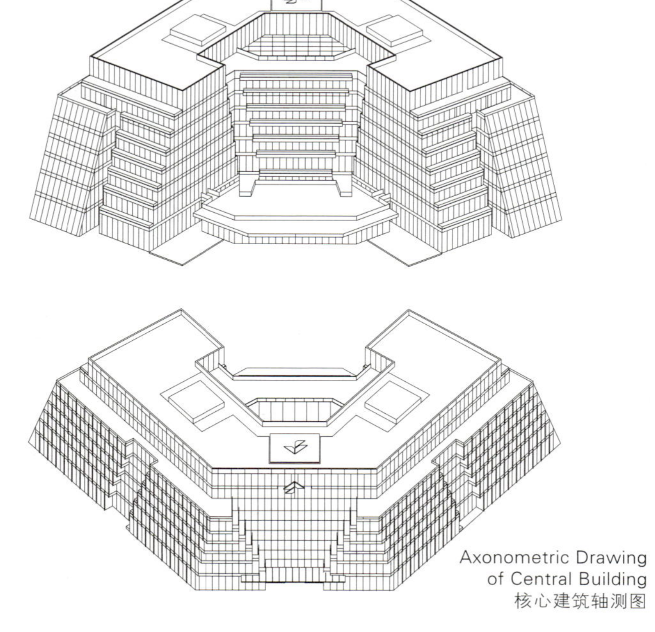

Axonometric Drawing of Central Building
核心建筑轴测图

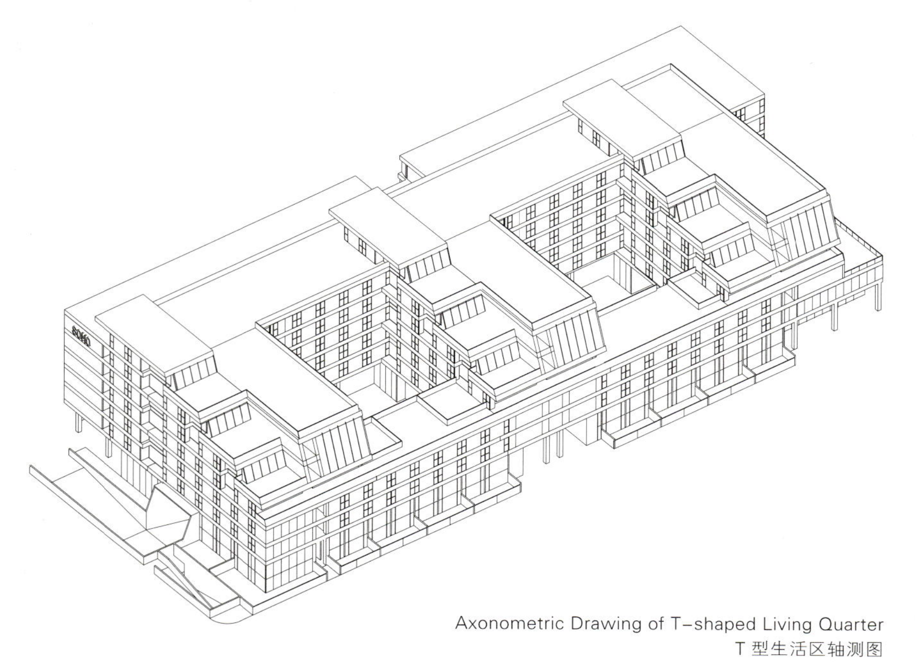

Axonometric Drawing of T-shaped Living Quarter
T 型生活区轴测图

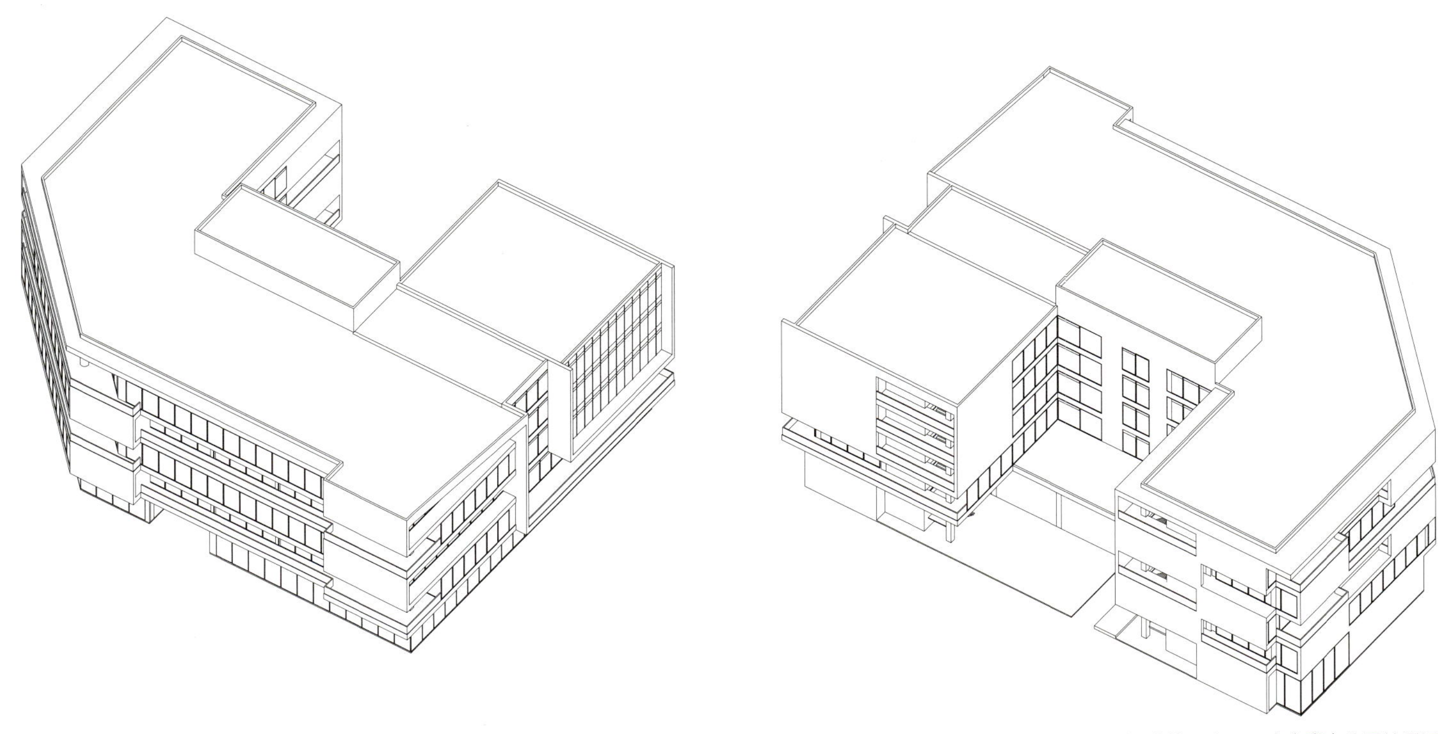

Axonometric Drawing of High-density Office Area 高密度办公区轴测图

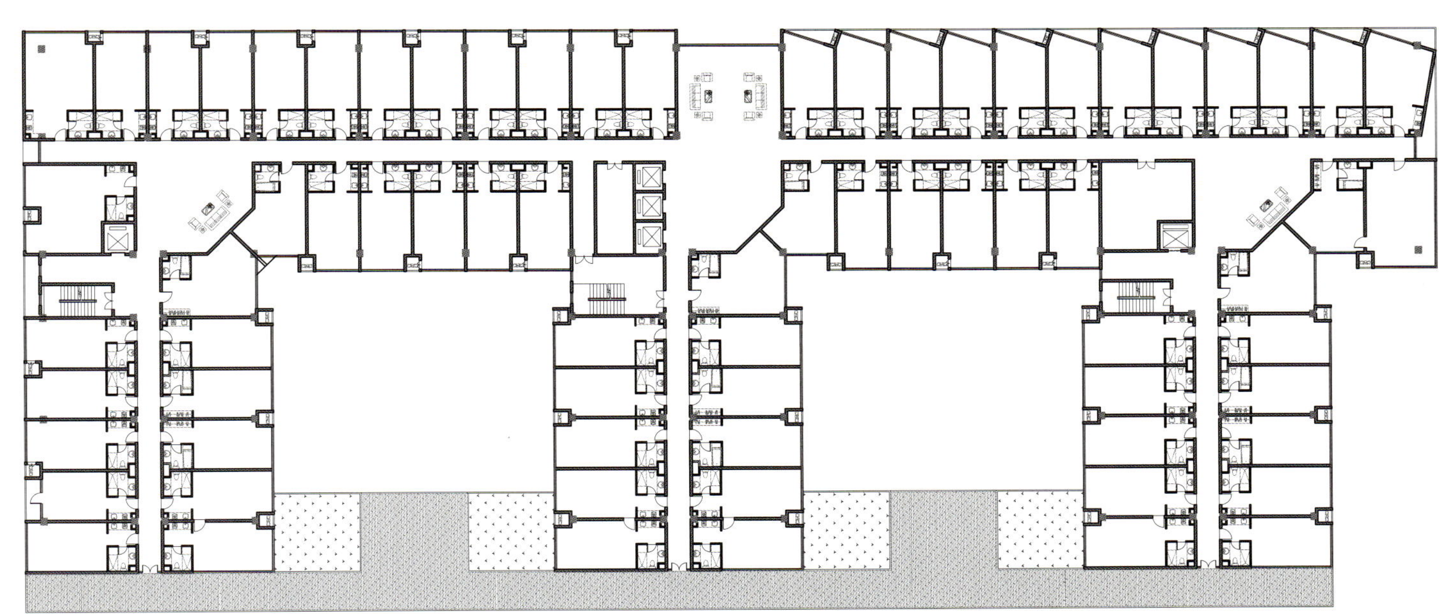

Plan of T-shaped Living Quarter T型生活区平面图

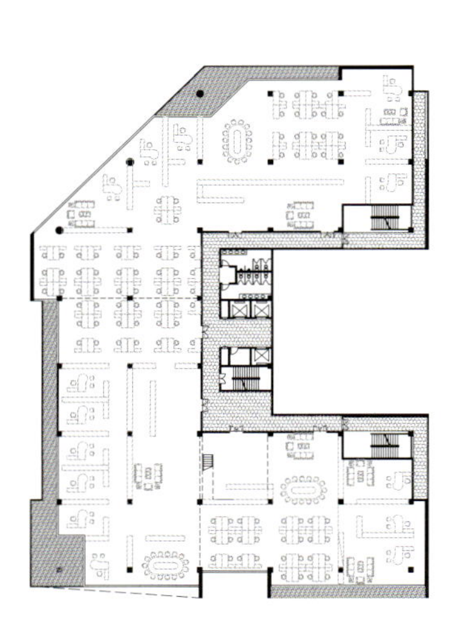

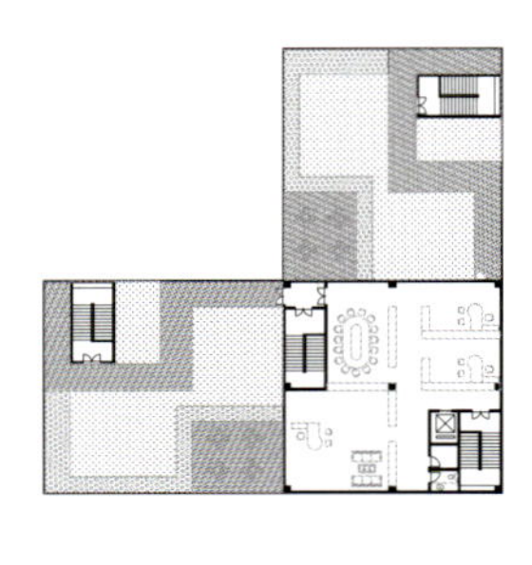

Plan of Low-density Office Area
低密度办公区平面图

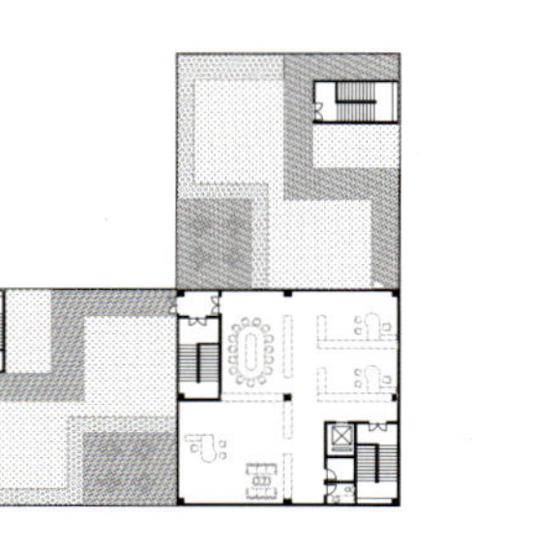

Plan of High-density Office Area
高密度办公区平面图

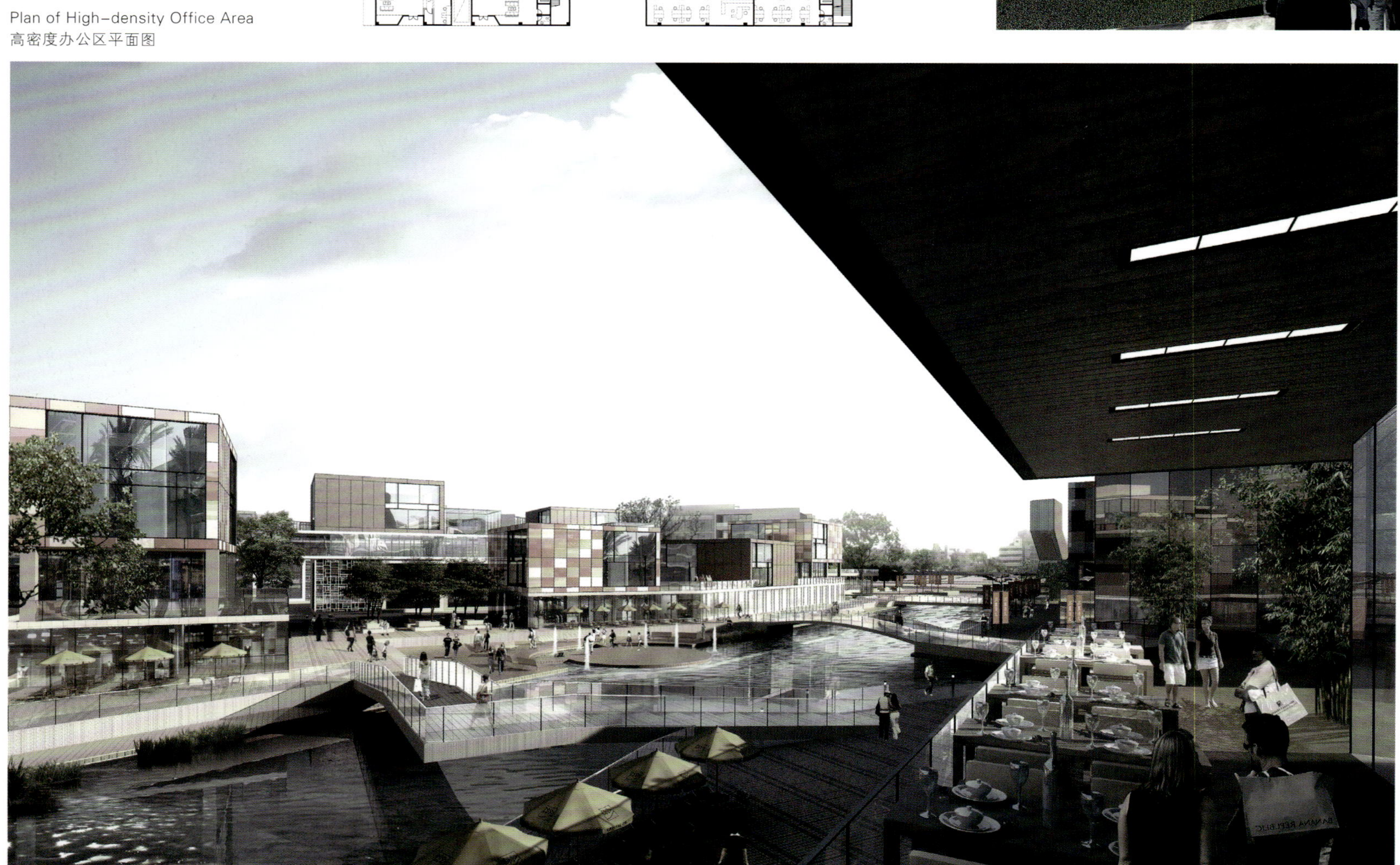

HOTEL
鸡鸣塘

Guangzhou Privately-owned Science and Technology Park

广州民营科技园

Features 项目亮点

According to the overall urban design, it takes advantage of the mature green technologies to create a sustainable and high-efficient organism.

项目综合考虑整体的城市设计，充分利用成熟的绿色技术，使之成为可持续发展和高效的有机整体。

Keywords 关键词

Convenient Transportation
交通便利

Green Science and Technology
绿色科技

Organic Integrity
有机整体

Location: Baiyun District, Guangzhou, Guangdong, China
Developer: KAISA Group
Architectural Design: AKG-chinagroup
Land Area: 282,563 m²
Floor Area: Approximately 530,000 m²

项目地点：中国广东省广州市白云区
开发商：佳兆业集团
建筑设计：德国 AKG-chinagroup 设计集团（深圳市凯筑建筑设计有限公司）
用地面积：282 563 m²
建筑面积：约 530 000 m²

Overview

Ideally located in the center of Baiyun District — the famous landscape area of Guangzhou City, sitting at the south foot of Maofeng Mountain Forest Park and on the south bank of Liuxi River, the Sci-tech Park is surrounded by mountains and waters, enjoying beautiful views and advantaged eco environment. It also boasts convenient transportation: it's only 15 km from downtown Guangzhou, 6 km from Guangzhou Baiyun Airport, and 23 km from the Huangpu Port. What's more, there is the Beijing-Zhuhai Expressway, the North 2nd Ring Road, the Airport Express-way, Guangzhou-Conghua First-class Highway and the 116 Provincial Road surrounding the park to provide great convenience.

项目概况

园区地理位置优越，地处广州市著名的山水城区——白云区中部，坐落在风景秀丽的帽峰山森林公园西麓、广州母亲河——流溪河南畔的平原，青山绿水环绕，风景秀丽，自然生态环境优越。园区交通便利，距广州市中心城区 15 km，距广州市白云机场 6 km，距黄埔港 23 km。京珠高速公路、北二环高速公路、机场高速公路、广从一级公路、省道 116 等在园区附近纵横交错，公路密度达到国际发达地区水平。

Position

As one of the five parks under Guangzhou Hi-tech Industrial Development Zone, it is an important national privately-owned sci-tech park contacted by the Ministry of Science and Technology of the People's Republic of China.

项目地位

项目是广州高新区五个园区之一，是国家科技部重点联系的国家级民营科技园。

Site Plan
总平面图

1# Plot (Culture Industrial Park) 1# 地块（时尚文化产业园）

2# Plot (Culture Industrial Park) 2# 地块（时尚文化产业园）

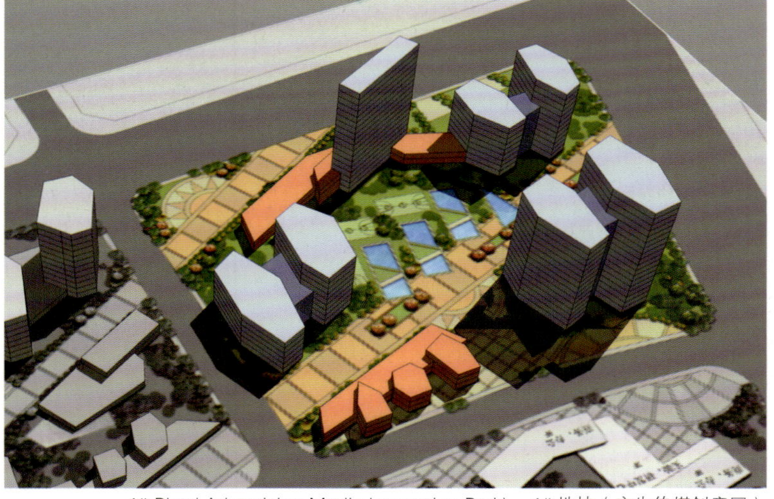

4# Plot (Advertising Media Innovation Park) 4# 地块（广告传媒创意园）

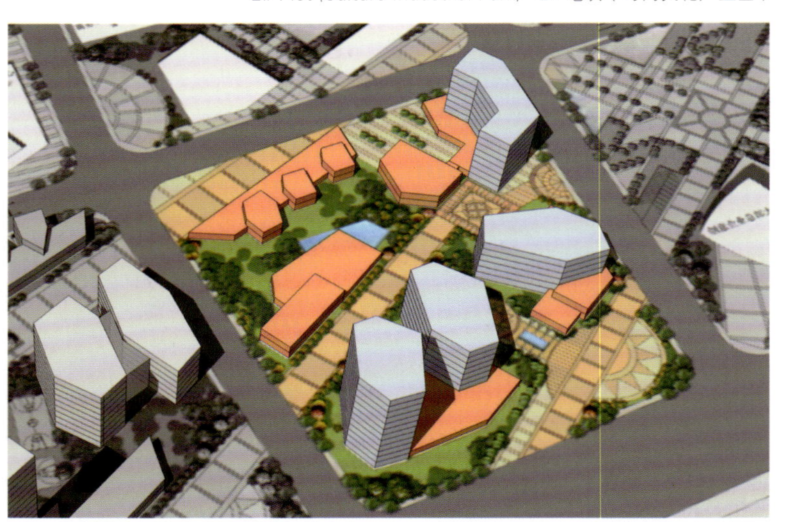

5# Plot (Advertising Media Innovation Park) 5# 地块（广告传媒创意园）

3# Plot (Residential) 3# 地块（住宅）

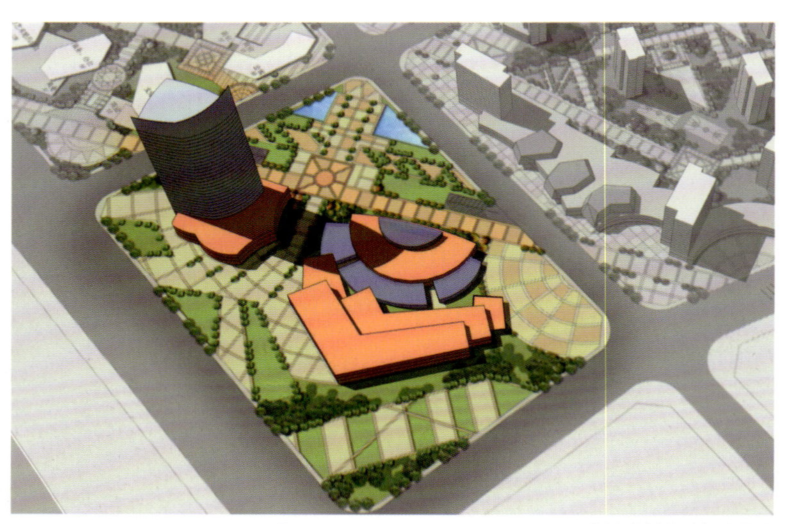

6# Plot (Innovation Headquarters) 6# 地块（创意总部大厦）

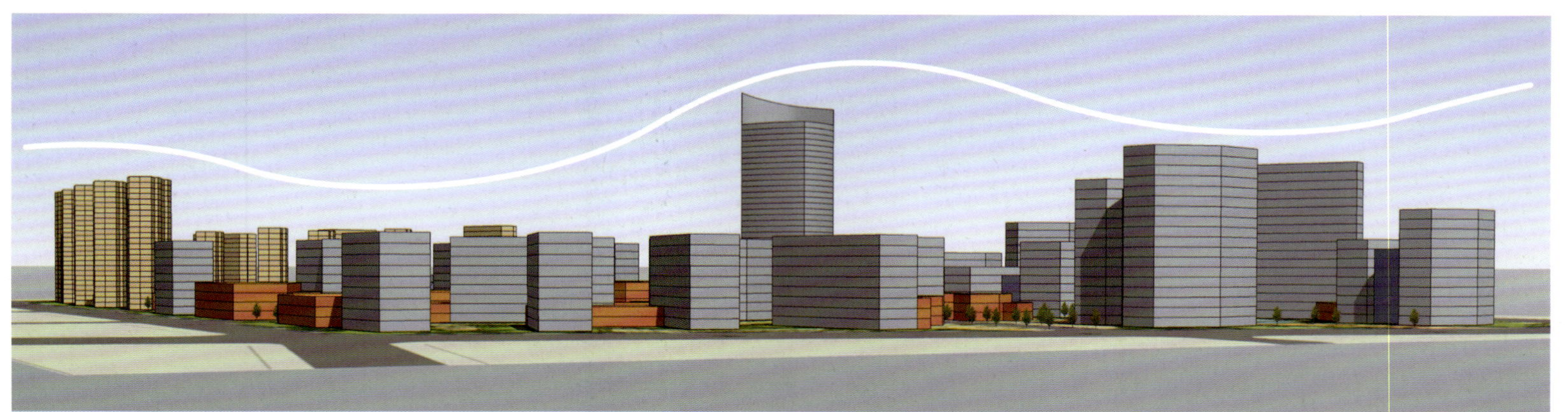

Skyline 天际线

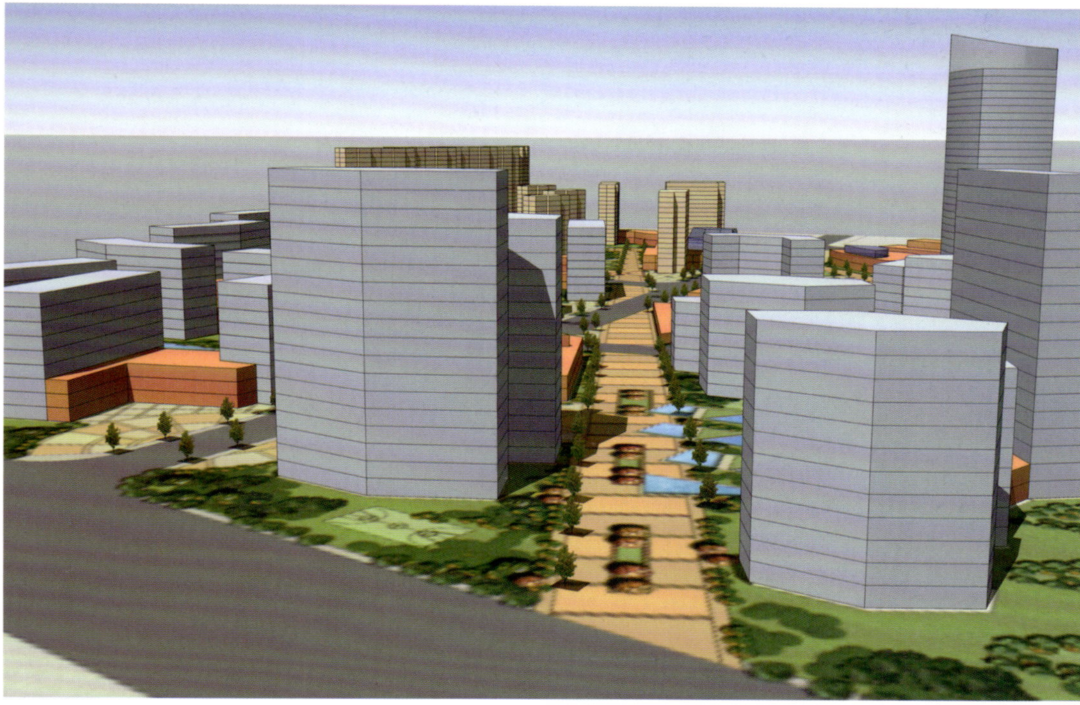

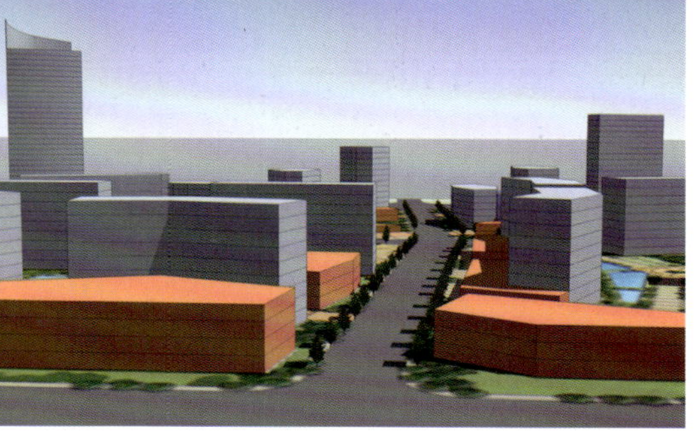

Dior
LOUIS VUITTON
Welcome

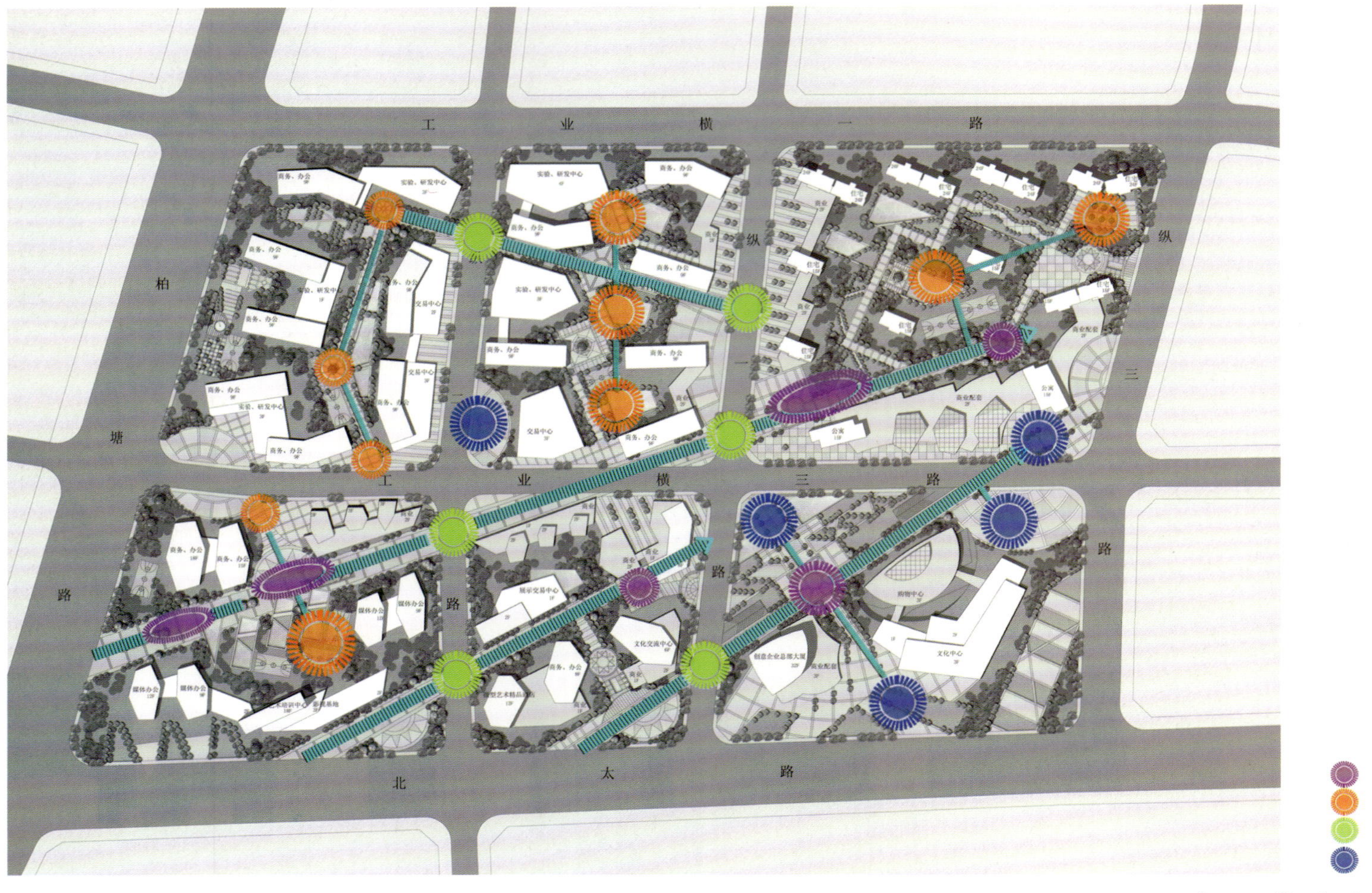

Structure Planning Drawing 规划结构图

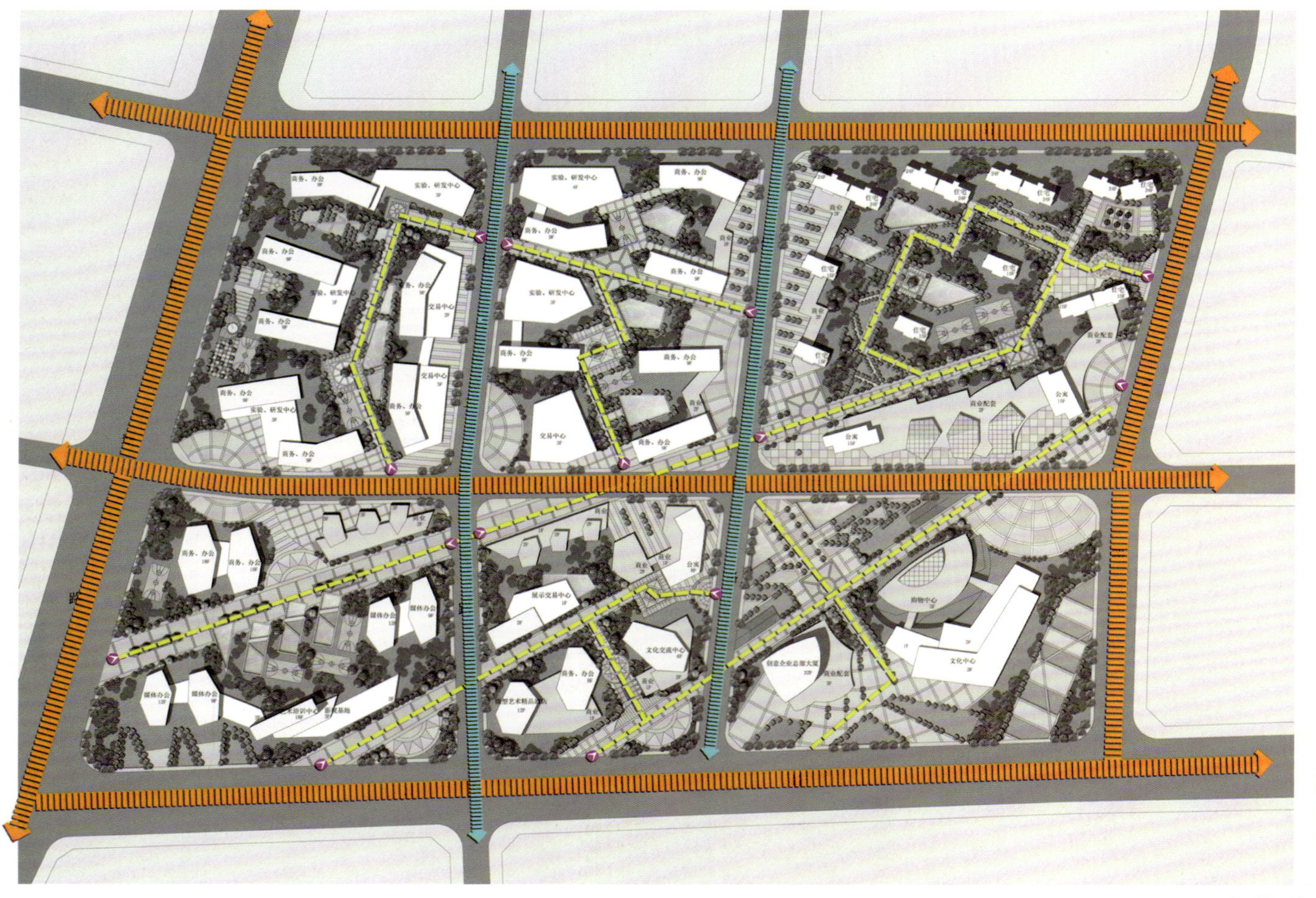

Traffic Network Drawing 交通流线图

Landscape Plan 景观意向图

功能：时尚文化创意园

用地面积：45 055 m²

总建筑面积：92 783 m²

容积率：2.06

总建筑面积包括以下建筑面积

商务、办公面积：76 654 m²

实验、研发中心：5 539 m²

交易中心：9 538 m²

时尚创意风情商业街：1 052 m²

地块区位示意图

Site Plan of Plot No.1 1# 地块平面图

功能：时尚文化创意园

用地面积：45 579 m²

总建筑面积：91 156 m²

容积率：2.00

总建筑面积包括以下建筑面积

商务、办公面积：63 198 m²

实验、研发中心：16 746 m²

交易中心：8 625 m²

时尚创意风情商业街：2 587 m²

地块区位示意图

Site Plan of Plot No.2 2# 地块平面图

功能：住宅

用地面积：67 059 m²

总建筑面积：134 247 m²

容积率：2.00

总建筑面积包括以下建筑面积

时尚创意风情商业街：63 198 m²

住宅面积：97 183 m²

配套公寓：20 772 m²

地块区位示意图

Site Plan of Plot No.3 3# 地块平面图

功能：广告传媒创意园

用地面积：45 814 m²

总建筑面积：94 257 m²

容积率：2.06

总建筑面积包括以下建筑面积

商务、办公面积：38 389 m²

时尚创意风情商业街：4 803 m²

媒体办公：34 046 m²

艺术培训中心：11 945 m²

影视基地：5 074 m²

地块区位示意图

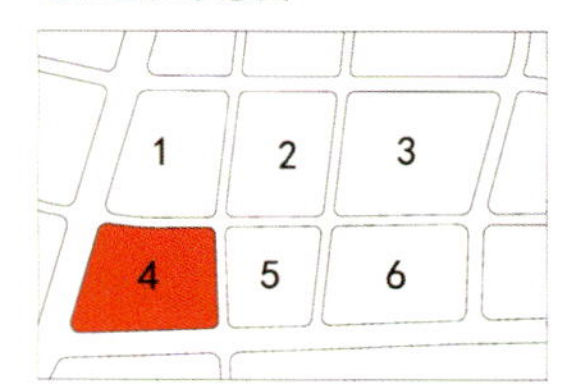

Site Plan of Plot No.4　4# 地块平面图

功能：广告传媒创意园

用地面积：31 376 m²

总建筑面积：45 128 m²

容积率：1.44

总建筑面积包括以下建筑面积

商务、办公面积：11 327 m²

时尚创意风情商业街：6 213 m²

微型艺术精品酒店：9 187 m²

文化交流中心：5 293 m²

展示交易中心：4 190 m²

配套公寓：8 918 m²

地块区位示意图

Site Plan of Plot No.5　5# 地块平面图

功能：创意总部大厦

用地面积：47 680 m²

总建筑面积：72 604 m²

容积率：1.52

总建筑面积包括以下建筑面积

商务、办公面积：48 725 m²

时尚创意风情商业街：11 546 m²

文化交流中心：12 333 m²

地块区位示意图

Site Plan of Plot No.6 6# 地块平面图

Design Idea

According to the overall urban design, the design for the park environment, the traffic system, the parking, the underground space as well as the interior and exterior public spaces has considered the electromechanical equipments, the intelligence and the future property management to provide basic functions and make full use of the resources. By taking advantage of the mature green technologies, it will become a sustainable and high-efficient organism.

设计理念

综合考虑整体的城市设计，在城市环境、交通、停车、地下空间和室内外公共空间等部分，结合机电设备、智能化和未来物业管理等因素，在满足功能要求的前提下，达到最佳配置和资源共享，充分利用成熟的绿色技术，使之成为可持续发展和高效的有机整体。

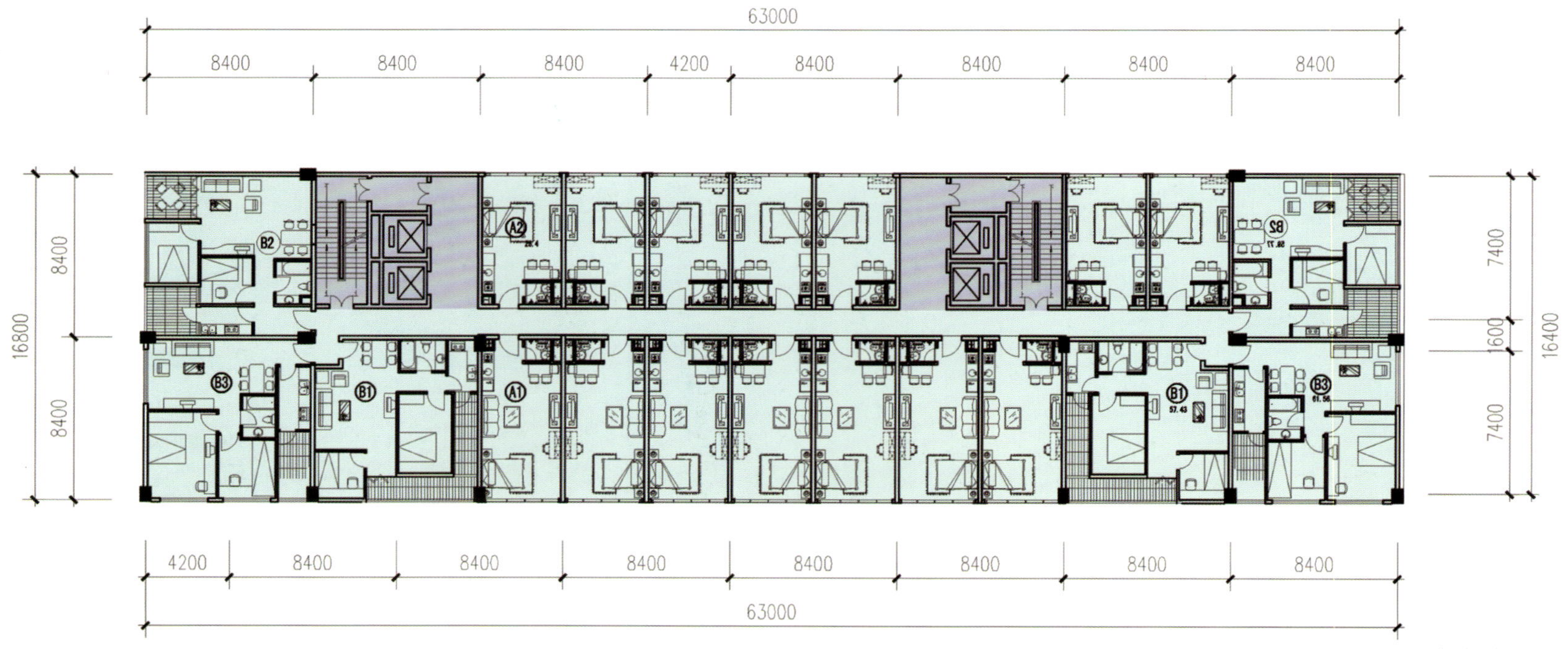

Micro Hotel Plan 微型酒店平面图

Functional Layout

The site is divided into six parts, accommodating the functions such as the cultural innovation and residences. The cultural innovative park is to the north of the 3rd Industrial Cross Road and to the west of the 1st Longitudinal Road, with the residential area on its east. To the south of the 3rd Industrial Cross Road and to the west of the 1st Longitudinal Road, it is also the cultural innovative park, to the east of which it is the creative headquarters building.

功能布局

该项目分为六个地块，具备文化创意及居住等功能，其中工业横三路以北、纵一路以西为时尚文化创意园，其东边为住宅区；工业横三路以南、纵一路以西为时尚文化创意园，其东边则为创意总部大厦。

Ninghe Ivy Science and Technology Park, Tianjin

天津宁河长春藤科技园

Features 项目亮点

The site is divided into three parts, of which the one in the middle serves as the main body. It includes a convention center, a comprehensive management and training center and an oval square which symbolizes a gem shining on the surroundings with the "light of science and technology".

地块分为三大块，中间地块为设计的主体，主要功能有商务会展中心、综合管理和培训中心，椭圆形广场象征璀璨的宝石并向周围散发科技之光。

Keywords 关键词

Shinning & Bright
璀璨亮丽

Multiple Functions
功能多样

Oval Square
椭圆形广场

Location: Tianjin, China
Architectural Design: KaziaLi Design Collaborative Inc.
Gross Land Area: 333,700 m²
Gross Floor Area: 447,999 m²
Floor Area Overground: 398,999 m²
Gross Floor Area Underground: 49,000 m²
Ground-breaking Time: April, 2011

项目地点：中国天津市
建筑设计：凯佳李建筑设计事务所
总用地面积：333 700 m²
总建筑面积：447 999 m²
地上建筑面积：398 999 m²
地下总建筑面积：49 000 m²
动工时间：2011 年 4 月

Overview

The project is located in Ninghe modern industrial park which is at the southwest of Ninghe town and with a planning land area of 40 km². This project is positioned as an innovative, displaying, growth-oriented and visitable technology incubator with the strongest industrial culture characteristics.

项目概况

本项目位于宁河县西南部的宁河现代产业区内，规划占地面积 40 km²。本项目定位为天津市最具有产业文化特色的创新型、展示型、成长型及可参观性的科技孵化基地。

Functional Layout

The design borrows advanced concepts, and focuses on the quality of surrounding environment. The whole project is divided into two categories: urban and community; the urban type contains garden-style offices, sci-tech venture trading center, courtyard-style plants, and urban convention center; the community type contains convenient services and blue-collar apartments.

功能布局

本项目设计吸取先进理念，注重周围环境质量。项目整体分为两大类：城市和社区；城市类包含花园式办公、科技创投交易中心、庭院式厂房、都市会展中心；社区类包括便捷服务、蓝领公寓。

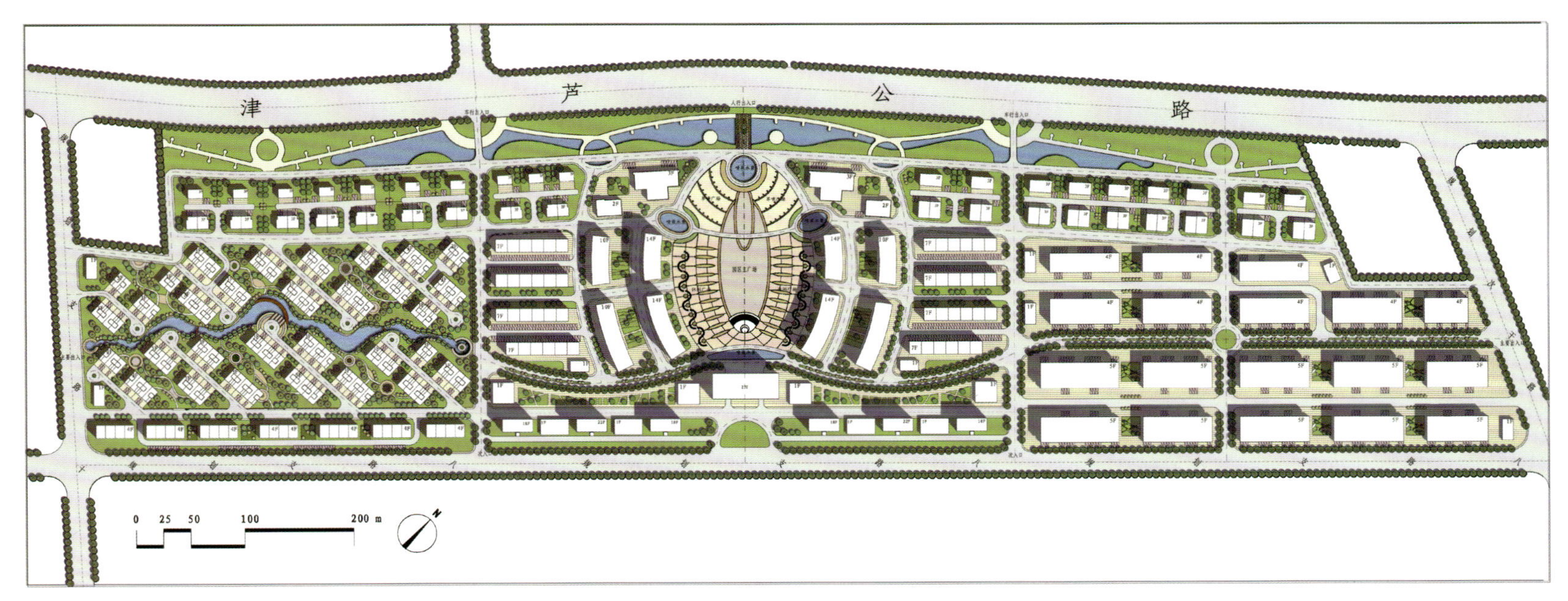

Site Plan
总平面图

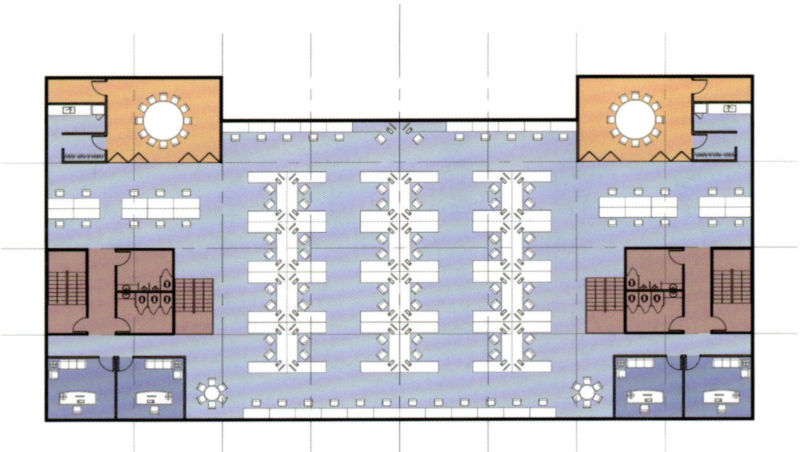

TYPICAL
full floor

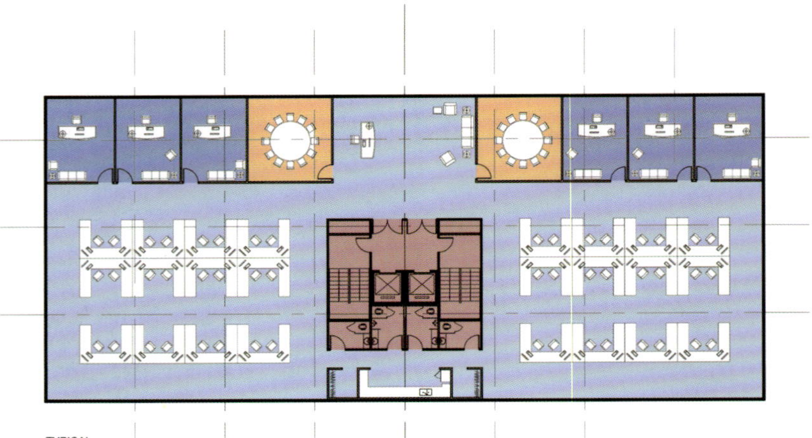

TYPICAL
full floor

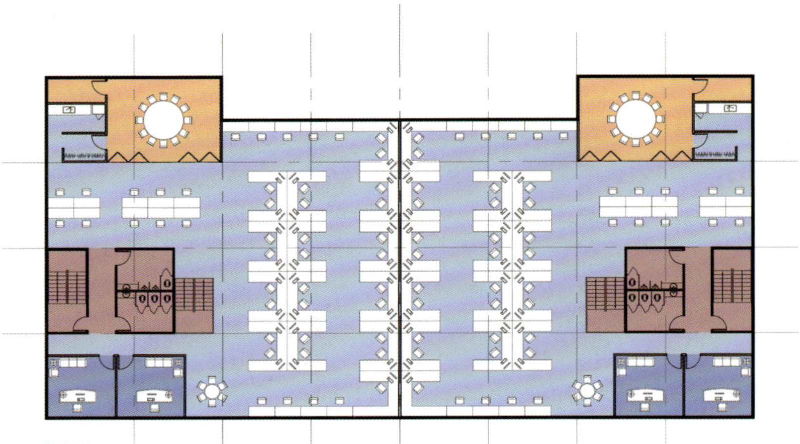

TYPICAL
half-floor

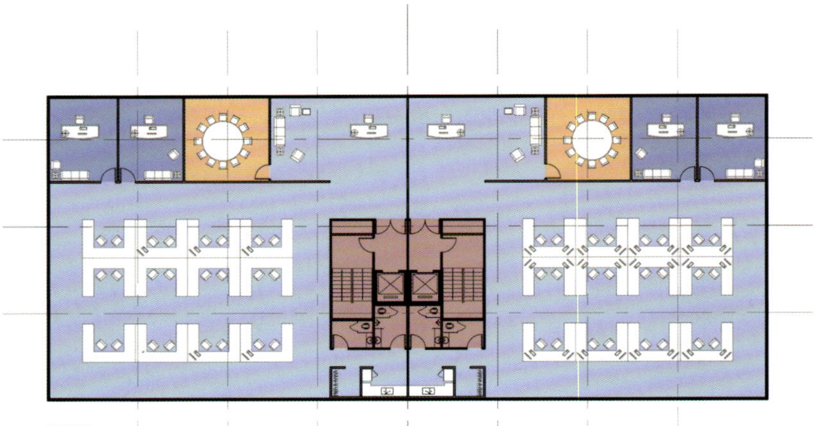

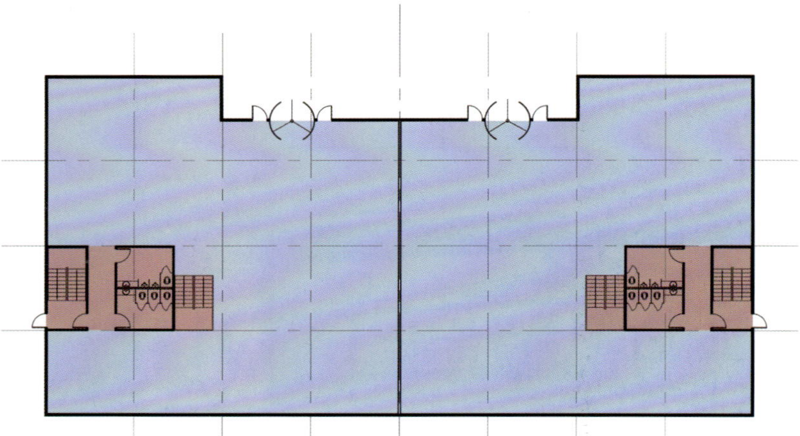

TYPICAL
half-floor

TYPICAL
first floor

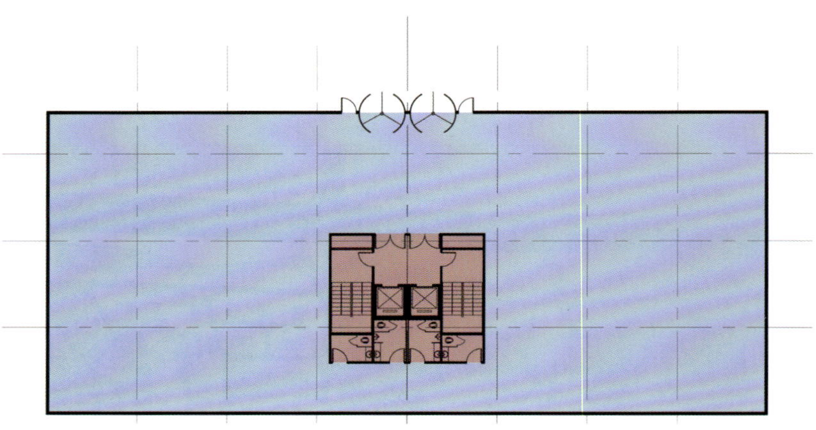

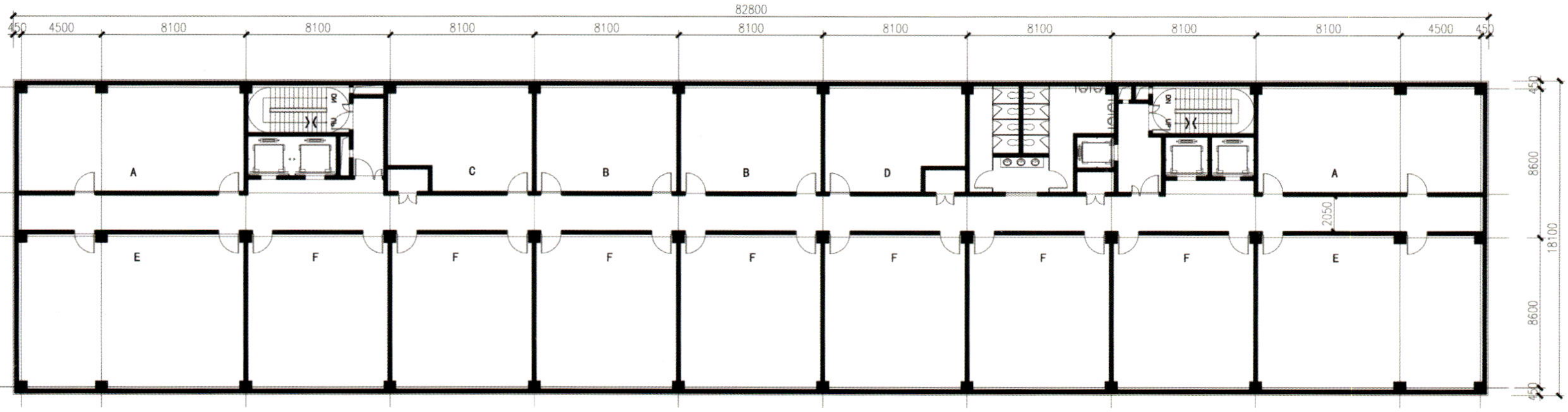

TYPICAL
first floor

Standard Floor Plan of Office Building 办公楼标准层平面图

Architectural Design

The site is divided into three parts, of which the one in the middle serves as the main body. It includes a convention center, a comprehensive management and training center and an oval square which symbolizes a gem shining on the surroundings with the "light of science and technology". The western plot is a strategic industrial incubation center and the eastern plot is an industrial processing base.

建筑设计

地块分为三大块，中间地块为设计的主体，主要功能有商务会展中心、综合管理和培训中心，椭圆形广场象征璀璨的宝石并向周围散发科技之光，地块西侧为战略产业孵化中心，地块东侧为产业加工基地。

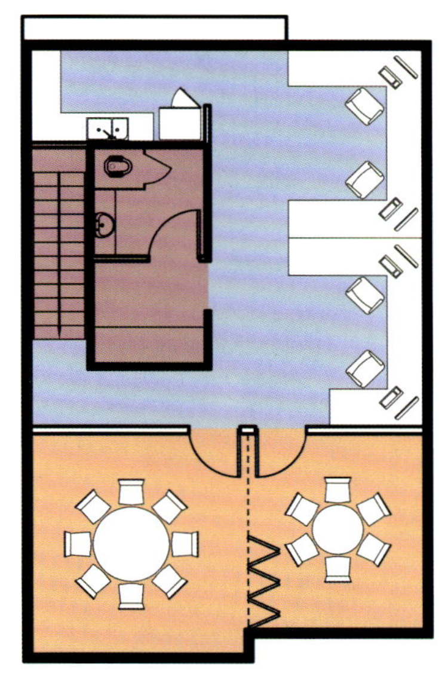

FP 03
conference
open office
break room

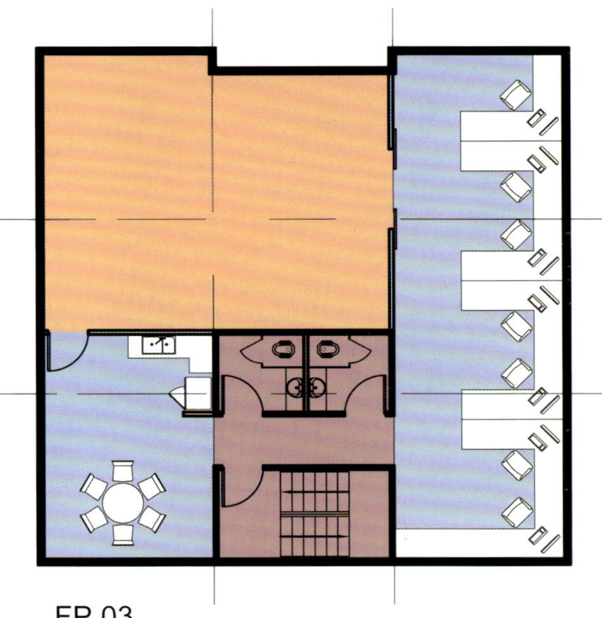

FP 03
conference
open office
break room

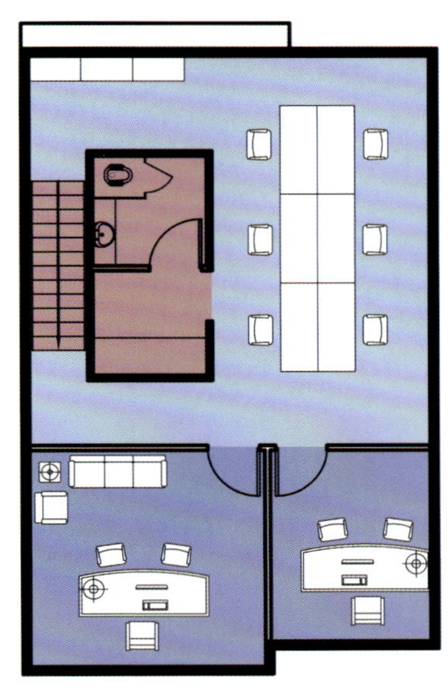

FP 02
offices
open office

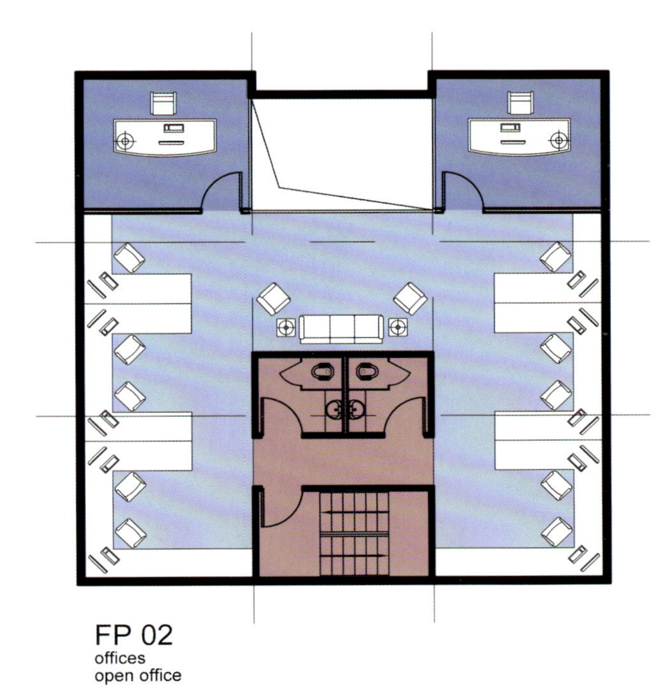

FP 02
offices
open office

FP 01
lobby
open office

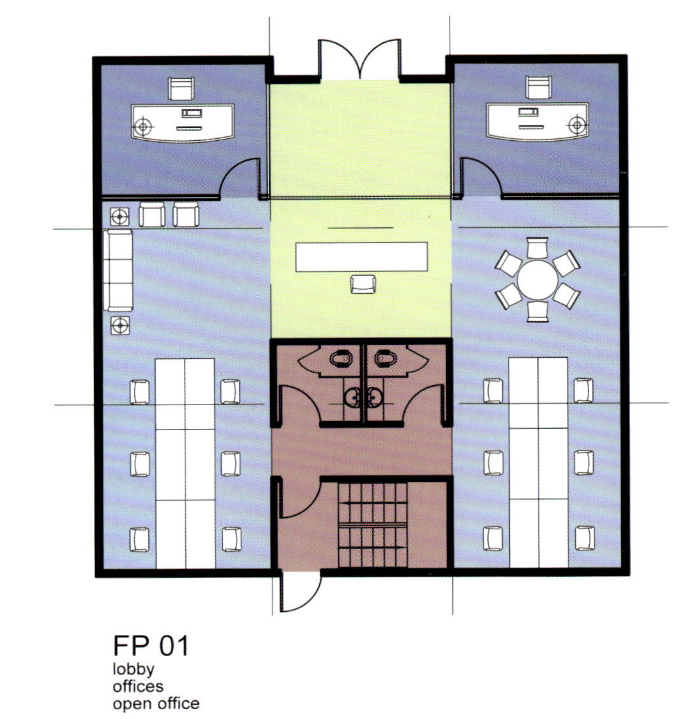

FP 01
lobby
offices
open office

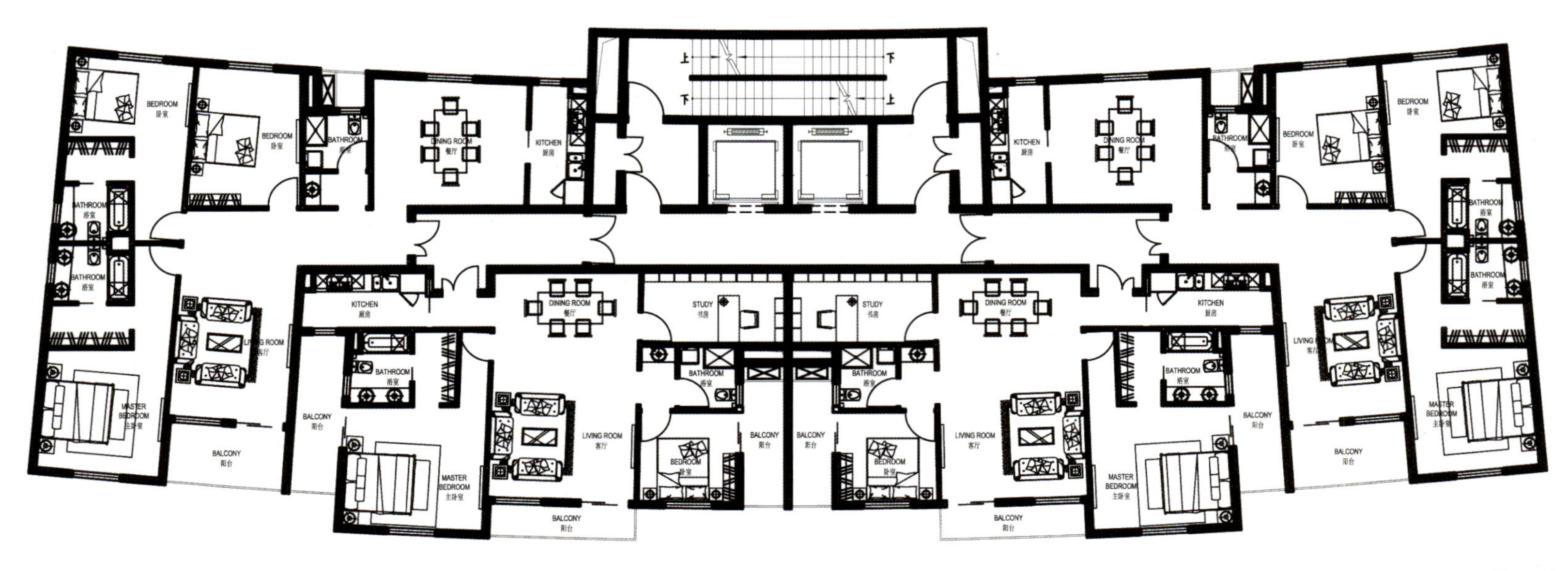

Floor Plan
平面图

Hong Kong Science Park Phase Ⅱ

香港科学园二期

Features 项目亮点

Science Park Phase Ⅱ in planar rectangular plan effectively creates landscape corridor, which extends to the adjacent harbor.

科学园二期以呈矩形的平面规划有效地营造景观长廊，一直伸延至毗邻的海港。

Keywords 关键词

The Central Axis
中轴线

Campus Tone
校园风格

Landscape Corridor
景观长廊

Location: Tai Po, New Territories, Hong Kong SAR China
Architecture Design: Leigh & Orange Architects
Land Area: 77,320 m^2
Gross Floor Area: 105,000 m^2
Completion Date: 2007

项目地点：中国香港特别行政区新界大埔区
建筑设计：利安顾问有限公司
用地面积：77 320 m^2
总建筑面积：105 000 m^2
竣工时间：2007 年

Overview

Hong Kong Science Park Phase Ⅱ is located in Pak Shek Kok, Tai Po, New Territories, Hong Kong, near the mountain and by the river, with pleasant scenery. Providing high quality research and development equipment, Science Park Phase II has eight high-tech buildings, of which two are energy towers and six are research offices, including two laboratories with special devices.

项目概况

科学园二期位于香港新界大埔白石角，风景宜人，依山傍水。科学园二期建设提供高质素科研及发展设备，园内迄立八幢高科技大楼；当中划分为两座能源塔及六座科研办公室（其中两座特设设备实验室）。

Design Concept

With “park” as the design concept, the entire project ingeniously arranges buildings by campus tone to enable the tenants and the general public to integrate with the natural landscape.

设计理念

整个项目以“园”为设计理念，以校园风格设计，将建筑物巧妙安排，令租户及大众融合于自然景致中。

Architectural Design

Science Park Phase Ⅱ in planar rectangular plan effectively creates landscape corridor, extending to the adjacent harbor. To organize multiple vast squares, the total plane even makes open space doubled by the concept of the underground garage, underground mechanical and electrical tunnels and energy towers, and it also makes repair, reorganization and extension of electrical and mechanical equipment become simple and feasible, and need not impede the normal operation of tenants.

Illustrative Master Layout Plan
总体规划示意图

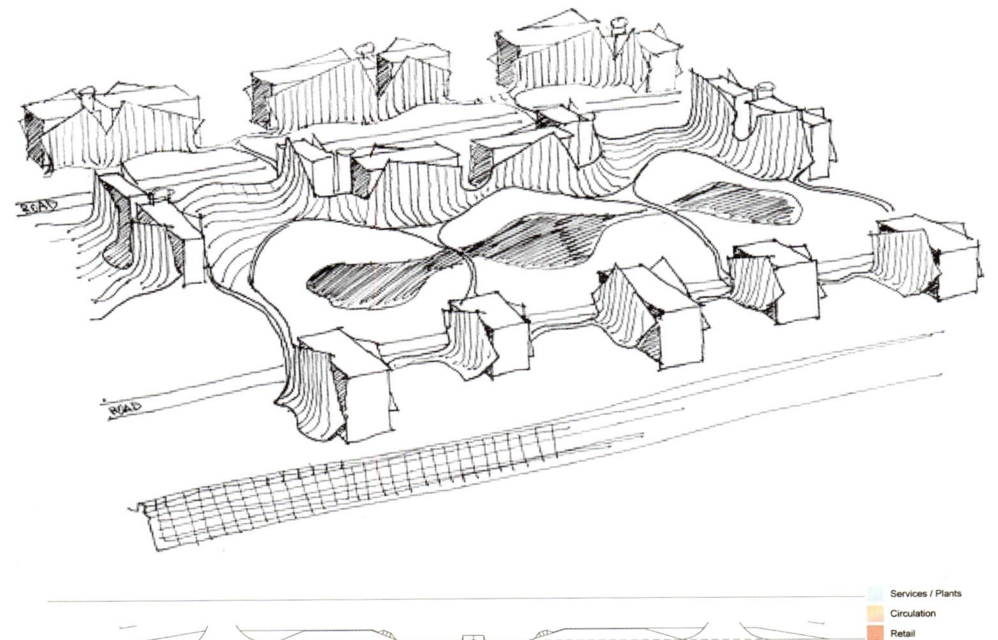

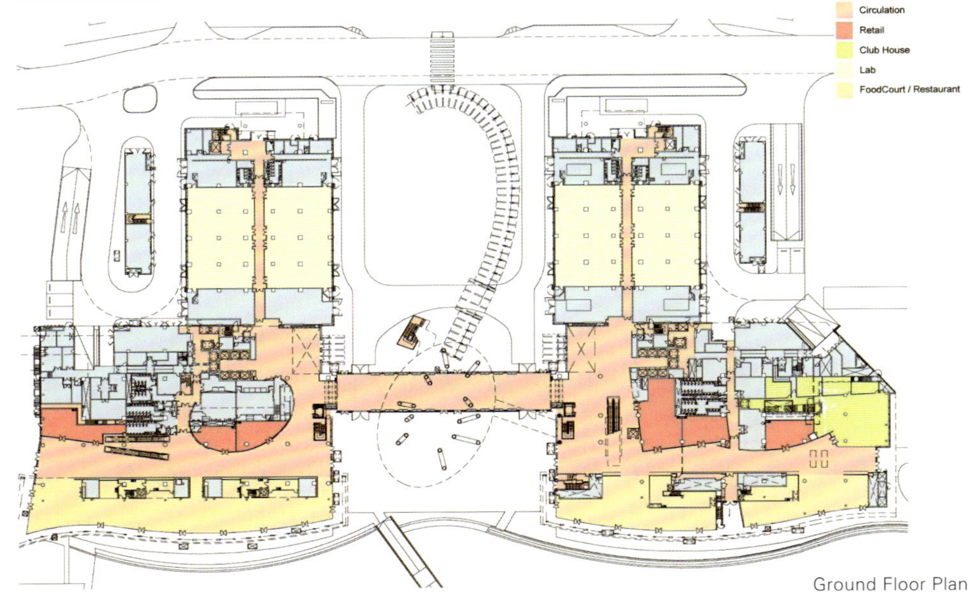

Ground Floor Plan
首层平面图

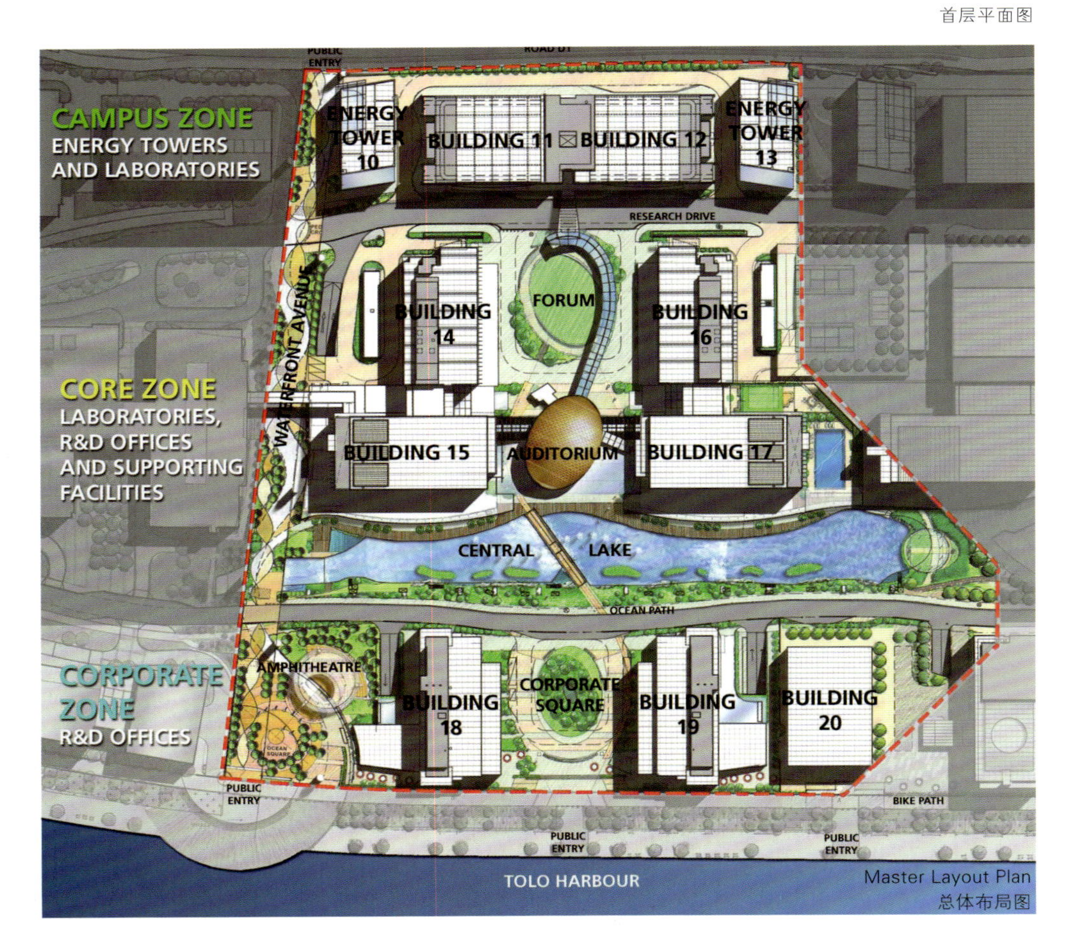

Master Layout Plan
总体布局图

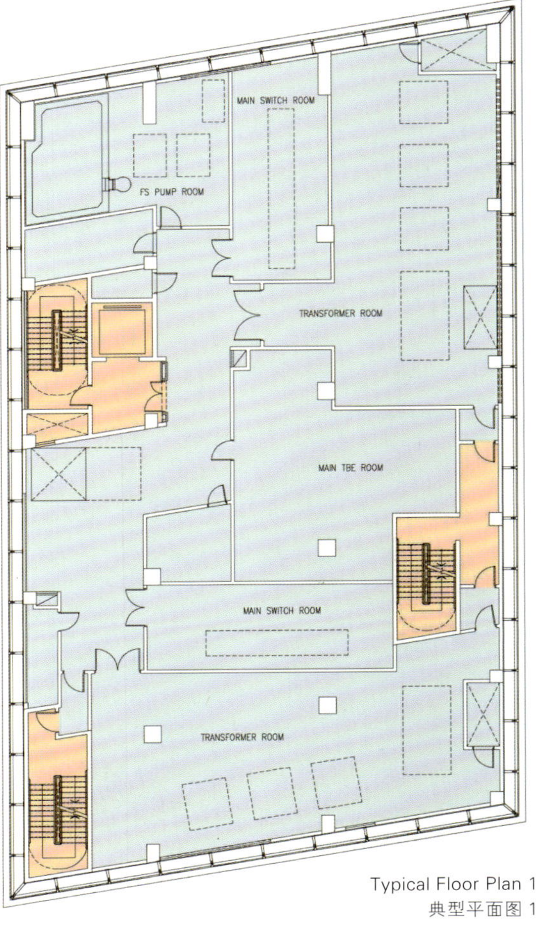

Typical Floor Plan 1
典型平面图 1

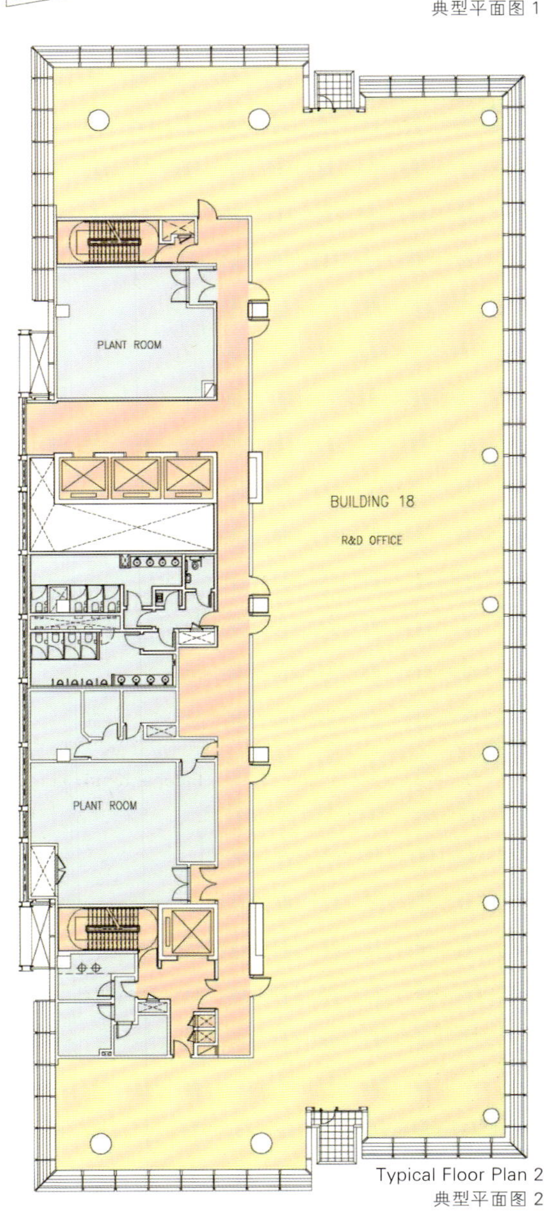

Typical Floor Plan 2
典型平面图 2

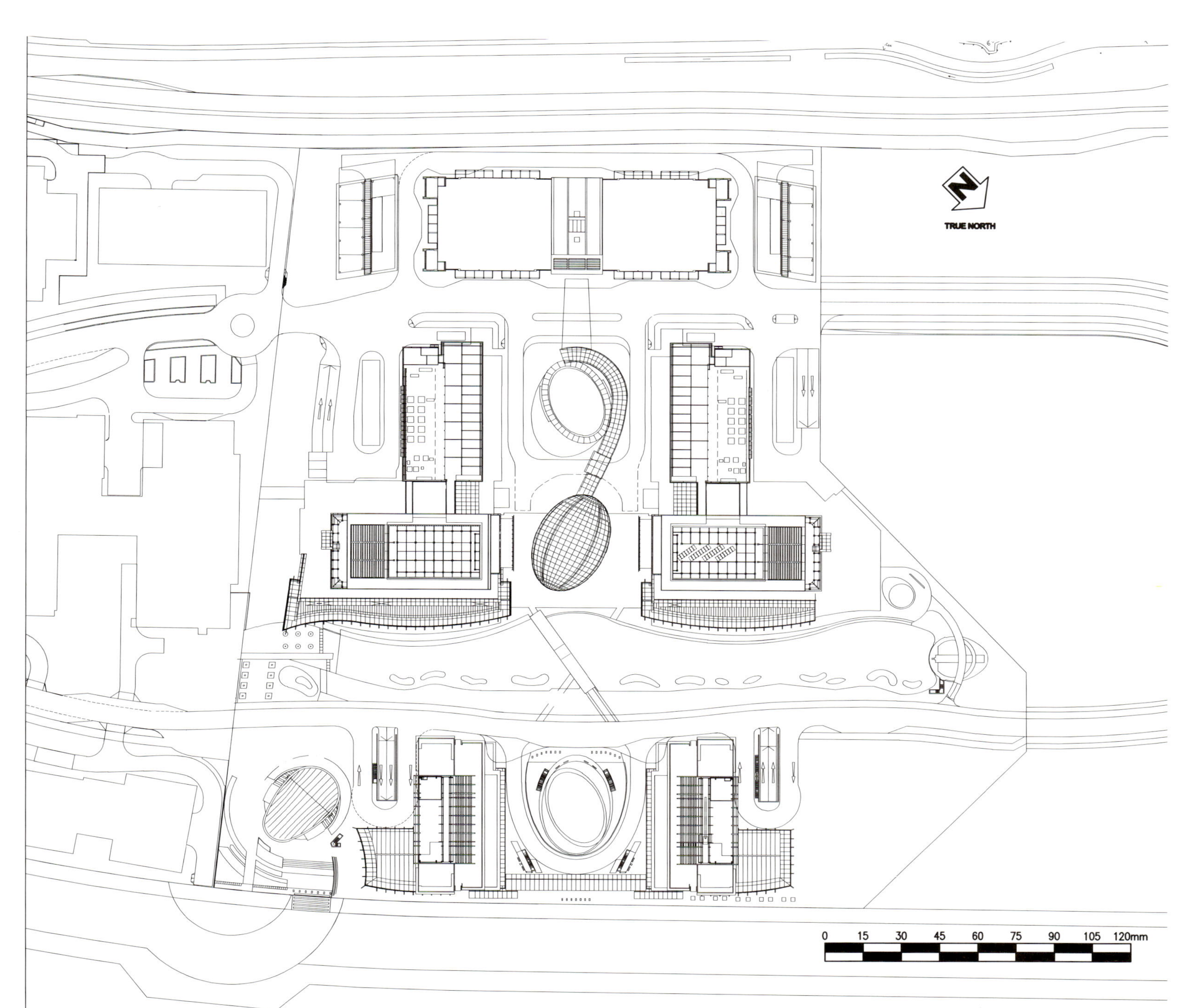

Floor Plan
平面图

The entire project begins with a central monitoring and control, and utilizes sophisticated building management, security and information system, fully reflecting the ever-changing high-tech mode of operation.

The central axis increases commercial taste for Phase II by different commercial spaces in their own type, including business center, conference center, shops and catering facilities, and then the club at the center region to provide users with convenient access to the channel, also effectively facilitate the exchange of user. Independently standing beside the lake with three hundred seats, the oval lecture hall is the focus of the Park. The striking landmark appearance also played a key role for the project.

建筑设计

科学园二期以呈矩形之平面规划有效地营造景观长廊，一直延伸至毗邻的海港。为规划多个辽阔广场，总平面更以地下车库、地底机电设备隧道及能源塔等概念令休憩空间倍增，更促使机电设备之维修、重组及延伸变得简易可行，同时无需阻碍租户的正常运作，一举两得。整个项目皆以中央监察及控制，采用精密的大厦管理、保安及信息系统，完全地体现日新月异的高科技运作模式，走在科技尖端。

中轴线以各适其式的商用空间为二期加添商业味道，当中包括商务中心、会议中心、商店及餐饮设施等，再以会所置于中心地域，为用户提供便捷的出入信道，亦有效促进用户交流。独立于湖畔并拥有三百个座位的椭圆形演讲厅更是园内的焦点所在，地标性的外形除引人注目外，更为项目起了点睛作用。

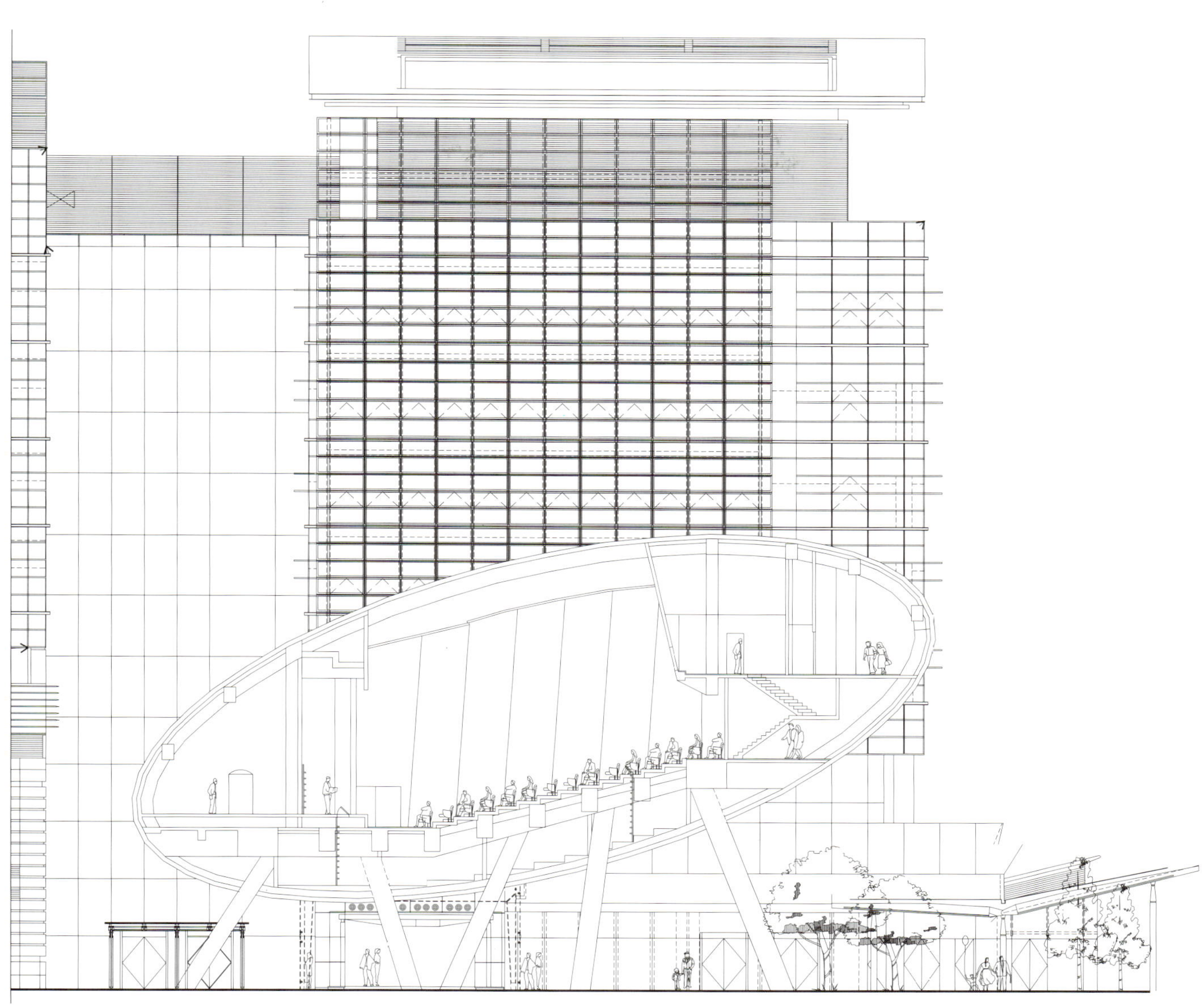
Section 1-1 1-1 剖面图

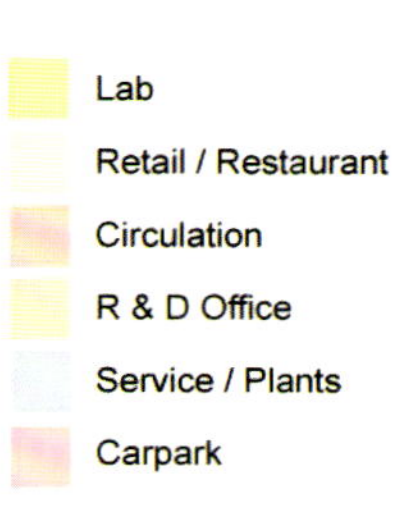

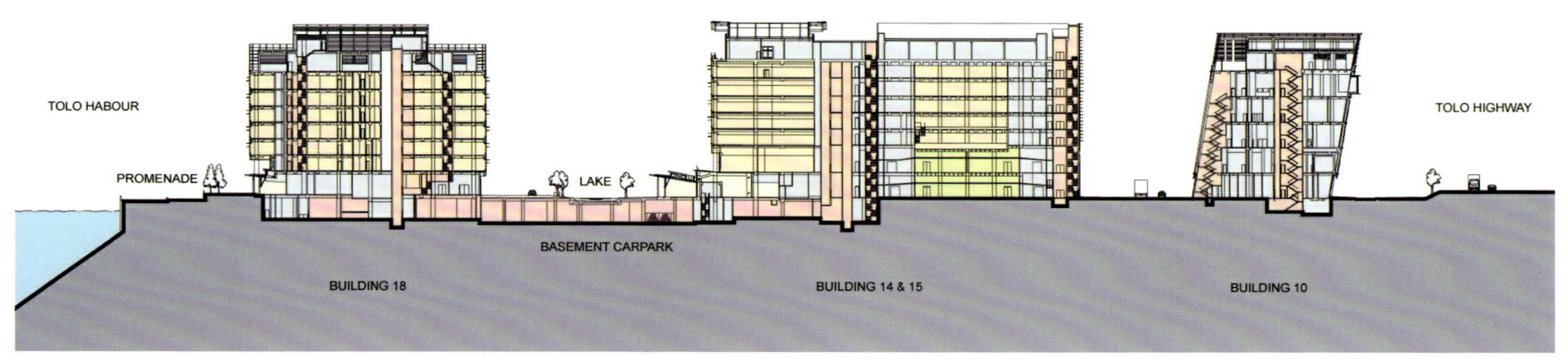

Section Across Phase 1
一期截面图

Interior Design

Laboratory's design concept of replicate unit fits for the flexible planar arrangement; research office relatively uses eccentric transport nuclear, provided with different area of the unit. This design not only has the possibility of restructuring the space, but also can save construction waste and resources, which shows evident environmental awareness.

室内设计

实验室以复制单元为设计概念，适用于灵活多变的平面安排；科研办公室相对地采用偏心交通核，同样地配合提供不同面积的单位。这样的设计既予以空间重组之可能性，同时能省却浪费建筑废料及资源的情况，环保意识可见一斑。

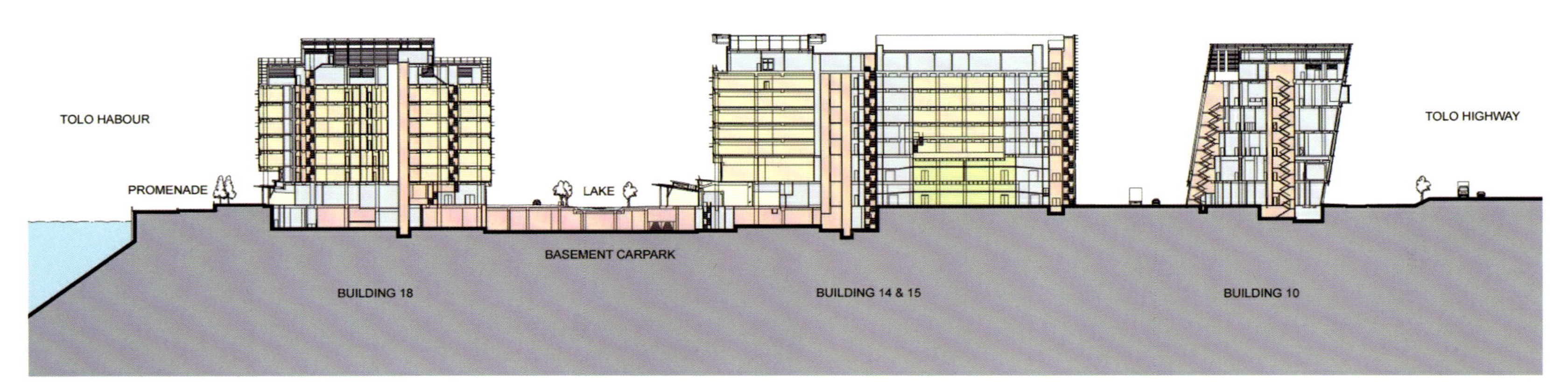

Section Across Phase 2
二期截面图

Zhongguancun Electronic Park West Zone A6

中关村电子园西区 A6 项目

Features 项目亮点

The office group in south employs the layout form of the middle low and sides high, creates a pile of group layout with shocking visual effects, and gives the innovative appearance by diagonal blocks interspersing with each other.

南侧办公组团采用中间低、两边高布局形式，创造出堆成群体布局，具有震撼性和视觉效果，并通过立面斜向体块穿插而赋予其创新精神。

Keywords 关键词

Group Layout
群体布局

High Building Volume
高建筑体量

Visual Impact
视觉冲击

Location: Chaoyang District, Beijing, China
Architecture Design: Beijing Victory Star Architectural & Civil Engineering Design Co., Ltd.
Land Area: 23,600 m^2
Gross Floor Area: 182,300 m^2
Completion Date: 2012

项目地点：中国北京市朝阳区
建筑设计：北京维拓时代建筑设计有限公司
用地面积：23 600 m^2
总建筑面积：182 300 m^2
竣工时间：2012 年

Overview

Zone A6 is located in Electronic City West of Zhongguancun Science Park, Chaoyang District of Beijing, bordering four roads on different directions: the Middle Road of Electronic City West is on its east, Zone A8 on the south, Guangying East Road on the west, and Electronic City West No. 1 Road on the north.

项目概况

A6 地块位于朝阳区中关村科技园电子城西区二期 A6 地块，用地四至为：东至电子城西区中路，南至电子城西区 A8 地块，西至来广营东路，北至电子城西区一号路。

Site Value

The built Metro Line 13 and the upcoming Line 14 provide multiple options for reaching the site. Moreover, Line 14 has even set its entry/exit at the southwest corner of the site, which will greatly benefit the project.

板块价值

已建的地铁 13 号线与即将建设的地铁 14 号线给地块提供了多种出行方式的可能性，其中地铁 14 号线更是直接在地块西南角设置地铁出入口，这也是该地块的最大优势之一。

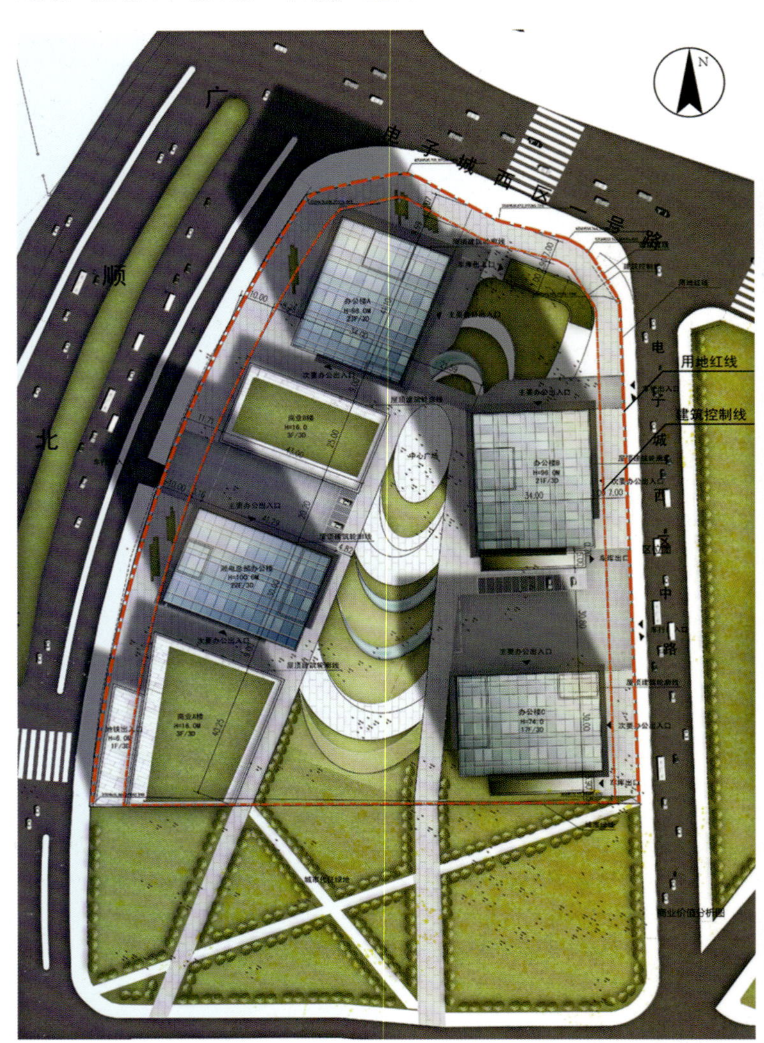
N
广
顺
北
电子城西区一号路
用地红线
建筑控制线
电
子
城
西
区
中
路
办公楼A
办公楼B
办公楼C
中心广场

Architectural Design

In view of the future users are well-known enterprises at home and abroad, the image of the building must be enough representative to reflect its corporate culture. 100 m high volume compared with the surrounding architectures, is easy to form landmark headquarter buildings, and focuses on building the street facade along the North Fifth Ring Road in the south side. The office group in south side employs the layout form of low middle and high sides, creates a pile of group layout with shocking and visual effects, and gives the innovative spirit by diagonal blocks interspersing through the facade.

建筑设计

鉴于未来使用者均是国内外知名企业，建筑必须有足够的形象代表性来体现其企业文化。与周边相比，其 100 m 高的建筑体量，易于形成标志性总部建筑群，并重点打造南侧对北五环路沿街立面。南侧办公组团采用中间低、两边高的布局形式，创造出群体布局，具有震撼性和视觉效果，并通过立面斜向体块穿插而赋予其创新精神。

China Mobile Phone Animation Base

中国移动手机动漫基地

Features 项目亮点

No.1 and No.2 Equipment Buildings stand randomly along the lake, forming a rhythmic building group and presenting a beautiful skyline.

沿湖的一号和二号机房楼，随意分布，形成富有韵律感的建筑群，提供了富有动感的沿湖建筑天际轮廓线。

Keywords 关键词

Randomly Scattered
随意分布

Dynamic
富有动感

Avant-garde Sense of Technology
前卫科技感

Location: Xiamen, Fujian, China
Architecture Design: Xiamen Hodor Design Group
Land Area: 133,239.5 m^2
Gross Floor Area: 250,000 m^2
Completion Date: 2010

项目地点：中国福建省厦门市
建筑设计：厦门合道工程设计集团有限公司
用地面积：133 239.5 m^2
总建筑面积：250 000 m^2
竣工时间：2010 年

» Overview

China mobile phone animation base officially opened in Xiamen on April 26, 2010, is one of China mobile nine value-added service bases, and also is another move following the music, reading, video and games in the field of digital culture of Chinese mobile services.

» 项目概况

中国移动手机动漫基地于 2010 年 4 月 26 日正式落户厦门，是中国移动九大增值业务基地之一，也是继音乐、阅读、视频和游戏之后中国移动服务数字文化领域的又一举措。

» Overall Planning and Layout

As a modification and deepening of the original, the program preserves the original idea of the overall layout: “one axis, one center, three squares and four Areas, phased construction can be expanded” and basic traffic organizational structure and takes interference ripples formed by waves radiation triggered by four “wave sources” as the source of the idea to reconstruct the axis frame of animation base park. Two “wave sources” formed the two central square areas, and the other two “wave sources” formed urban space near lake and street. All secondary road structure of the original program has been retained, such as the middle square, circular road and the curved viaduct across the two sections.

» 总体规划布局

作为原方案的修改和深化，本方案尽量保留了原方案的总体布局构思。保留了原方案“一轴线，一中心，三广场，四片区，分期建设可扩容”的总体规划布局和基本的交通组织结构，对原先的总体构思概念进行了重新梳理，用 4 个“波源”引发的电波辐射形成的干涉波纹作为构思来源，重新建构动漫基地园区的轴心构架。两个“波源”形成了两个中心广场区域，另两个“波源”则形成临街和临湖的城市空间。原方案的次级道路结构全部得到了保留，比如，中间广场圆形道路以及跨越两个地段的弧形高架桥等。

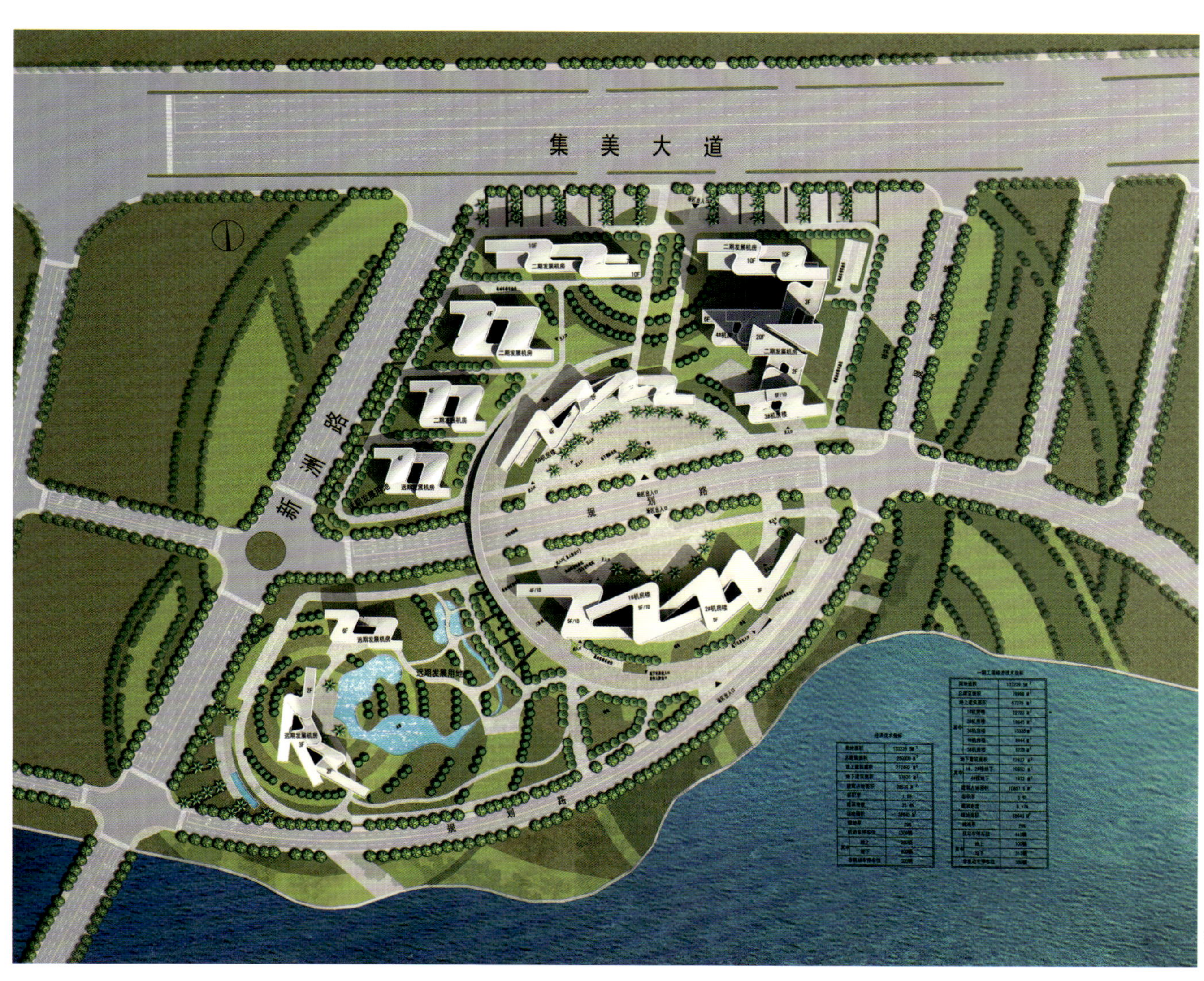

Site Plan
总平面图

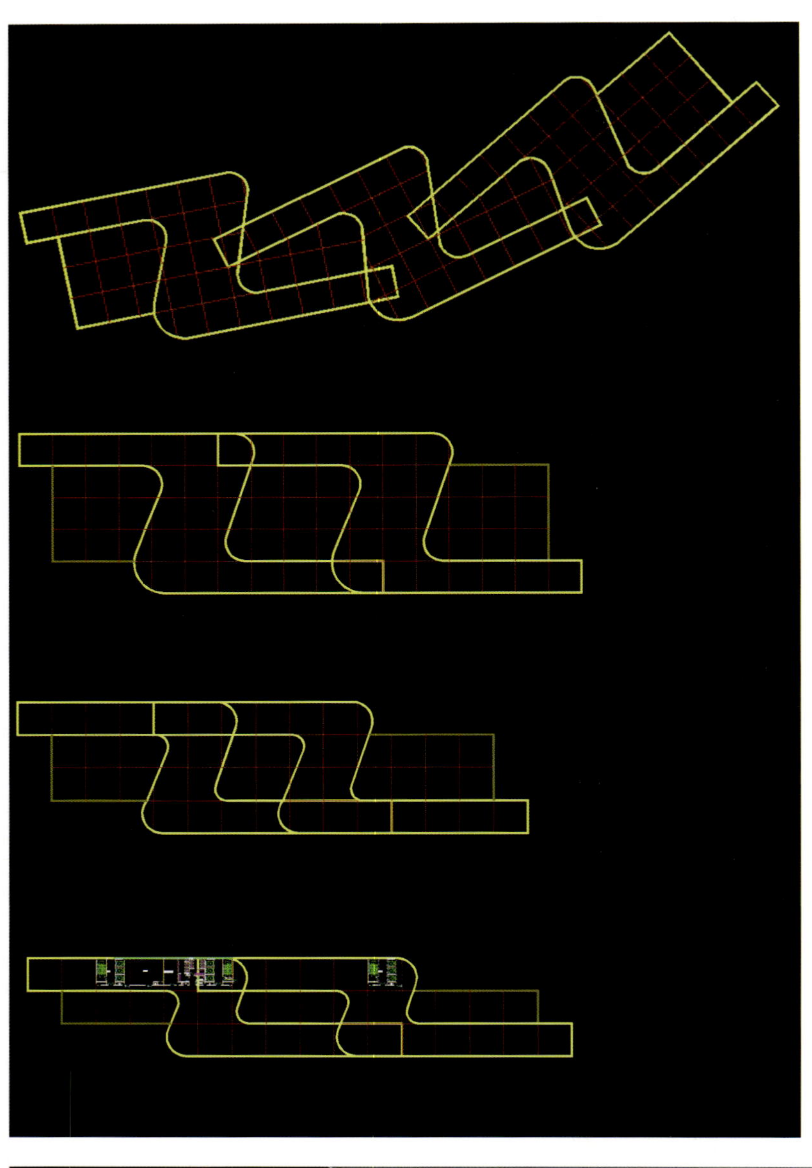

Design Idea

The project is largely inspired by the interference of electromagnetic wave and the animated movie — TRON: Legacy which has greatly influenced the design style of many industrial products, including the mobile phones, computer and accessories, and even the shape of buses. The smooth and dreamlike waves in the movie reflect the relationship between the virtual and physical world, bringing great inspirations to the architectural design of China Mobile Phone Animation Base.

设计理念

项目深受电磁波干涉和动画电影《创·战纪》的启发。《创·战纪》所描绘的电子游戏世界中的虚拟形象影响了大量工业产品的设计风格，如手机、电脑周边，甚至公共汽车造型等，其流畅而梦幻的波动线条反映了虚拟与物理之间联系，成为手机动漫建筑群绝好的建筑构思来源，非常符合中国移动手机动漫基地产业园的建筑性质和个性。

Architectural Design

Facade's windows employ Z-shaped lines under orthogonal grid control. All belt-type windows correspond to storey height to reduce the implementation cost. Under the premise of in ensuring adequate rate of open windows belt-type windows can be flexibly arranged according to the different needs of internal functions, creating rich facade texture changes, which provides a unified overall and smooth texture without repeat throughout the park. No. 1 and No.2 Equipment Buildings stand randomly along the lake, forming a rhythmic building group and presenting a beautiful skyline for the lakeside area and the nearby Xinglin Garden Show Park.

建筑设计

立面的开窗方式采用了正交网格控制下的Z字形线条。所有带型窗与层高相对应，降低了实施造价。在保证各层充分的开窗率前提下，可以根据内部功能的不同需求进行带型窗的灵活分布，产生丰富的立面纹理变化，为整个园区提供了总体统一但又不重复的流畅肌理。沿湖的一号和二号机房楼，高低错落，形成富有韵律感的建筑群，为紧邻湖边和杏林园博园的这个重要地段，提供了富有动感的沿湖建筑天际轮廓线。

Functional Layout

After the analysis on the basic form of the interference ripples, the architectural form employed Z-shaped unit plane layout, through different rotation, stitching and dislocation formed dynamic sense similar to the undulating wave crest and trough.

Undulating buildings achieved through completely regular architectural form by certain regular rotation and transformation of four kinds of basic unit forms, whose shape composition are based on 8.4 × 8.4m orthogonal column grid. Building depth has two spans (16.8 m), three spans (25.2 m) and four spans (33.6 m) to meet the functional requirements for the telecommunications room in original plan. Curved building volume is not composed of building column grid, but of the Z-shaped unit form after rotated, providing good operability and enforceability.

The monomer form forms the overall image like continuous mountains by alternation of ups and downs of the rising computer room layer and equipment layer and each monomer's interleaving, and more like the mapping radio waves presenting China Mobile Phone Animation Park's unique enterprise connotation and image.

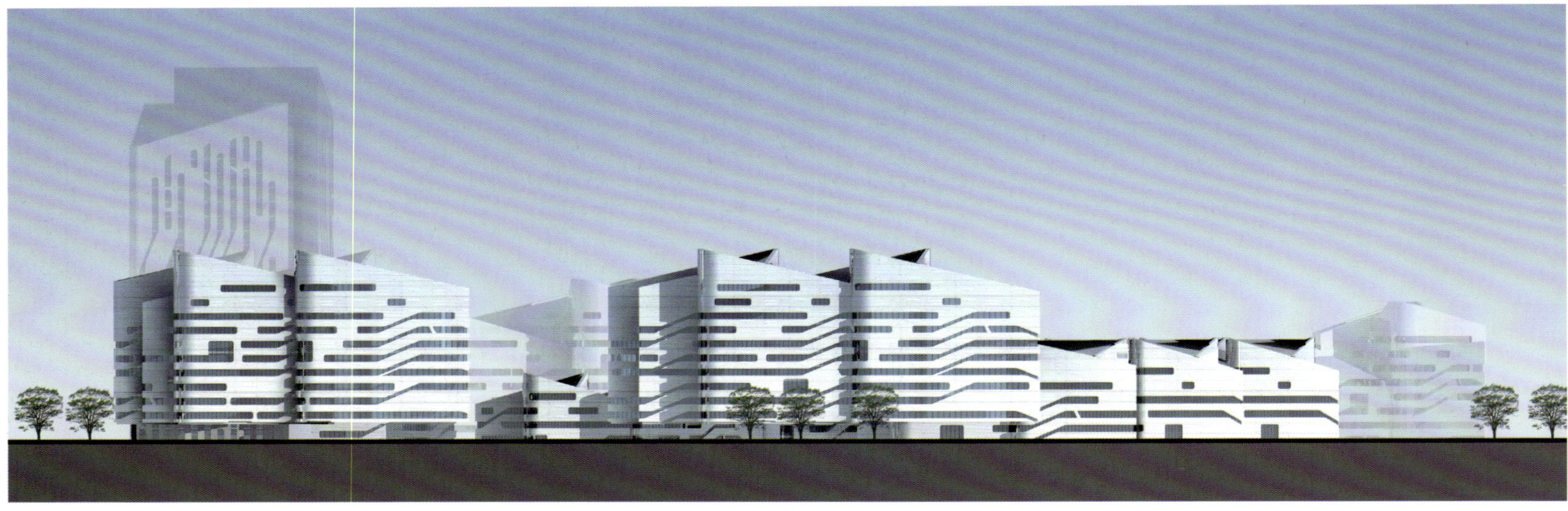

功能布局

经过对干涉波纹基本形态的抽象和分析，建筑形态采用了 Z 形的单元平面布局，通过不同的旋转、拼接和错动形成类似于干涉波的波峰和波谷交错起伏的动态感。

起伏的建筑群由完全规整的建筑形体以 4 种基本单元形体经过具有一定规律的旋转和变换得到，各种单元形体的形态构成都基于 8.4 × 8.4 m 的正交柱网，建筑物进深有两跨（16.8 m）、三跨（25.2 m）、四跨（33.6 m），以满足原方案对电信机房的功能性要求。弧形的建筑体量并不是由曲线的建筑柱网组成，而是经过旋转的 Z 形单元体组成，提供了良好的操作性和可实施性。

单体形态由高起的屋顶机房层和设备层交替起伏，各单体的交错形成了如连绵山峰般的总体形象，更如同映射的电波，传递中国移动手机动漫园的独特企业内涵和企业形象。

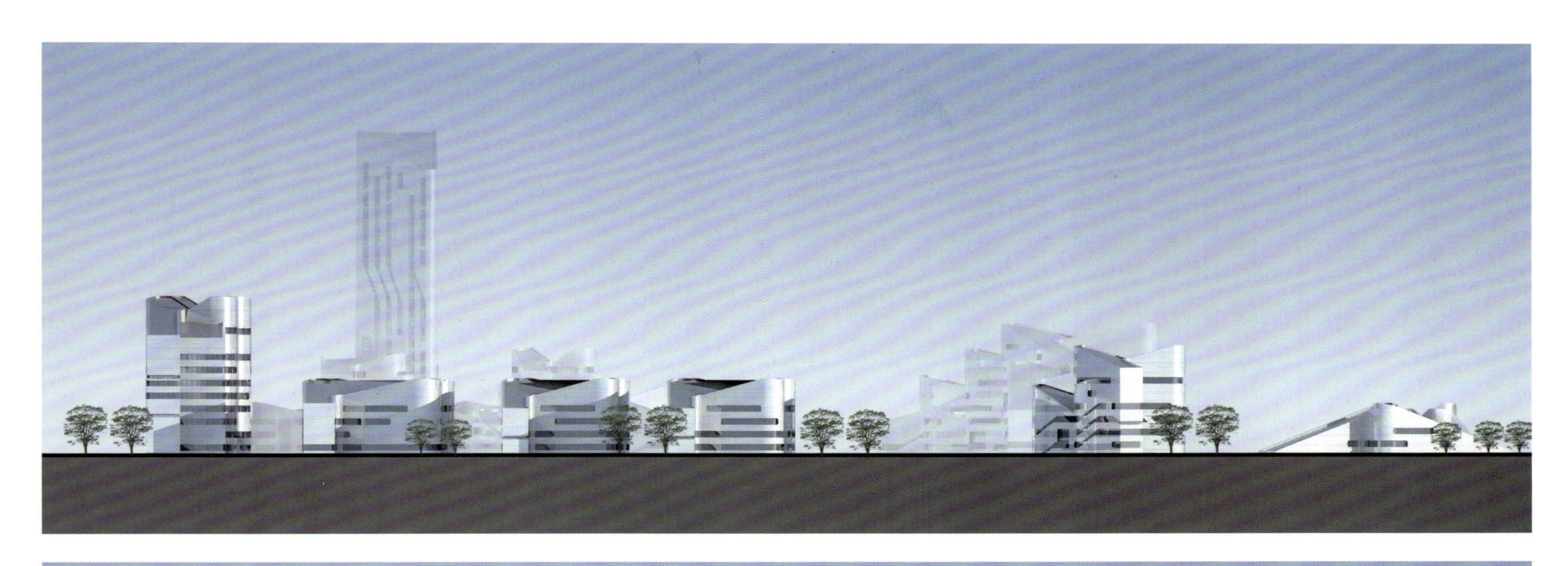

Beijing Olympics Sports Culture and Business Park

北京奥体文化商务园区

Features 项目亮点

Elegant roof style not only highlights the focal point but also becomes the landmark of this area to improve the cultural value of the project.

飘逸的顶盖造型不仅在视觉上强化焦点性，更形成地标，提升项目文化价值。

Keywords 关键词

Elegant Styling
飘逸造型

Landmark Building
地标建筑

Grand and Concise
大气简洁

Location: Beijing, China
Architecture Design: Gensler
Cooperative Design: Beijing Victory Star Architectural & Civil Engineering Design Co., Ltd.
Construction Unit: Beijing Tian Yuan Xiang Tai Properties Limited
Land Area: 20,021.3 m^2
Floor Area: 202,600 m^2
Plot Ratio: 6.6
Green Rate: 15%
Design Date: 2012

项目地点：中国北京市
建筑设计：Gensler
合作设计：北京维拓时代建筑设计有限公司
建设单位：北京天圆祥泰置业有限公司
用地面积：20 021.3 m^2
建筑面积：202 600 m^2
容积率：6.6
绿地率：15%
设计时间：2012 年

Overview

The project is positioned as an international top office, composed of two 147.15 m office buildings that are respectively located in the corner of northeast and southwest to ensure nice sight when overlooking Olympic Centre and Asian Games venues.

项目概况

本项目定位为国际顶级写字楼，场地中地面建筑由两栋 147.15 m 高的办公楼组成。分别位于东北角和西南角，保证其能很好地远眺到奥林匹克中心和亚运会场馆区。

Design Concept

From the functional plane to architectural design, designs have fully considered "dual cores" characteristics of this project, reaching visual and inner balance through uniform distribution of "public and private" spaces. The design has followed the current domestic norms to make the buildings look concise, classic, unique, innovative and elegant.

设计理念

从功能平面规划到建筑设计，设计都充分考虑本项目的“双核心”特质，均匀分布“公共与私有”空间，达到视觉和内在的双重平衡。该项目设计充分结合国内现行规范，打造都市化和经典化大气简洁风范，同时表现出新颖、独特、高档、典雅的气质。

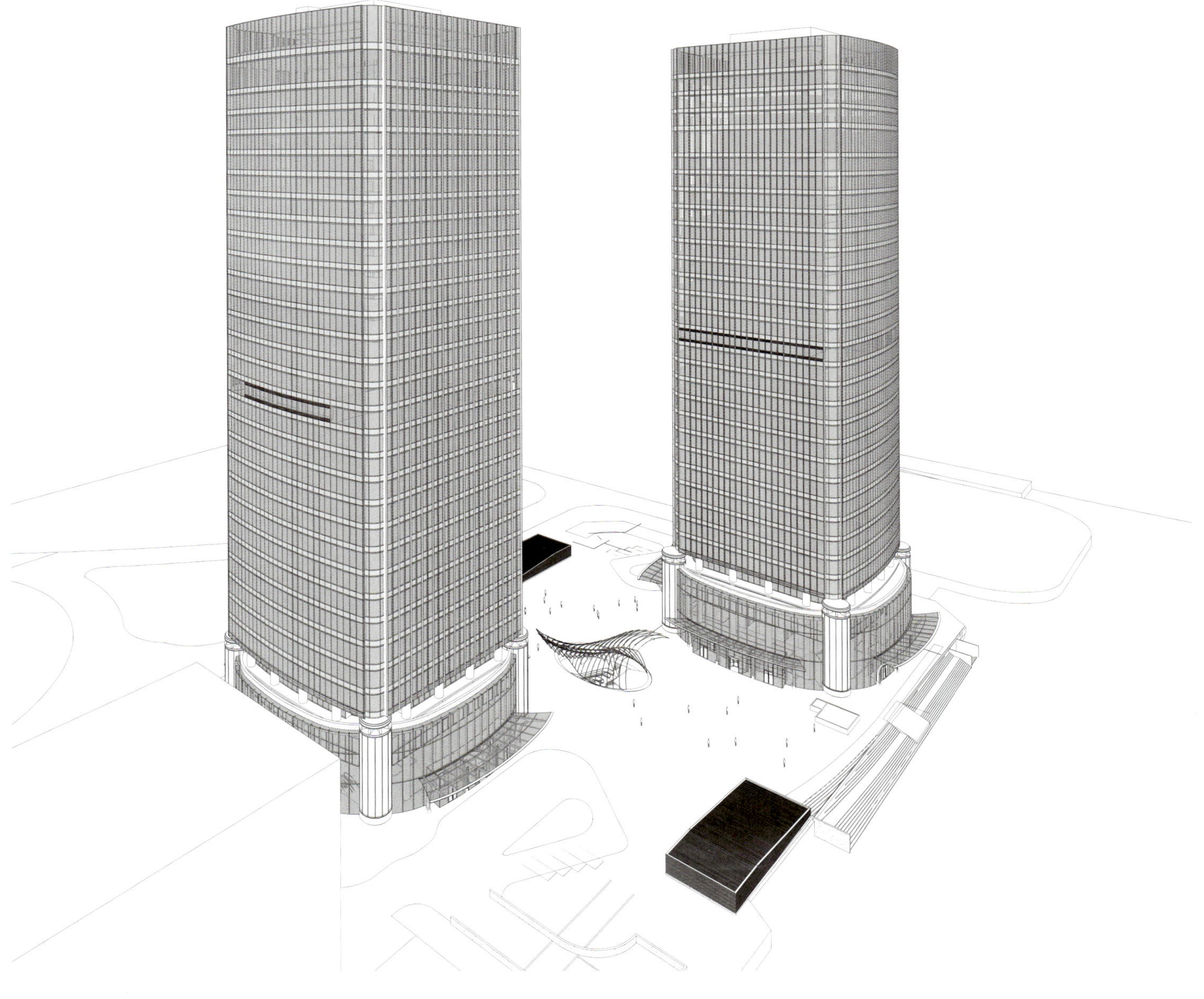

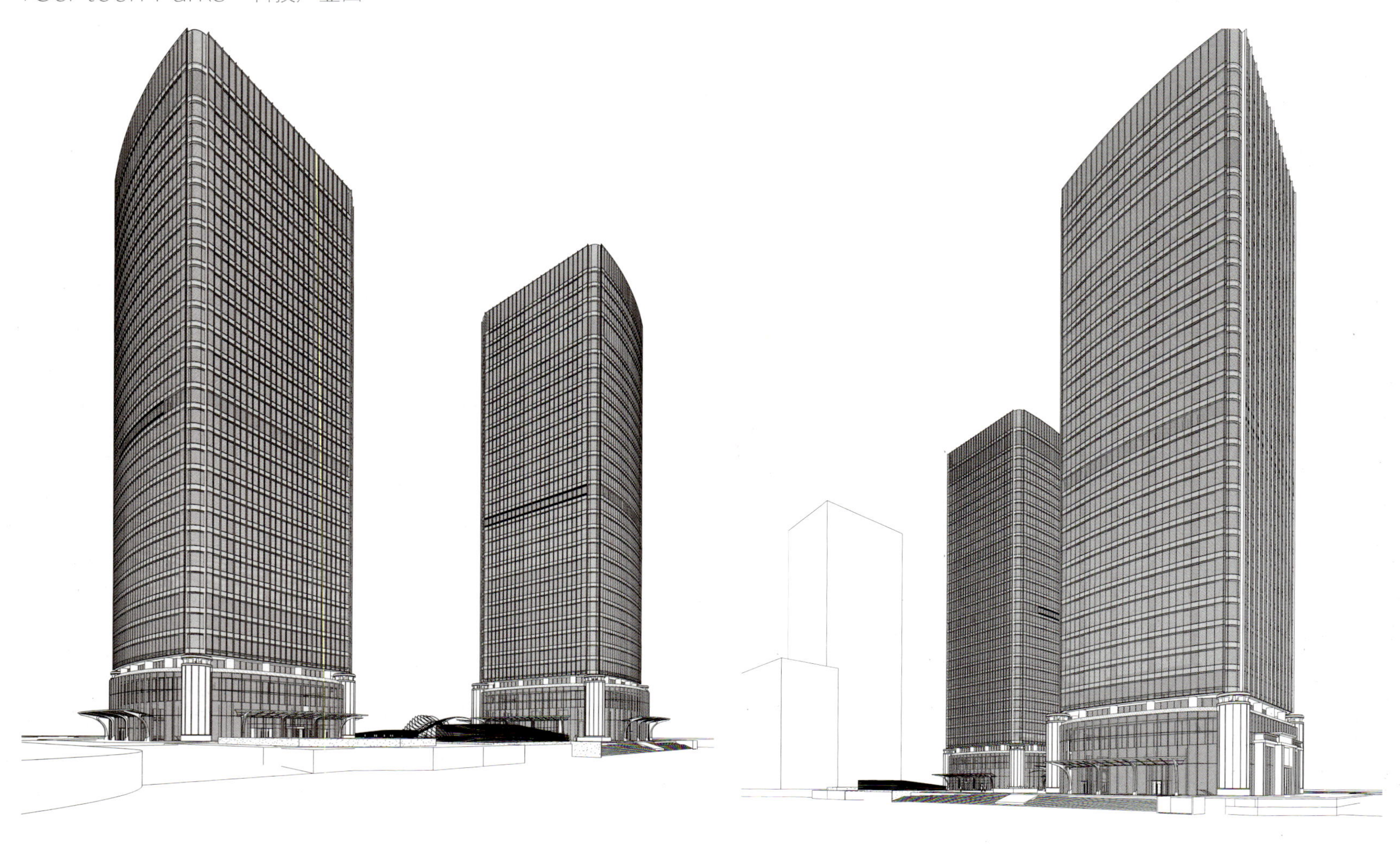

Architectural Design

Elegant roof style not only highlights the focal point but also becomes the landmark of this area to improve the cultural value of the project. Dominated by the "Bird's Nest" and the "Water Cube", the urban space in the South Olympic Area is complemented by buildings with different proportions, making the building complex within Plot OS-2 full of historical sense and cultural atmosphere.

建筑设计

飘逸的顶盖造型不仅在视觉上强化焦点性，更形成地标，提升项目文化价值。在保持奥南地区以"鸟巢""水立方"为中心的城市空间主次关系的基础上，尽量强化都市尺度的通畅感，同时建立起"5000 m－500 m－50 m"的多层次建筑机理，使整个OS-2地块建筑群更具历史感和文化包容感。

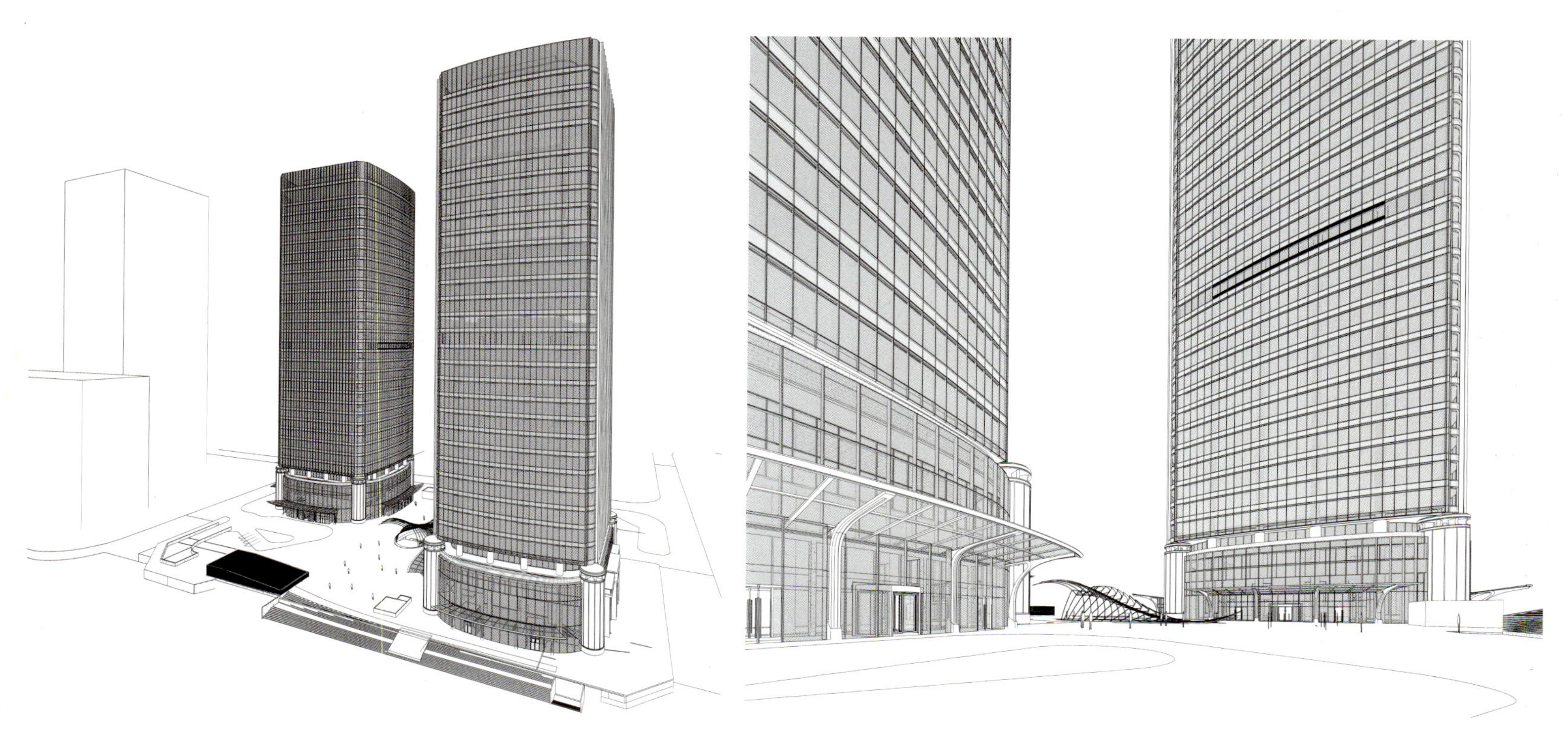

Commercial Space Design

Integrating comprehensive considerations of Olympic Sports Culture and Business Park on the underground business groups, project design created new commercial market space, multi-dimensionally organized the flow of people, thus enable the full integration of business office, commercial entertainment and other functions.

商业空间设计

项目设计结合奥体文化商务园区对地下一层集合式商业群的整体考量，塑造新生代的商业市场空间，立体化地组织人流，从而实现商务办公和商业娱乐等多功能的充分融合。

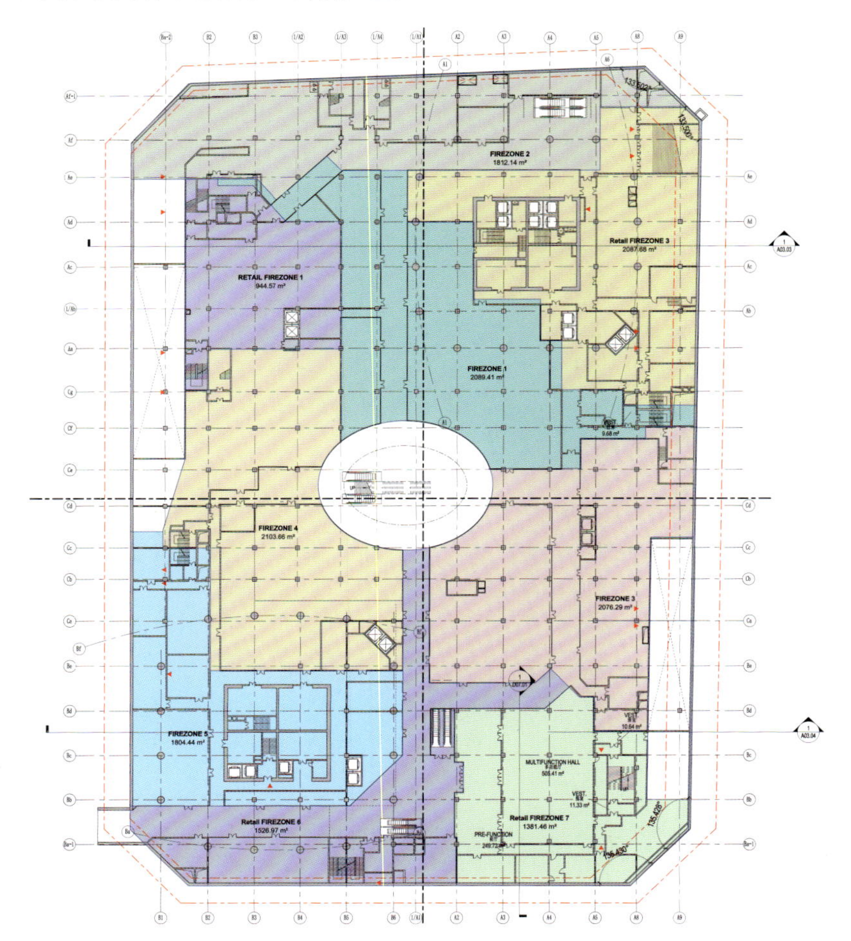

Basement Floor 1, Fire Compartment Plan 地下一层防火分区平面

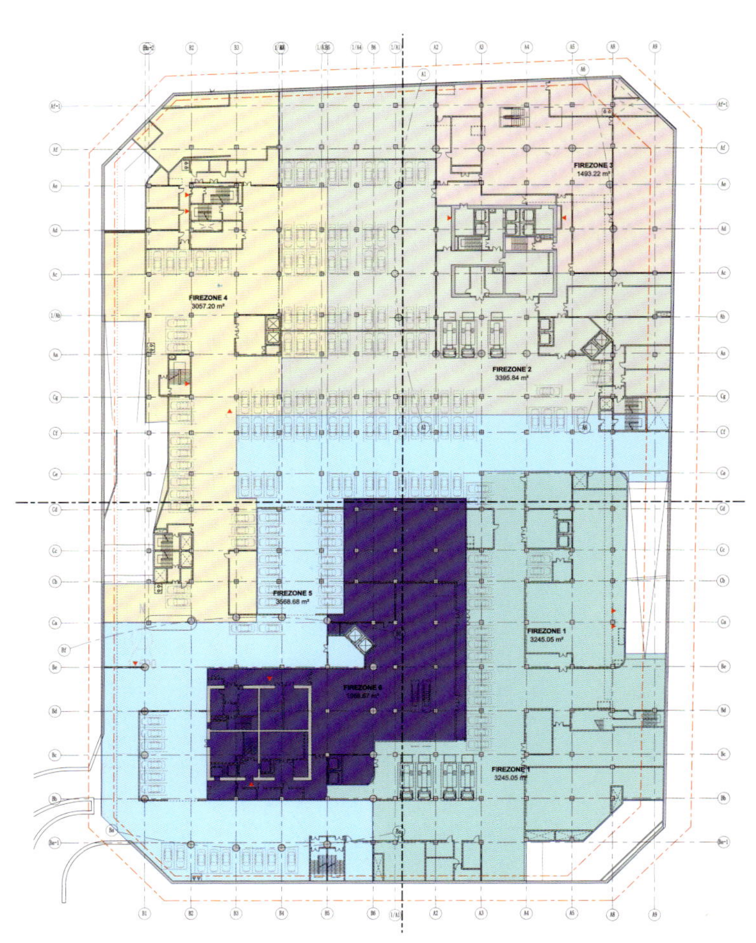

Basement Floor 2, Fire Compartment Plan 地下二层防火分区平面

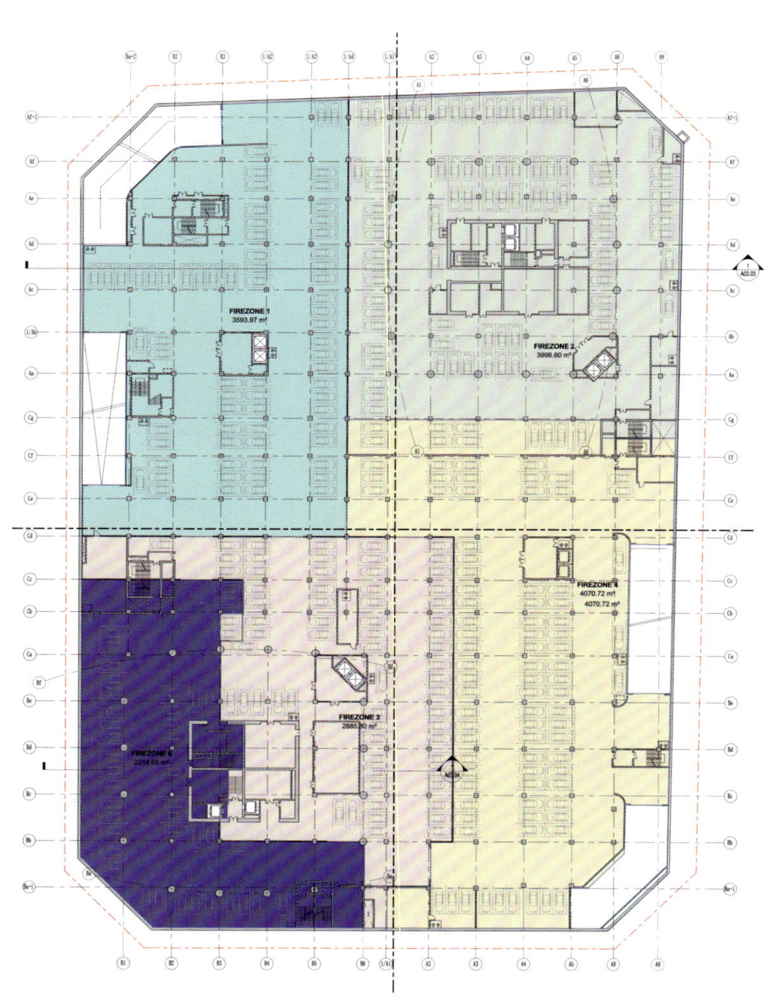

Basement Floor 3, Fire Compartment Plan 地下三层防火分区平面

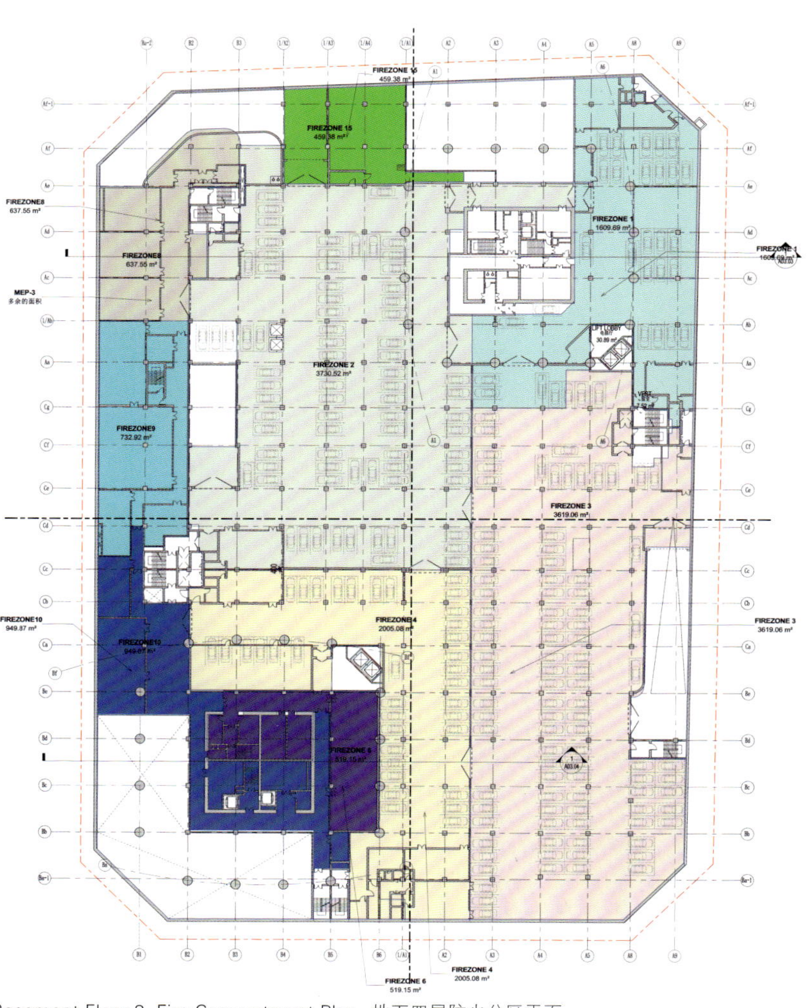

Basement Floor 3, Fire Compartment Plan 地下四层防火分区平面

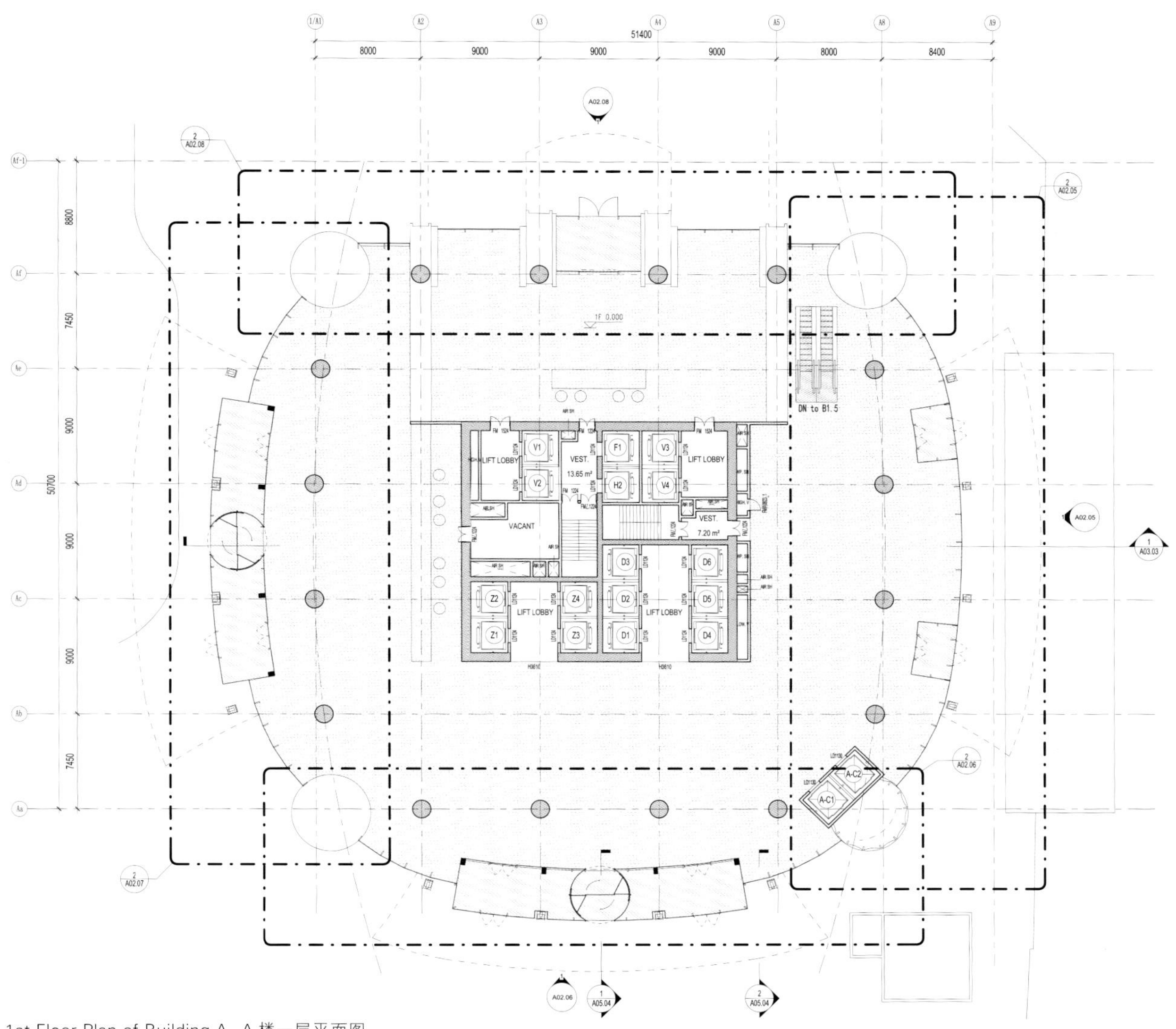

1st Floor Plan of Building A　A楼一层平面图

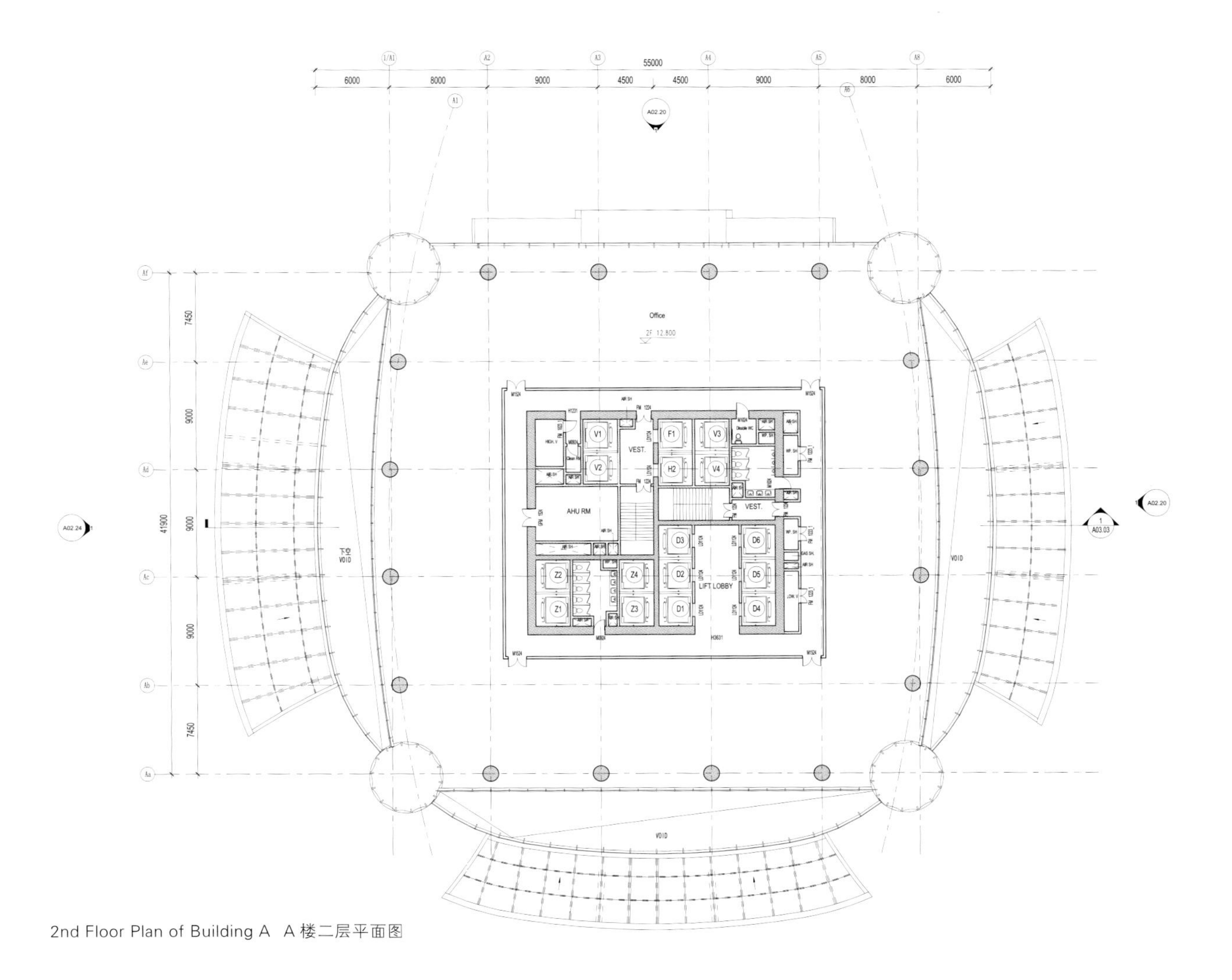

2nd Floor Plan of Building A　A楼二层平面图

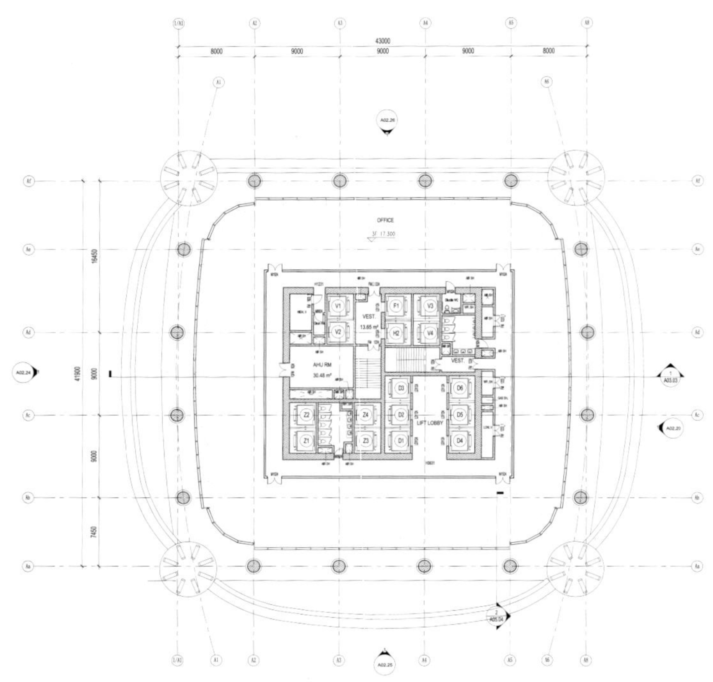

3rd Floor Plan of Building A A楼三层平面图

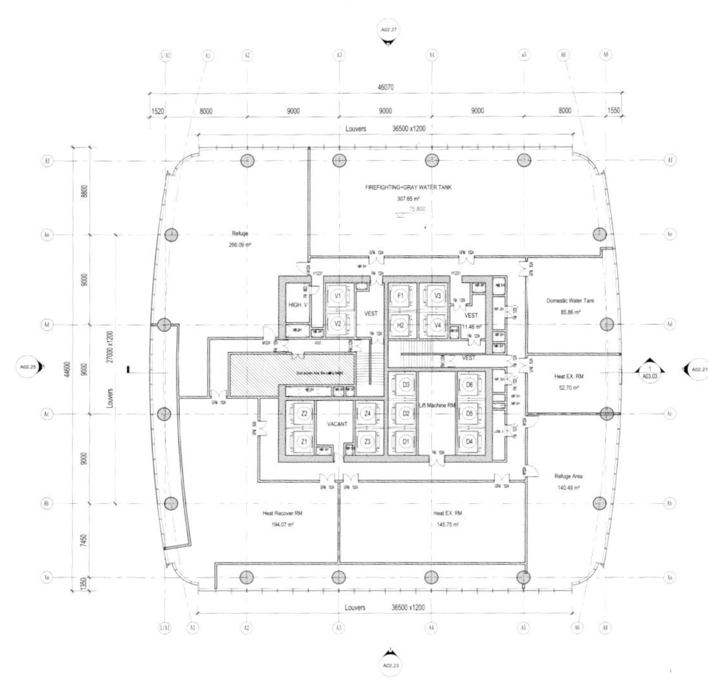

16th Floor Plan of Refuge 十六层避难层平面图

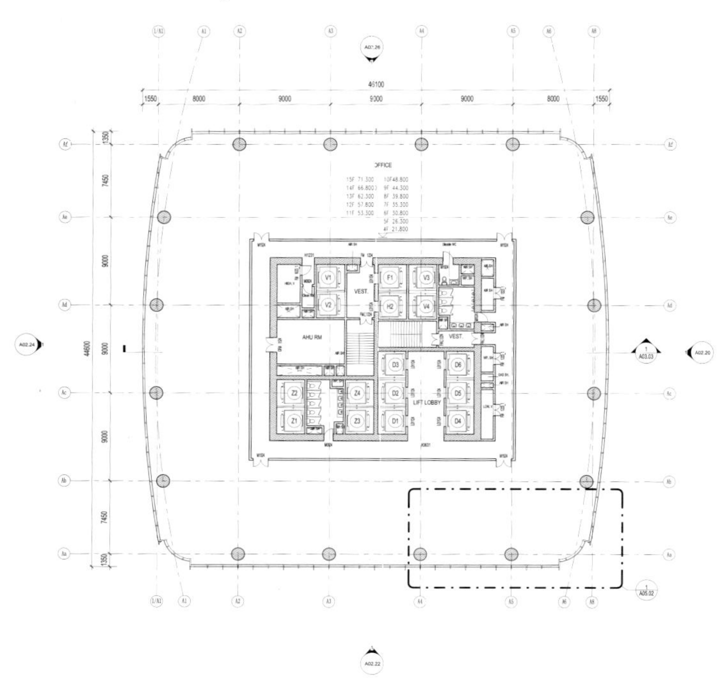

4th ~ 15th Floor Plan of Building A A楼四 – 十五层平面图

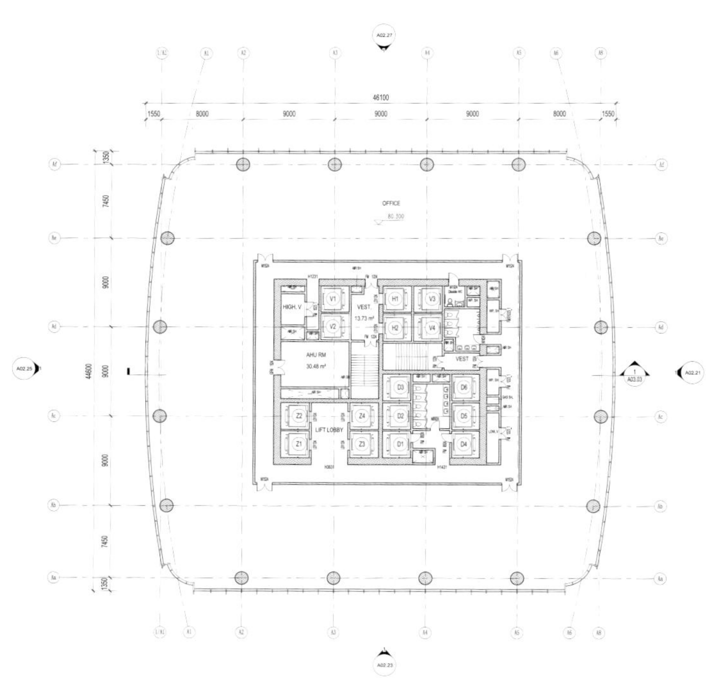

17th Floor Plan of Building A A楼十七层平面图

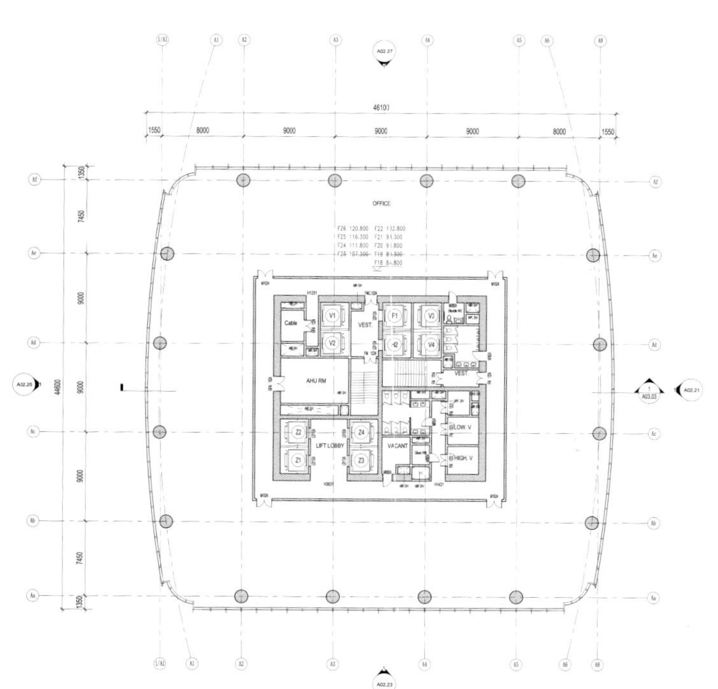

High Floor Plan of Building A A楼高区平面图

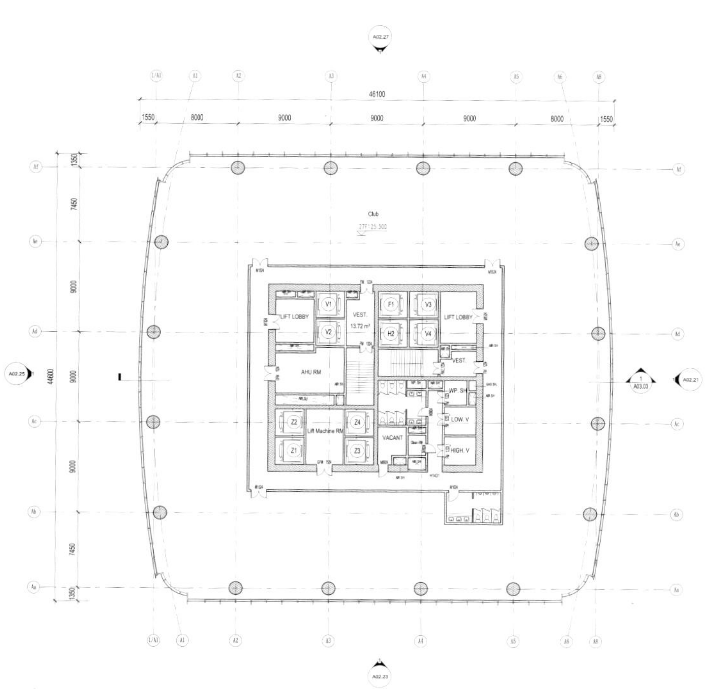

Club Plan of Building A A楼会所平面图

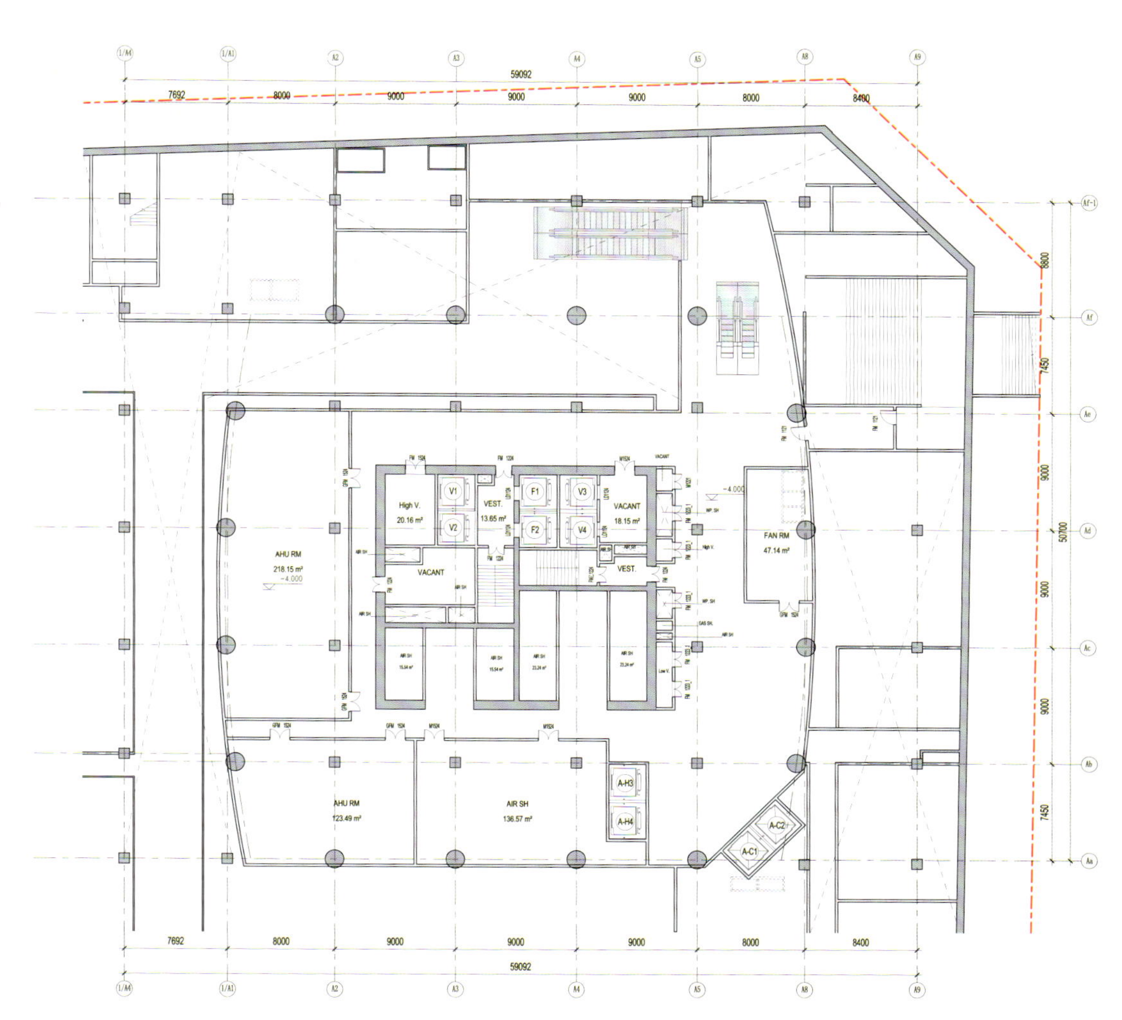

Mezzanine Plan 1　夹层平面图 1

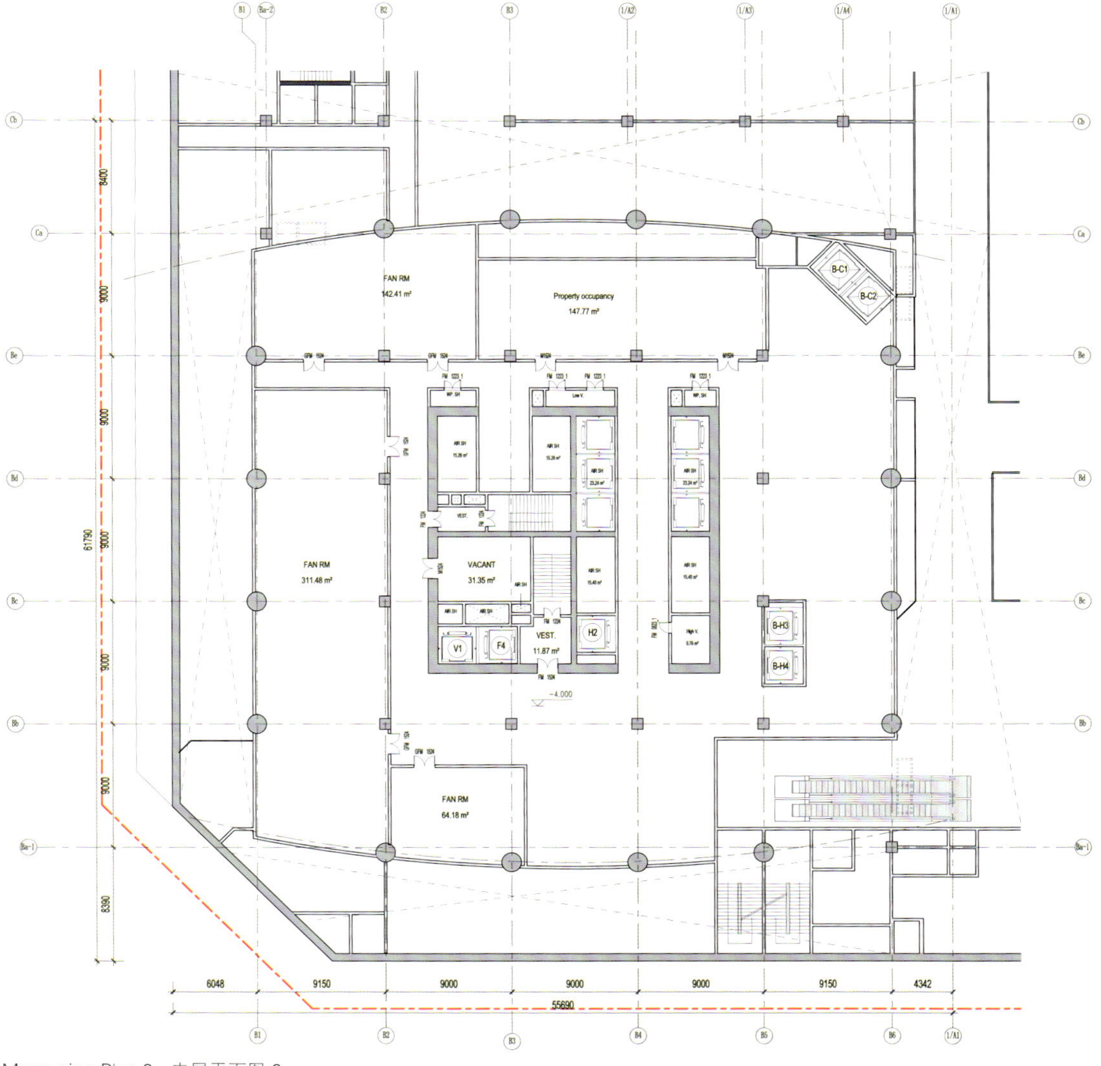

Mezzanine Plan 2　夹层平面图 2

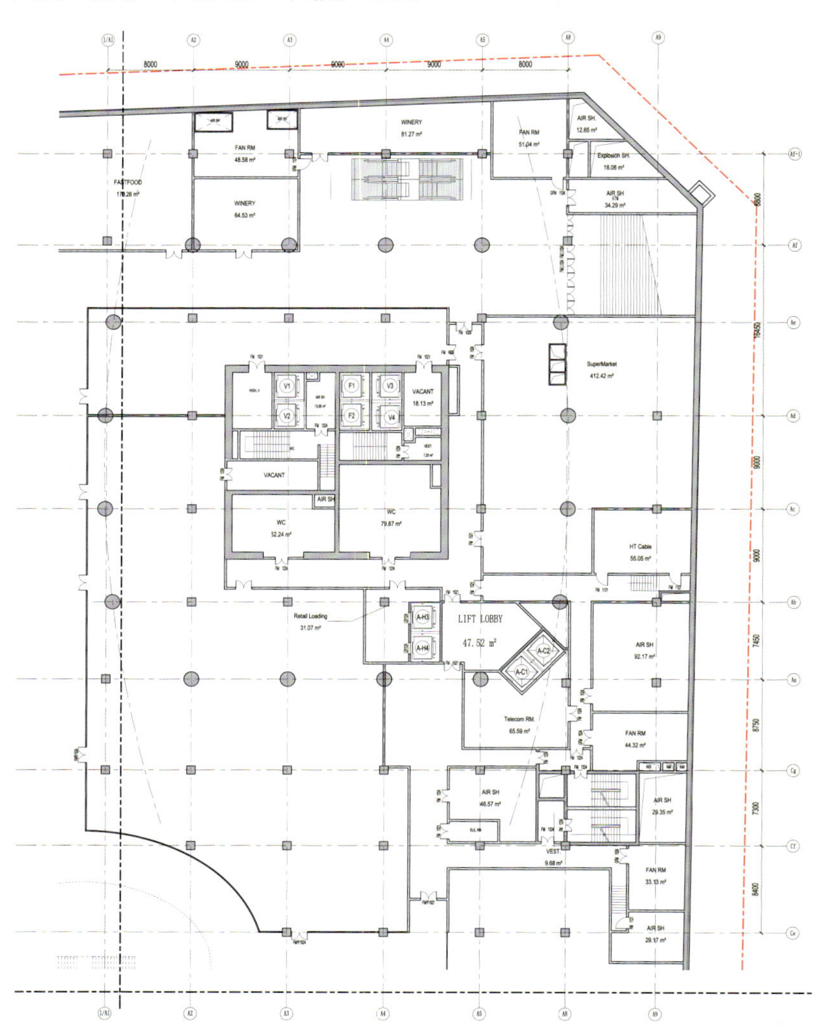

Plan for Basement One Floor 1 地下一层平面图 1

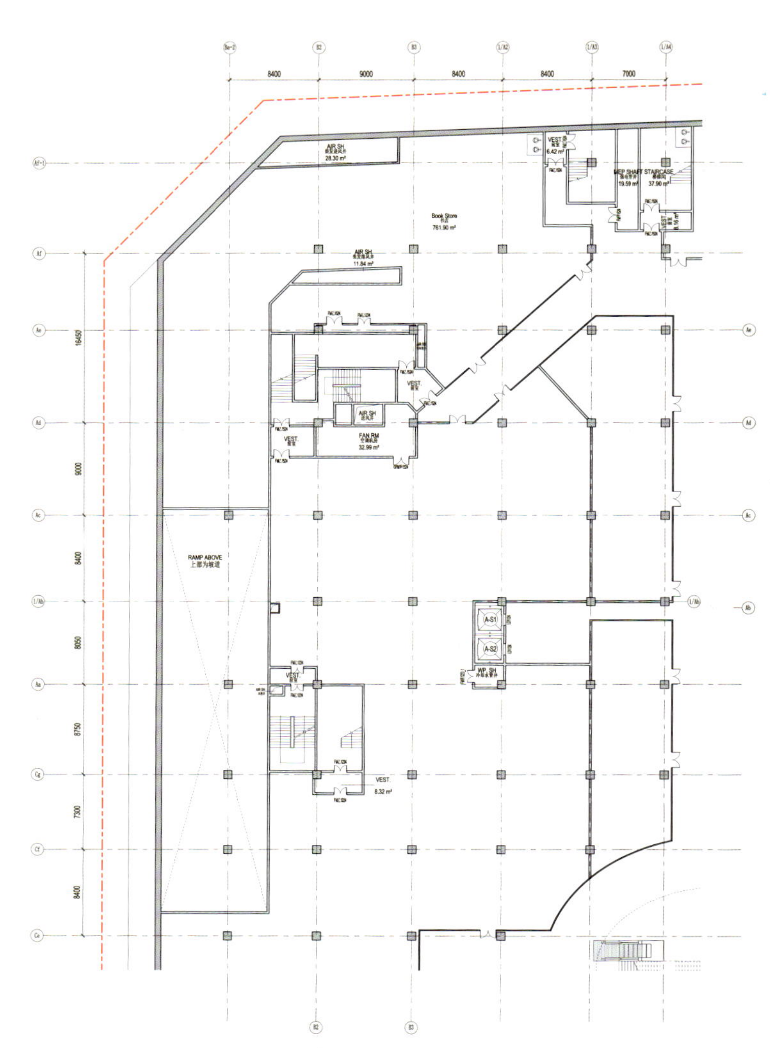

Plan for Basement One Floor 1 地下一层平面图 1

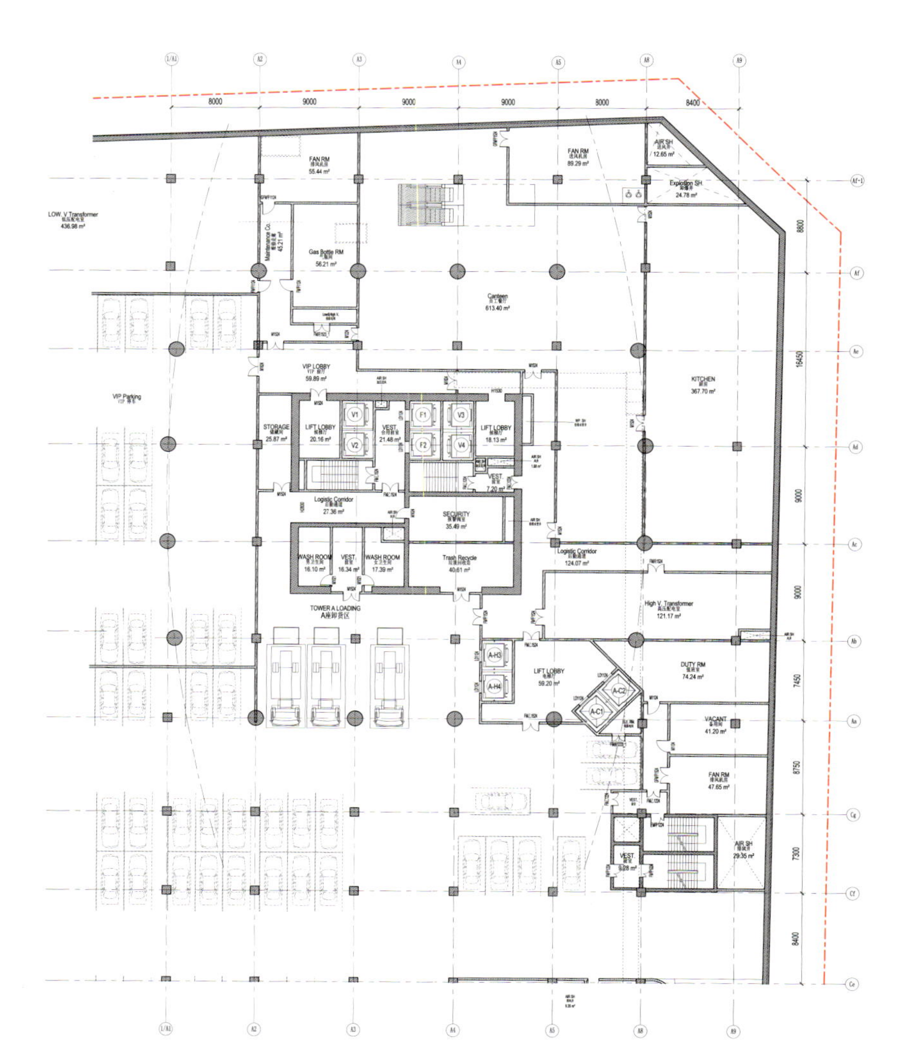

Plan for Basement Two Floor 1 地下二层平面图 1

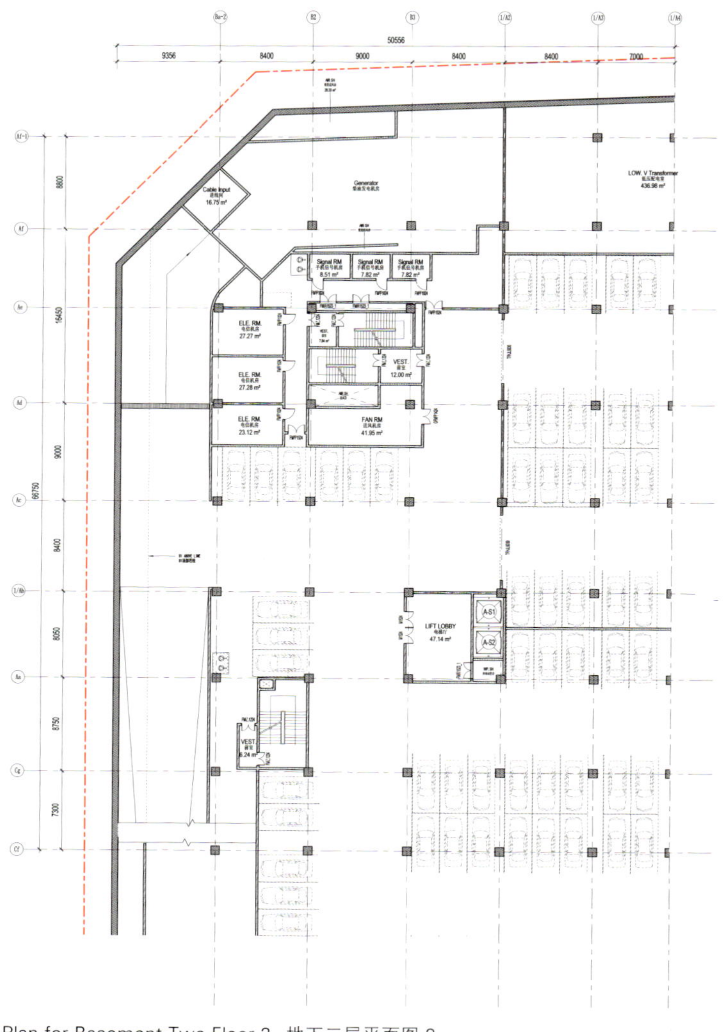

Plan for Basement Two Floor 2 地下二层平面图 2

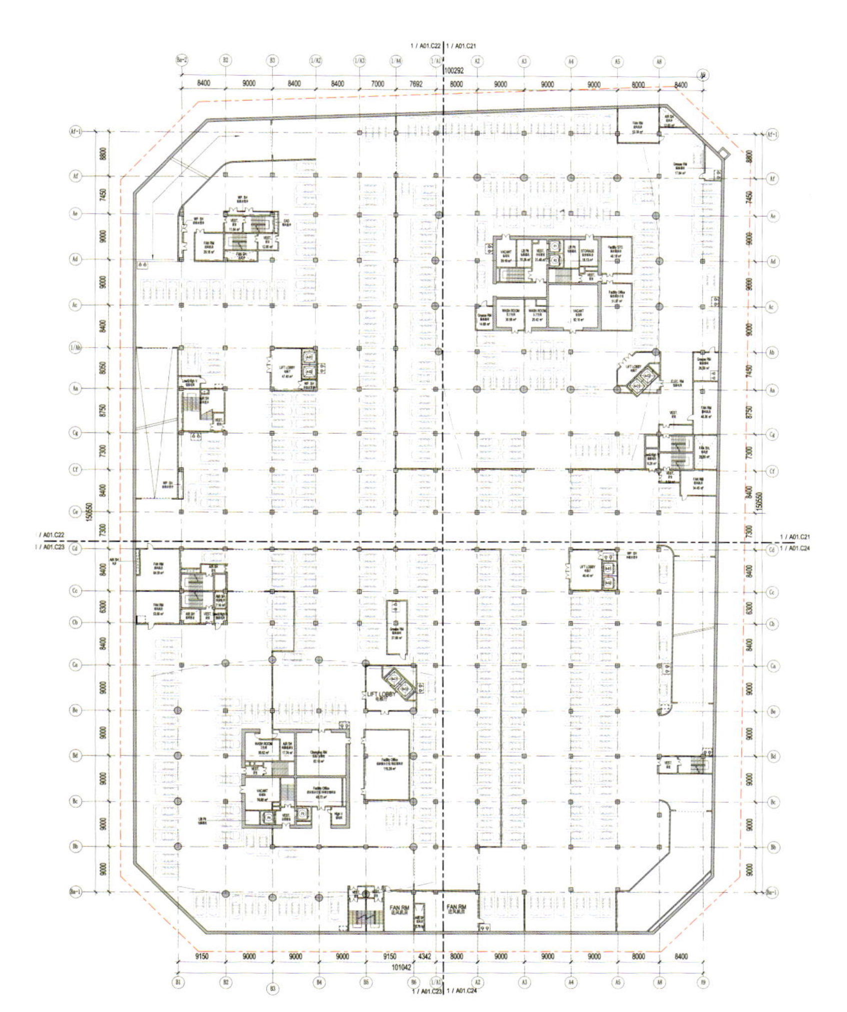

Plan for Basement Three Floor 地下三层平面图

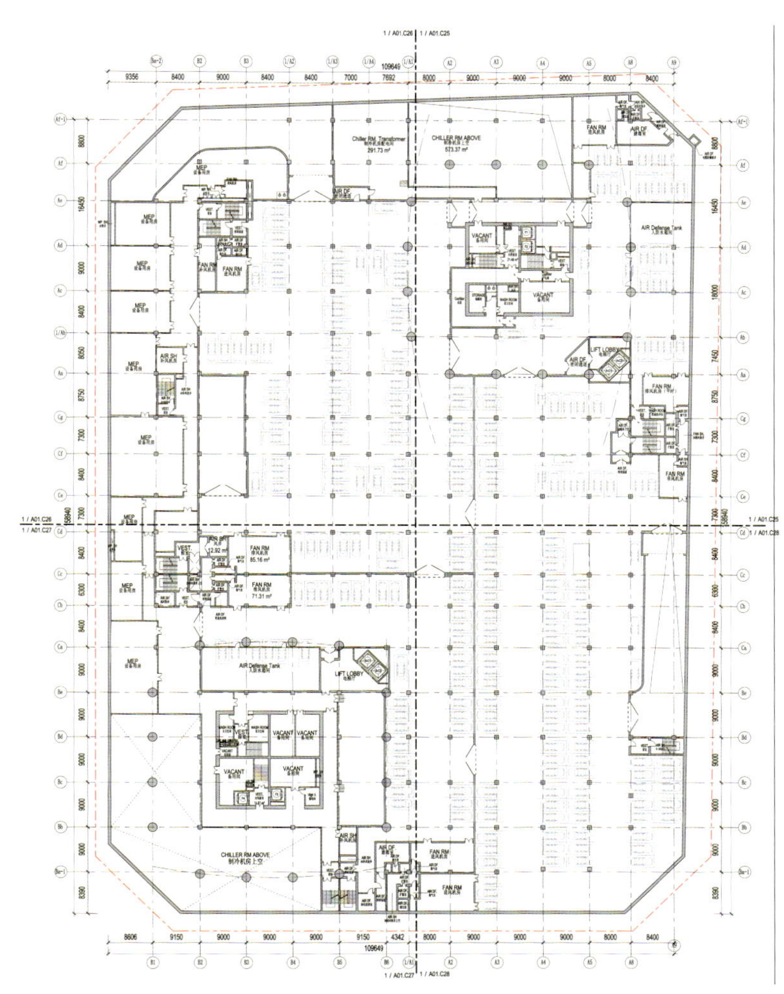

Plan for Basement Four Floor 地下四层平面图

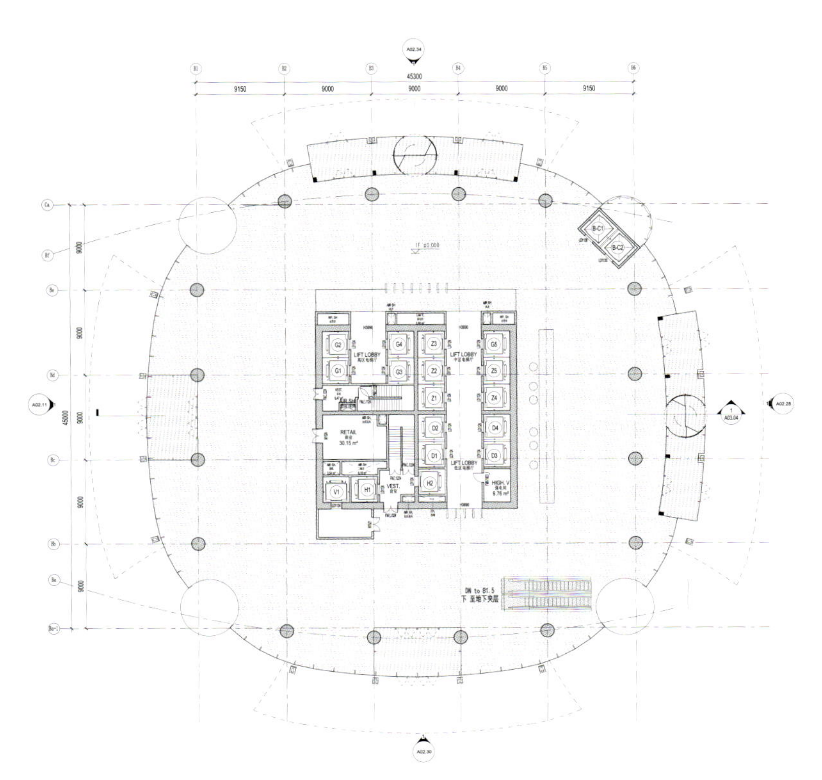

1st Floor Plan of Building B B 座一层平面图

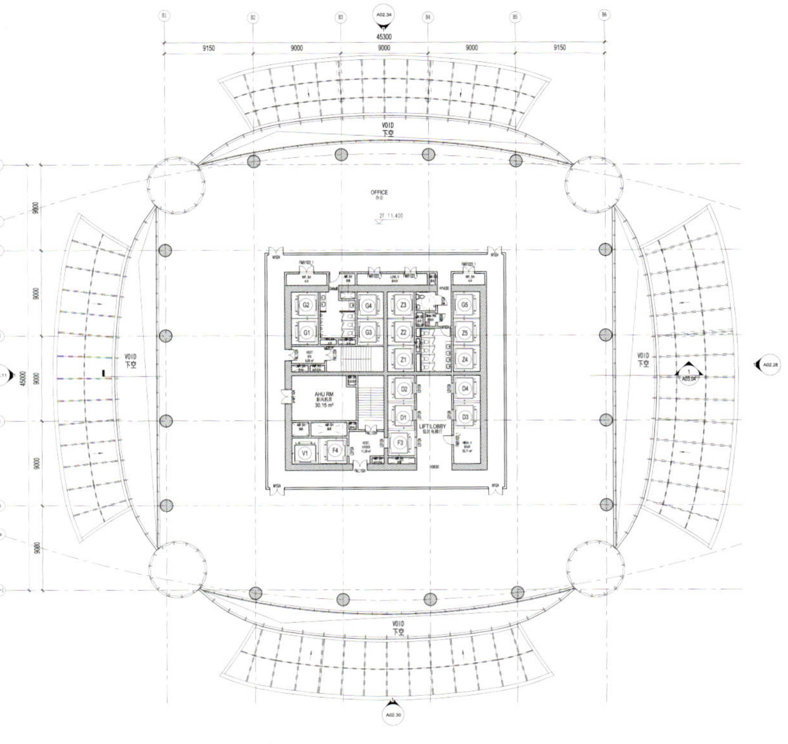

2nd Floor Plan of Building B B 座二层平面图

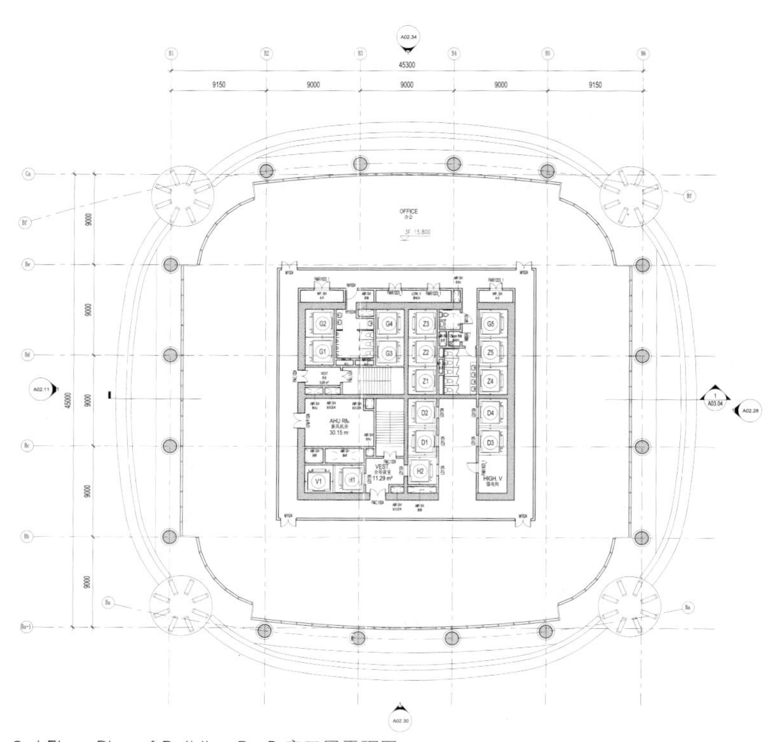
3rd Floor Plan of Building B B座三层平面图

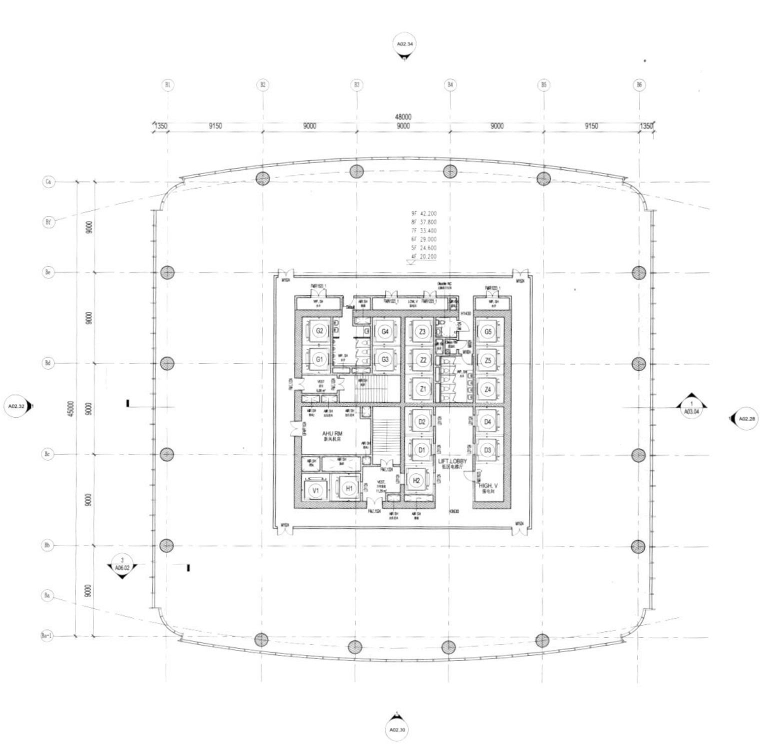
4th ~ 9th Floor Plan of Building B B座四～九层平面图

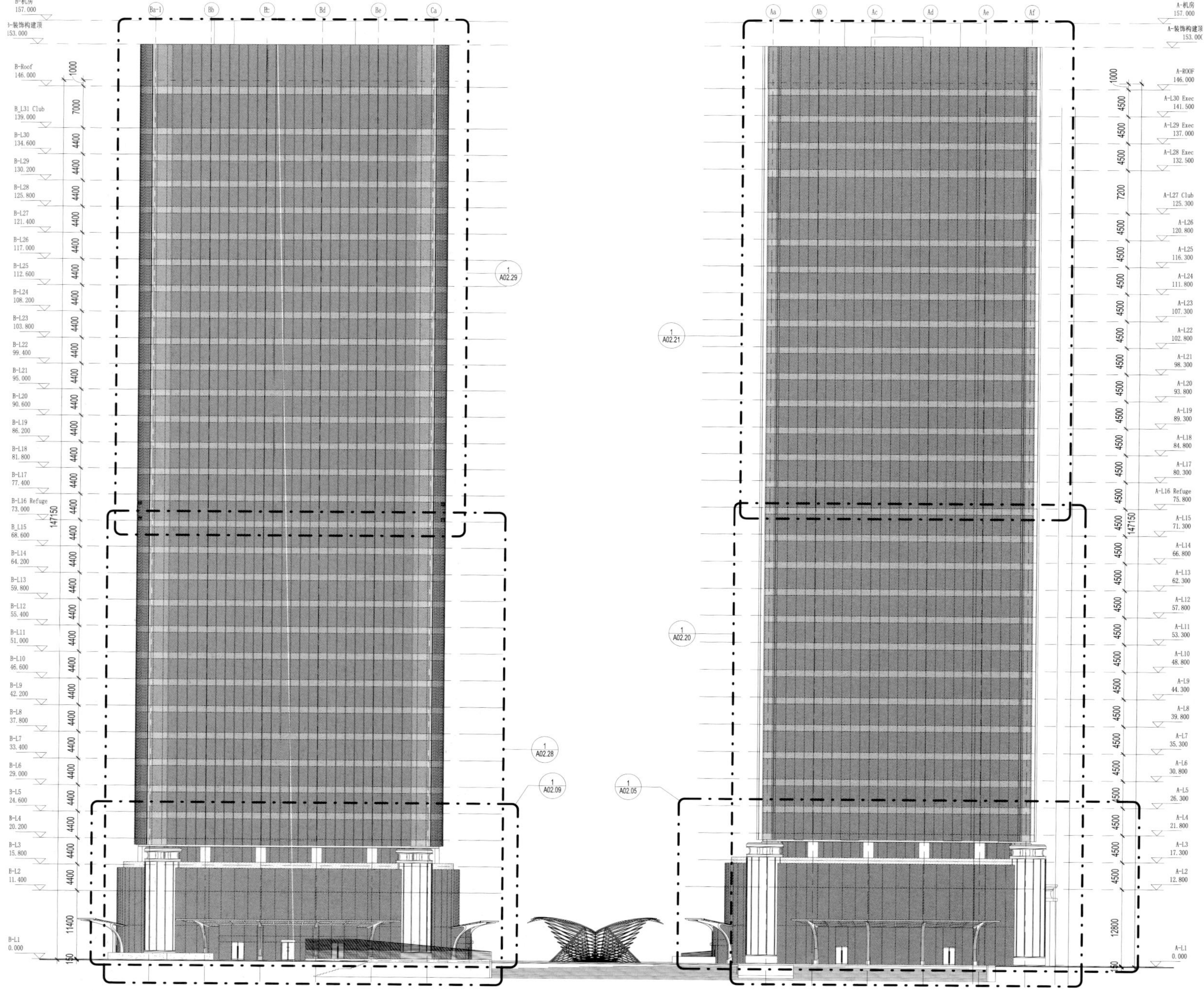
East Elevation 东立面图

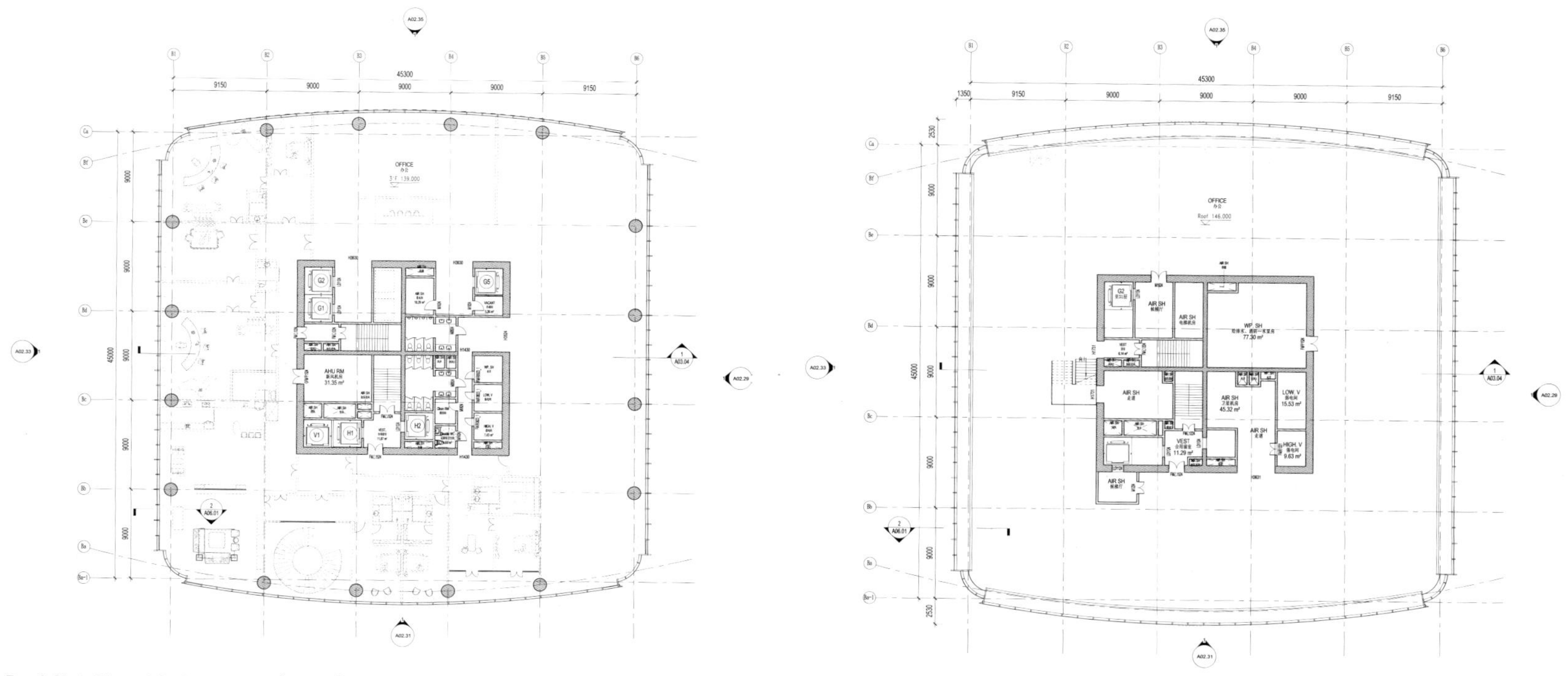

Roof Club Plan of Building B　B座顶层会所平面图

Roof Plan of Building B　B座屋顶平面图

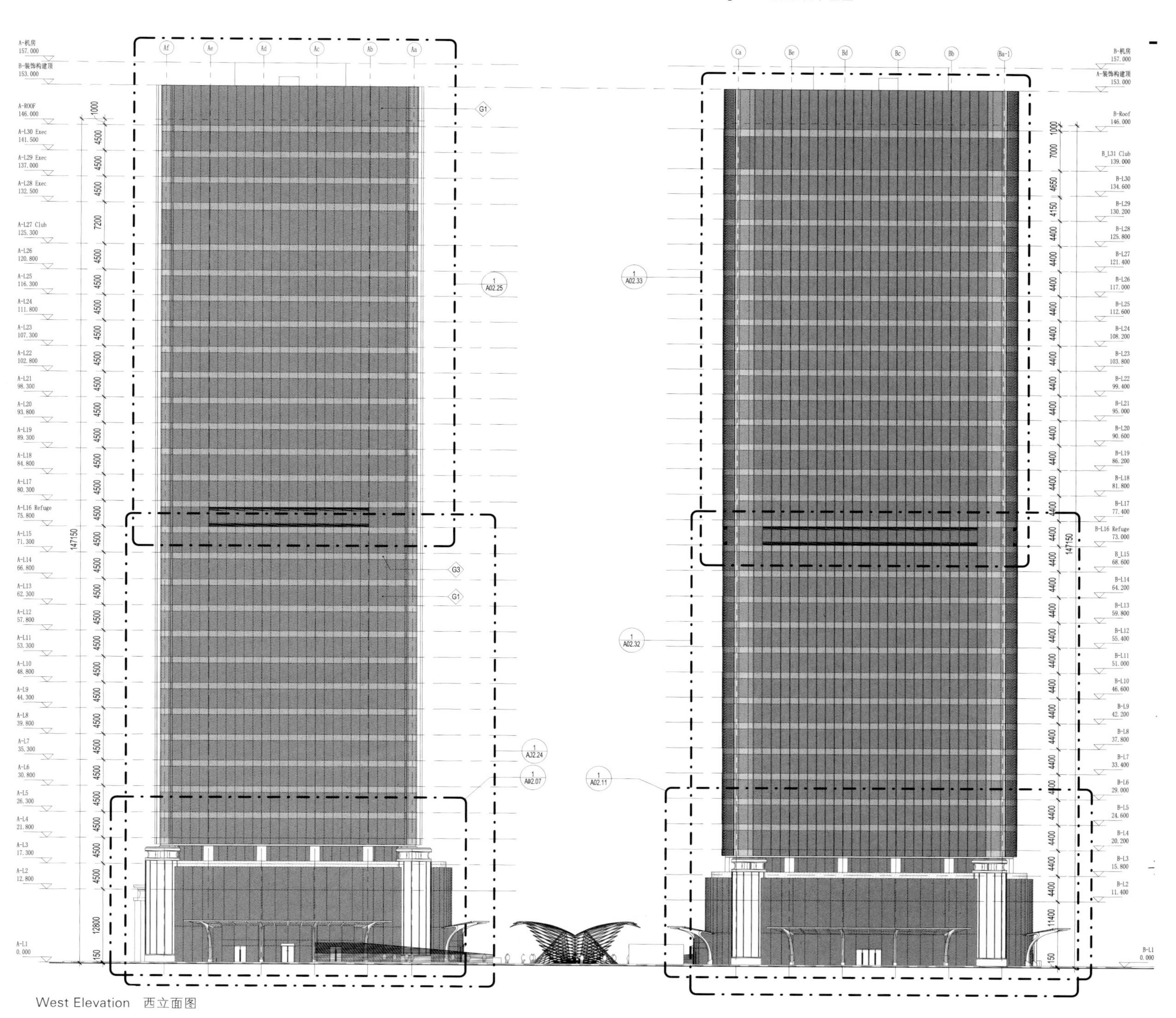

West Elevation　西立面图

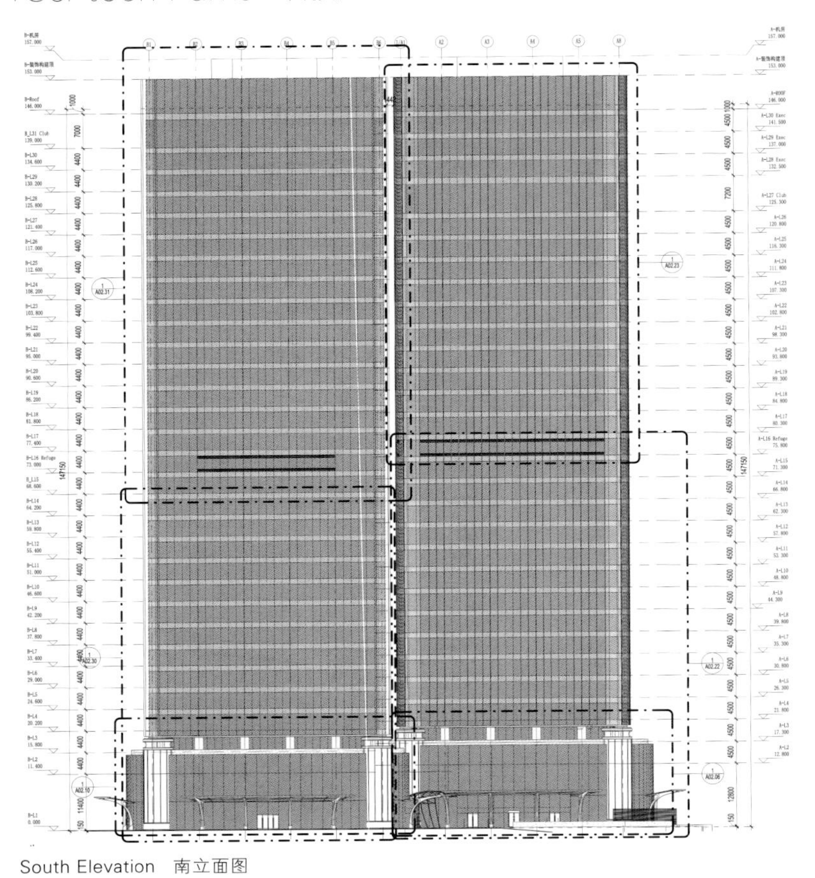

South Elevation 南立面图

North Elevation 北立面图

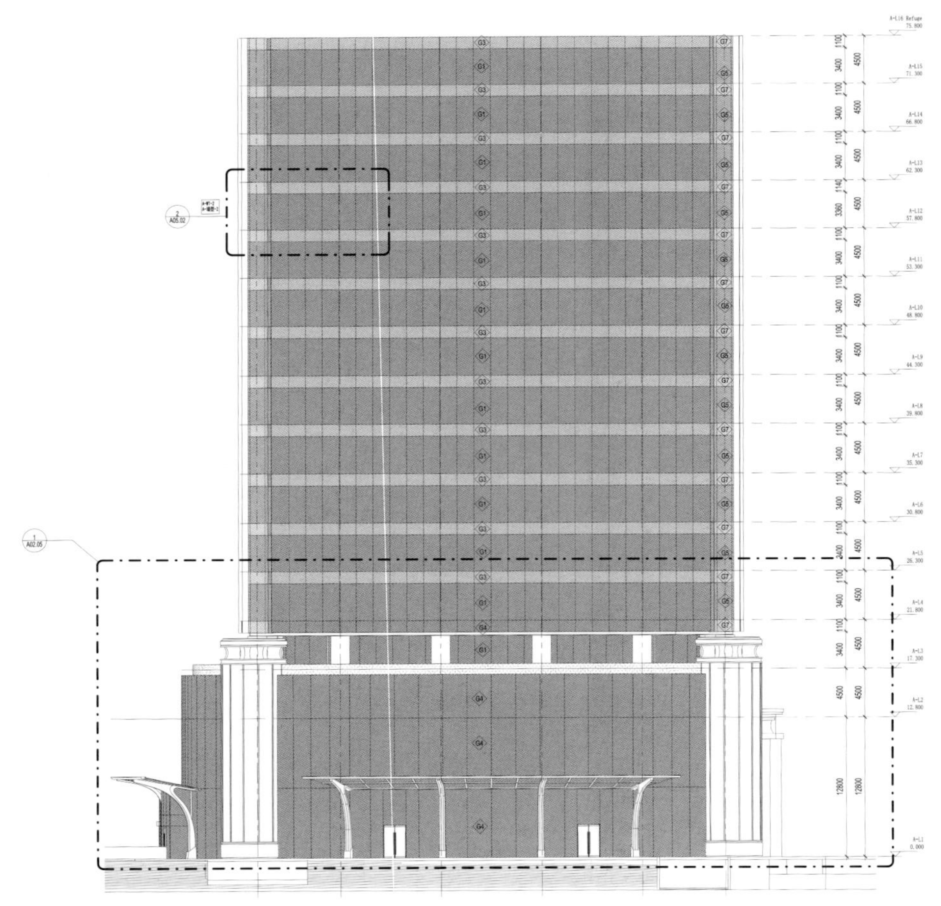

Tower A East Elevation 1 A座东立面图 1

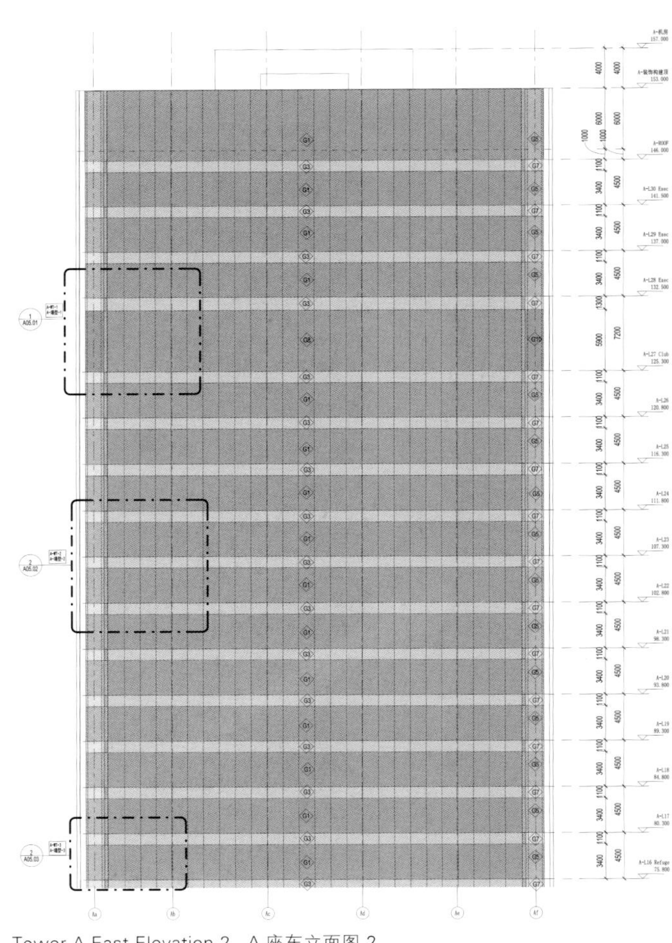

Tower A East Elevation 2 A座东立面图 2

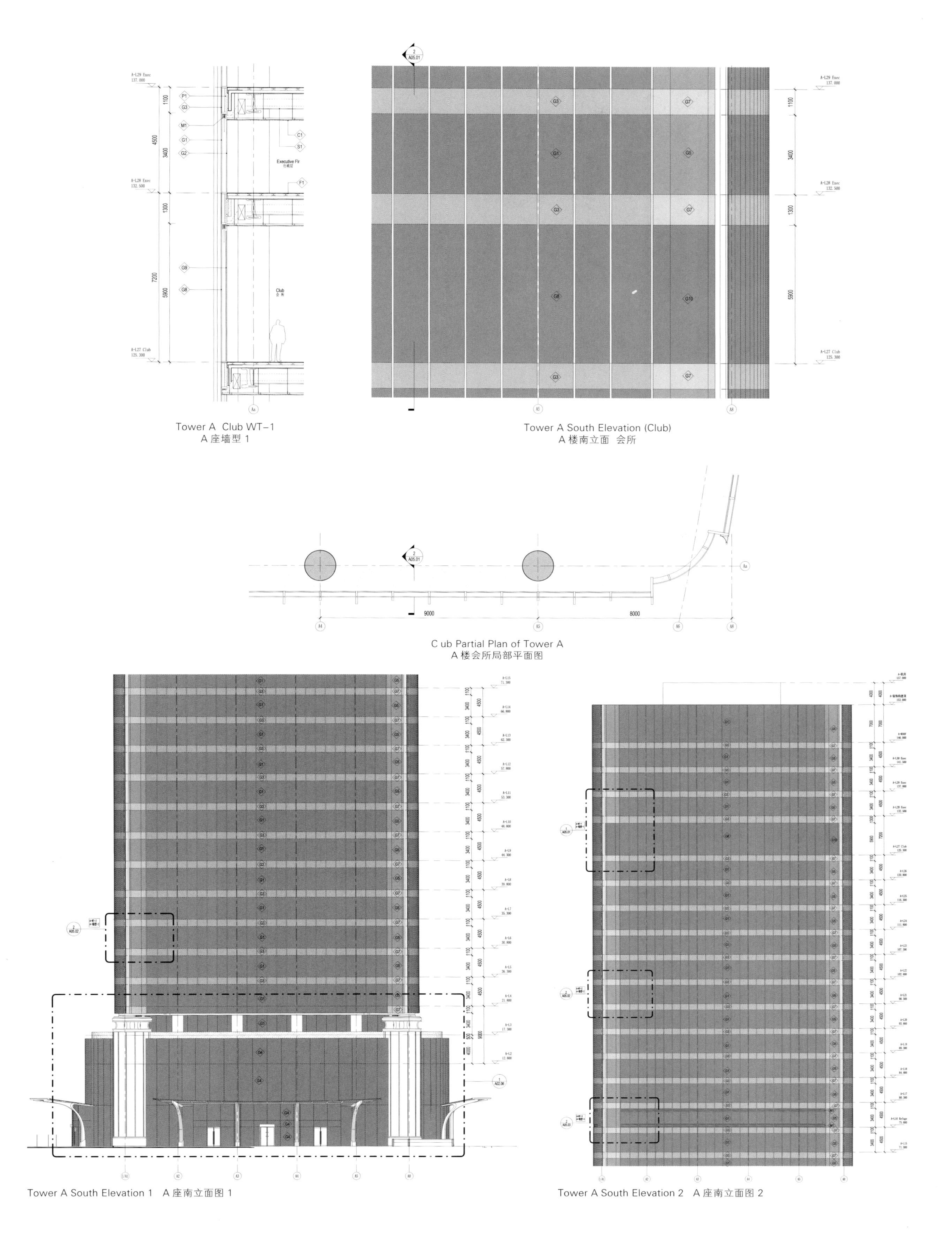

Tower A Club WT-1
A座墙型 1

Tower A South Elevation (Club)
A楼南立面 会所

C ub Partial Plan of Tower A
A楼会所局部平面图

Tower A South Elevation 1 A座南立面图 1

Tower A South Elevation 2 A座南立面图 2

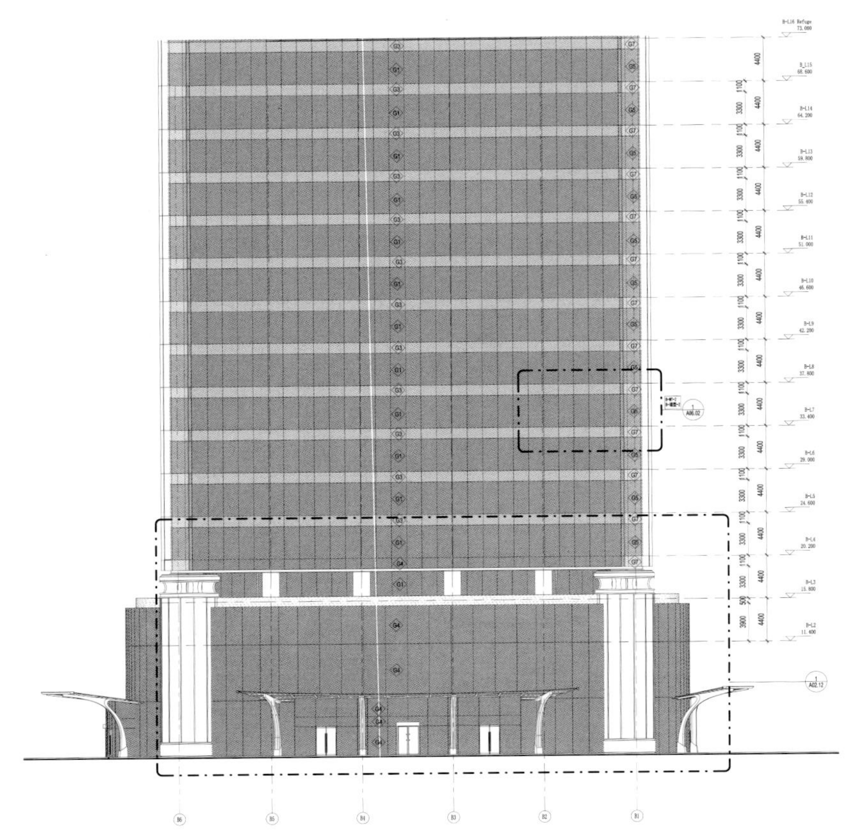

Tower B North Elevation 1 B 座北立面图 1

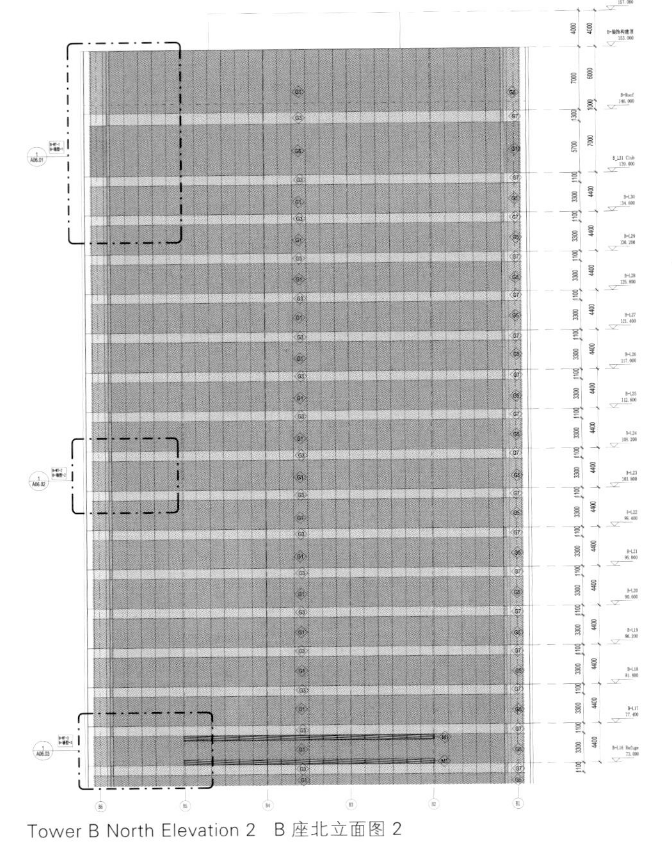

Tower B North Elevation 2 B 座北立面图 2

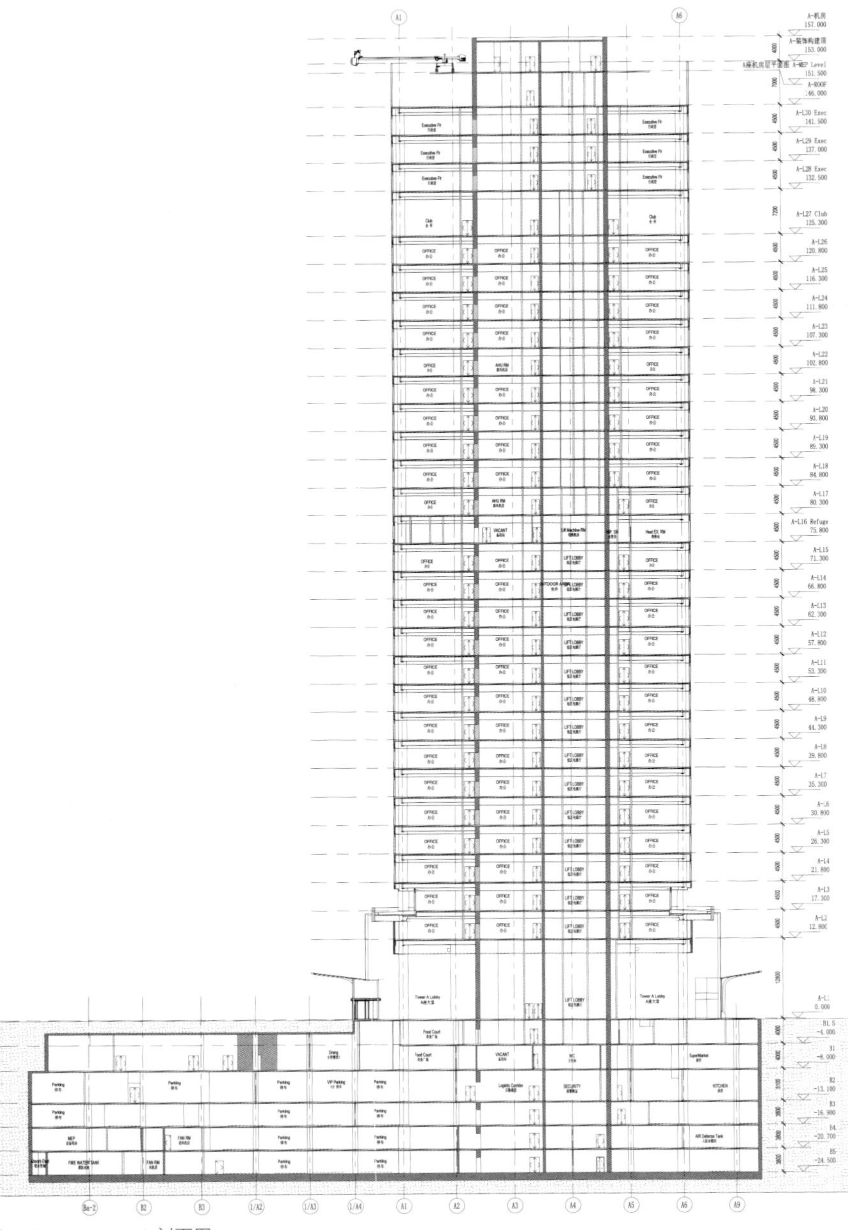

Section 1-1 1-1 剖面图

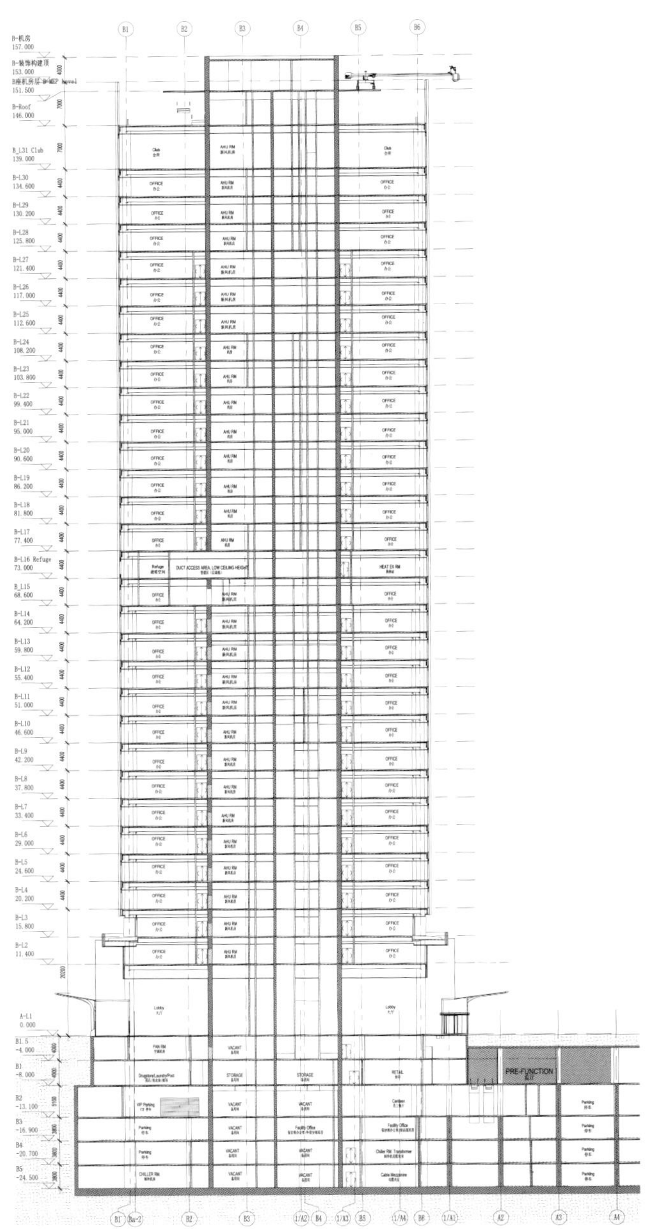

Section 2-2 2-2 剖面图

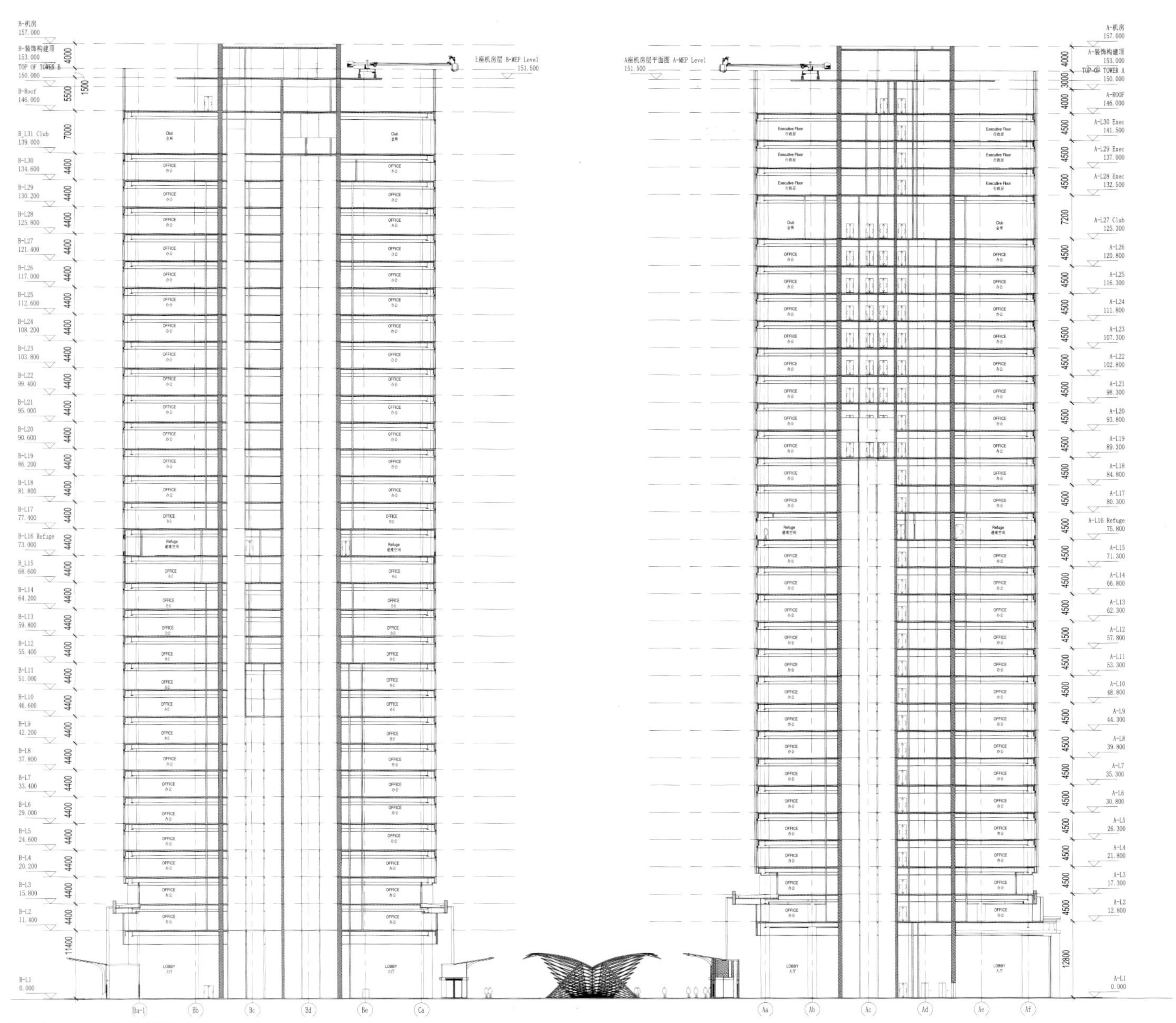
Section 3-3　3-3 剖面图

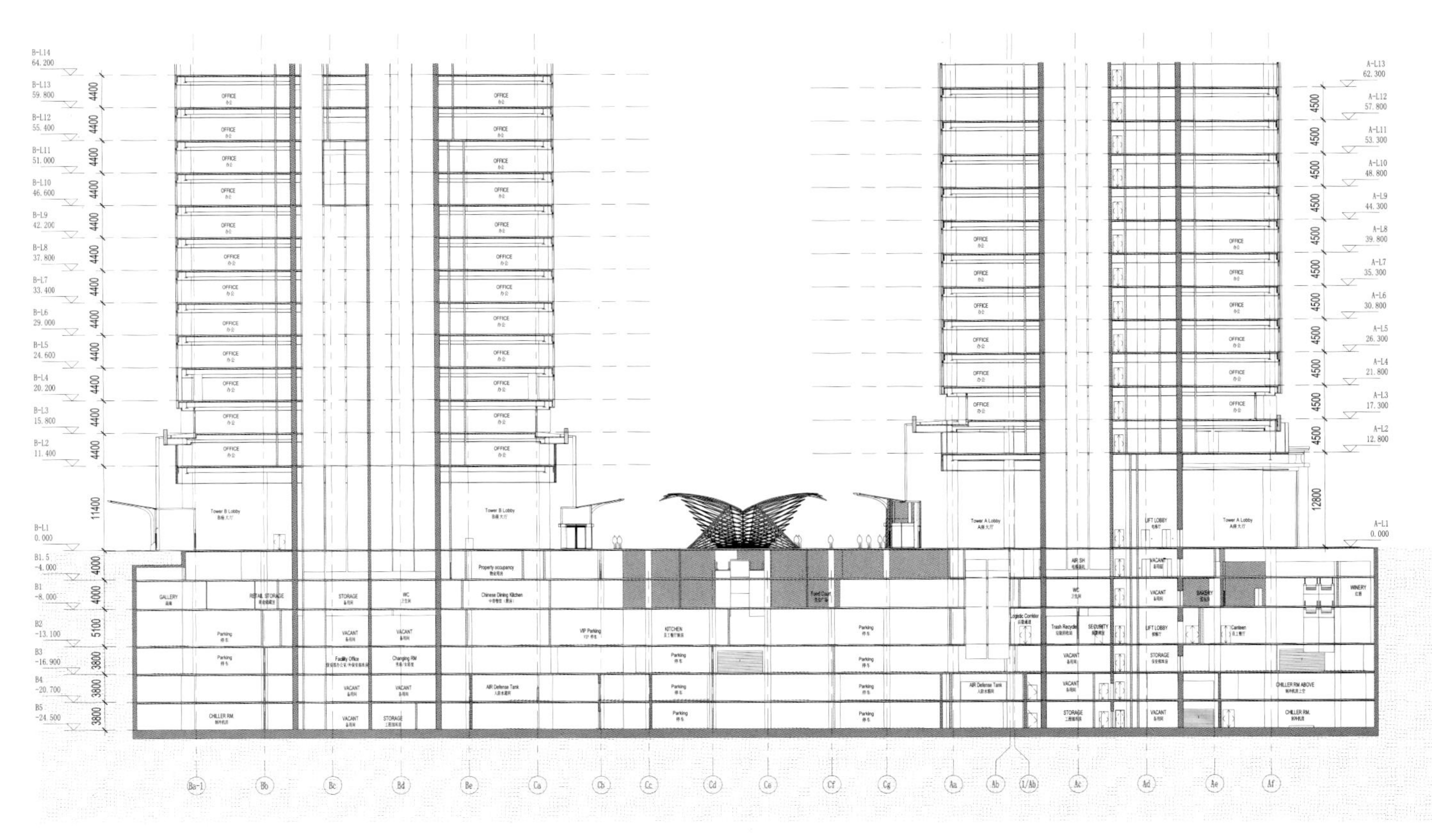
Section 4-4　4-4 剖面图

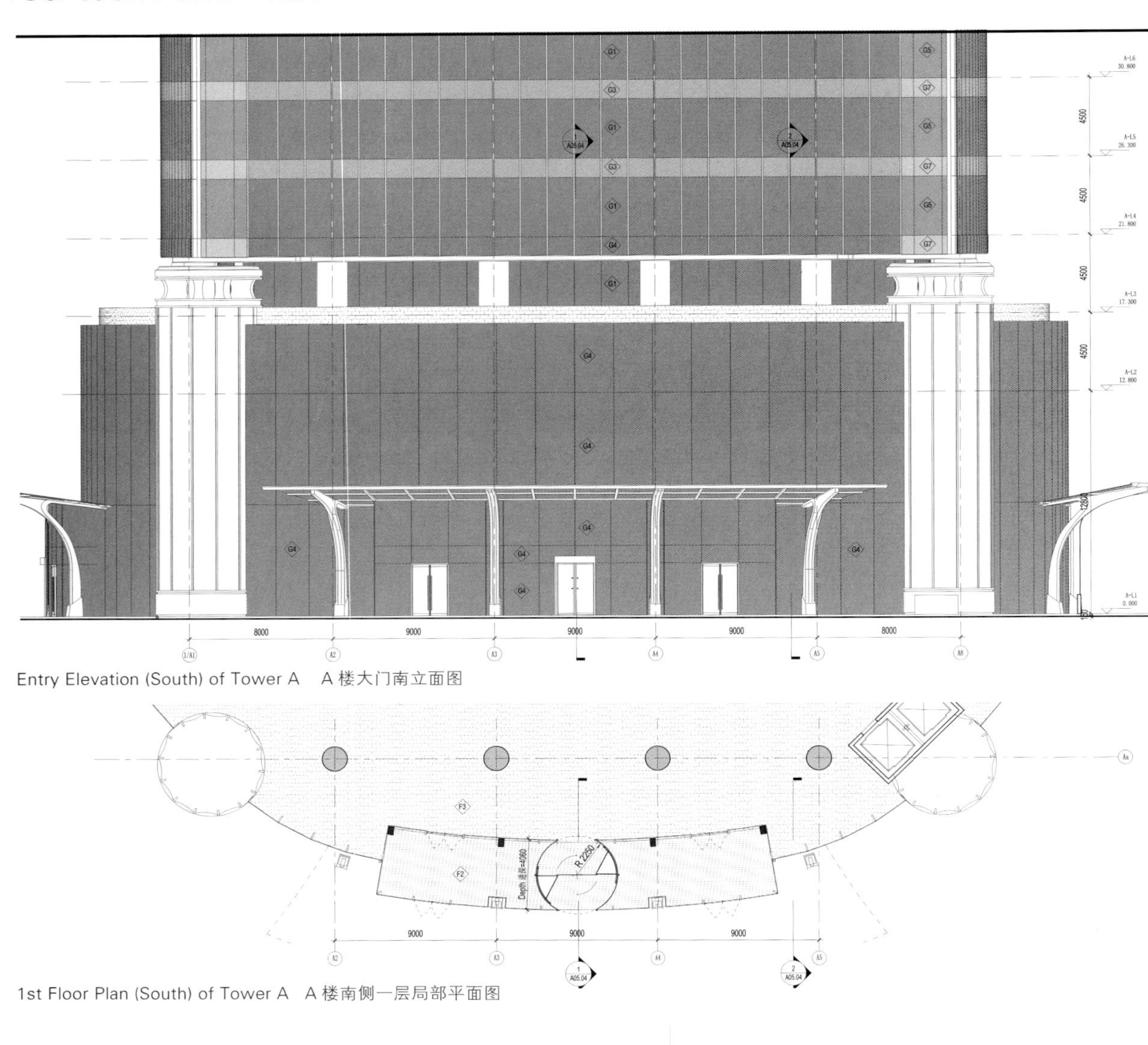

Entry Elevation (South) of Tower A　A 楼大门南立面图

1st Floor Plan (South) of Tower A　A 楼南侧一层局部平面图

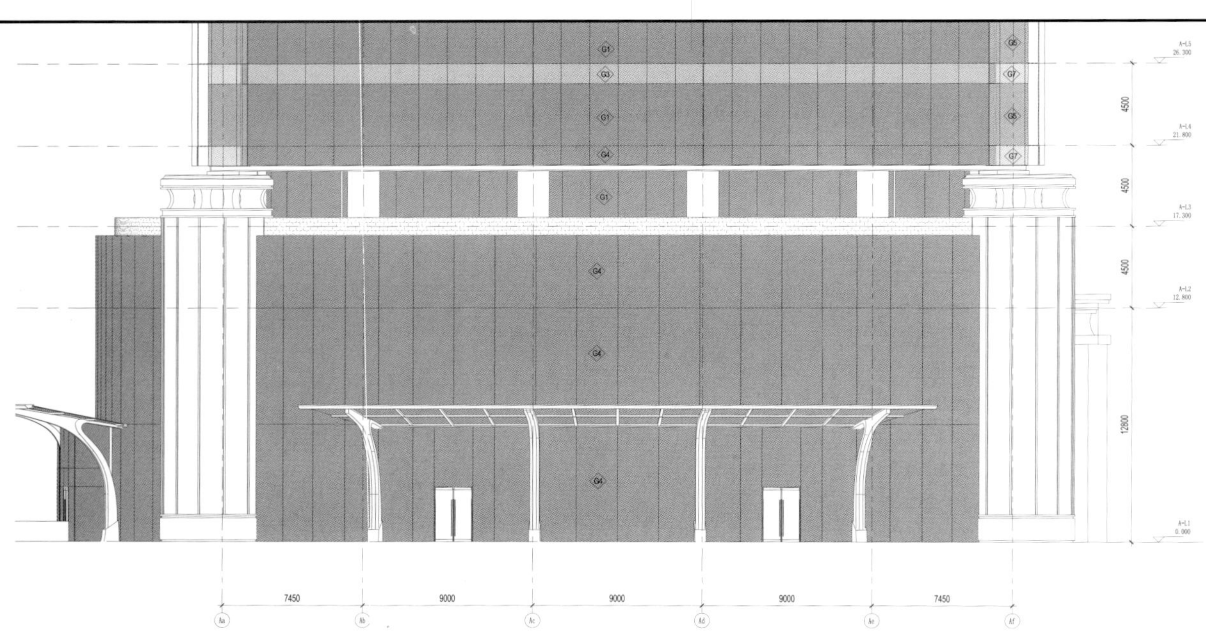

Entry Elevation (East) of Building A　A 楼大门东立面图

1st Floor Plan (East) of Building A　A 楼东侧一层局部平面图

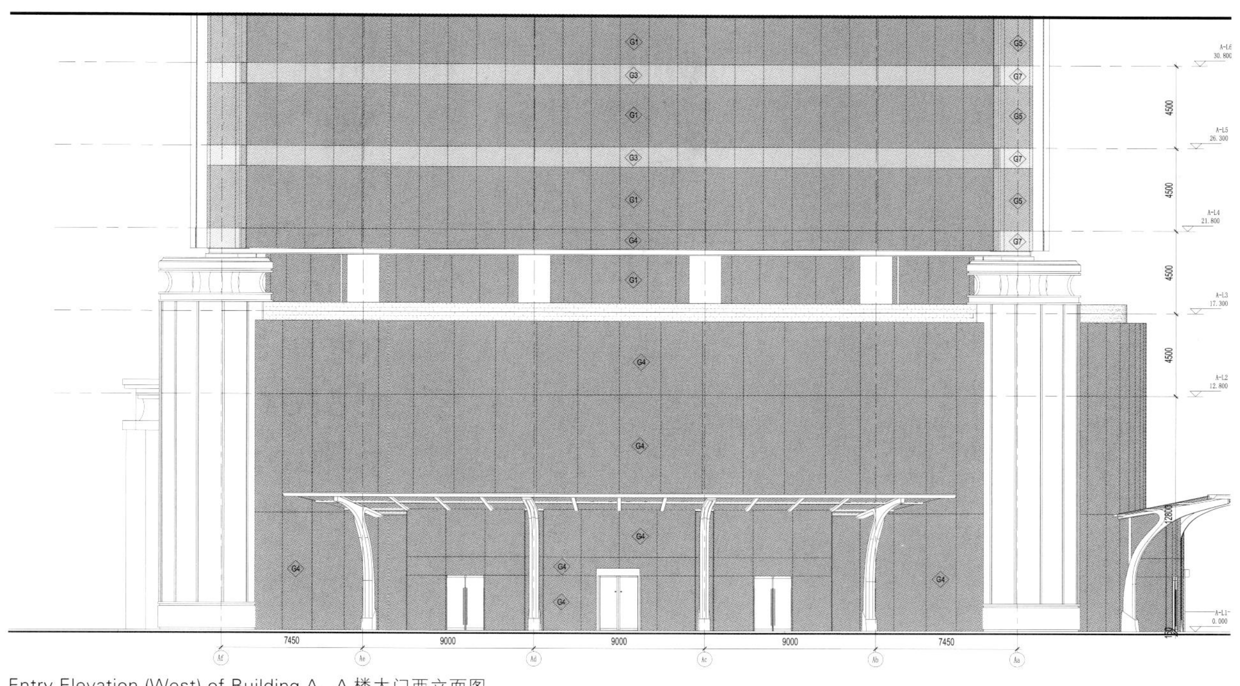

Entry Elevation (West) of Building A　A楼大门西立面图

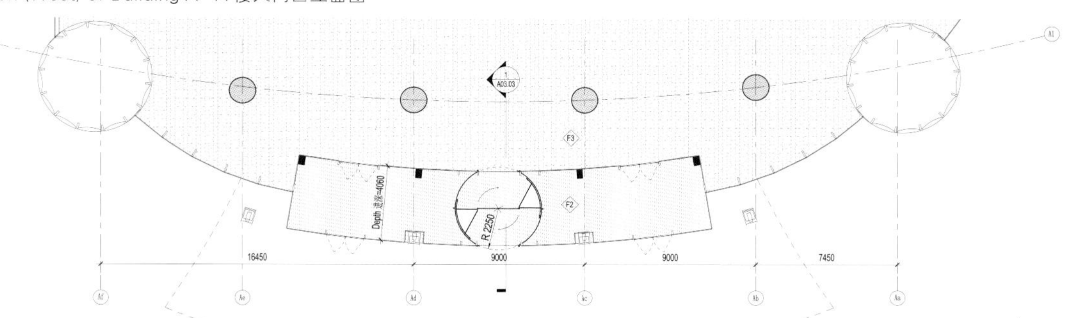

1st Floor Plan (West) of Building A　A楼西侧一层局部平面图

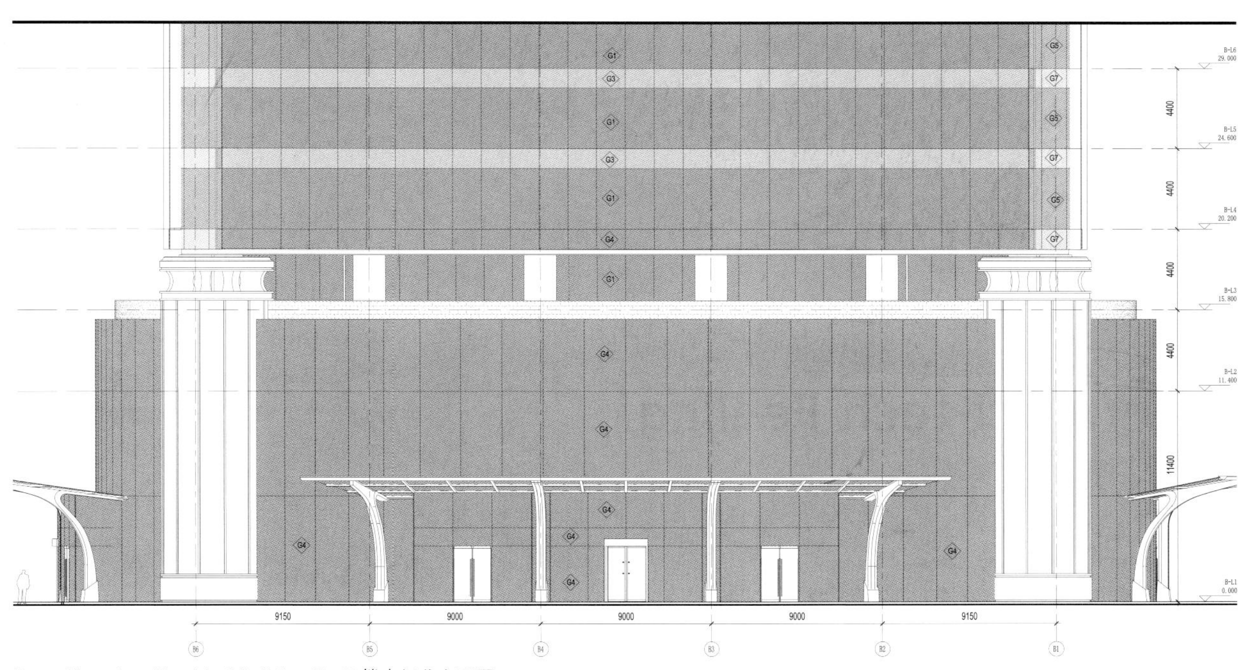

Entry Elevation (North) of Building B　B楼大门北立面图

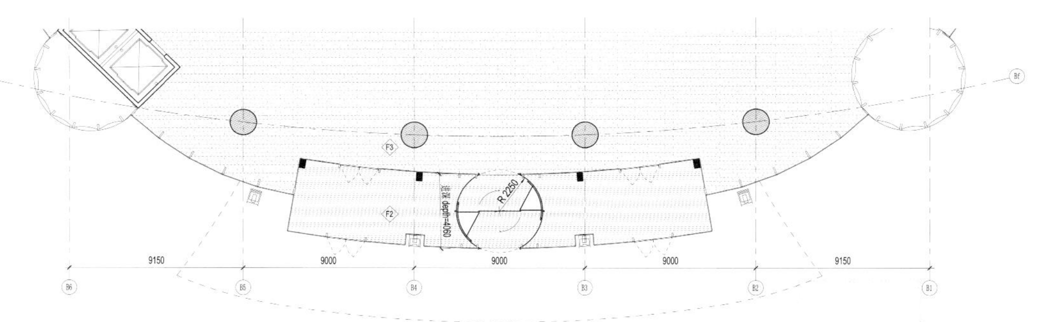

1st Floor Plan (North) of Building B　B楼北侧一层局部平面图

National Digital Electronic Product Testing Center (NETC)

国家数字电子产品质量监督检验中心

Features 项目亮点

It pays attention to create a modern sci–tech and humanized working environment by using innovative forms and advanced building materials, highlighting the great influence of modern science and technology.

项目设计注重营造富有现代科技感又具有人性化的办公环境，采用新颖的外形、先进的建筑材料，使得建筑充分体现了现代科技的影响力。

Keywords 关键词

Public Atrium
共享中庭

Communication Space
交往空间

Modern Sci-tech Feeling
现代科技感

Location: Shenzhen, Guangdong, China
Developer: Shenzhen Academy of Metrology & Quality Inspection
Architectural Design: CAPOL
Land Area: Approximately 12,000 m^2
Gross Floor Area: 29,000 m^2
Plot Ratio: 2.05
Design Date: 2008
Completion Date: 2013

项目地点：中国广东省深圳市
建设单位：深圳市计量质量检测研究院
建筑设计：华阳国际设计集团
用地面积：约 1.2 万 m^2
总建筑面积：2.9 万 m^2
容积率：2.05
设计时间：2008 年
竣工时间：2013 年

» Overview

The project is located in Xili Area of Nanshan District, Shenzhen, with Tongfa Road to its west and the Nanshan Construction Bureau close to its south. To the south and west not far from the testing center, it is a big mountain to provide great landscape views. Surrounded by the residential area on the southwest, Shenzhen Experimental School on the northwest, the reserved green land on the east, and the mountain area on the north, the project enjoys an advantaged natural environment.

» 项目概况

项目位于深圳南山西丽片区，项目用地西边为同发路，南部紧临南山建设局，南边与西边不远处都是大片山体，有一定的外部景观资源；西南为现有住宅区；西北面为深圳实验学校；东边为预留绿地；北部为保留山体，周边自然山体较多，自然条件优越。

» Project Value

Upon completion, the project has contributed a lot to the independent innovation, transformation and upgrading, quality improvement and international market expansion in the national digital industry. It will make great significance in making Shenzhen a national innovative city and in building its modern industrial system.

» 项目价值

项目的建成为深圳乃至全国数字电子产业自主创新、转型升级、品质提升、国内外市场开拓等方面作出杰出贡献，并将为深圳国家创新型城市建设和现代产业体系构建产生深远意义。

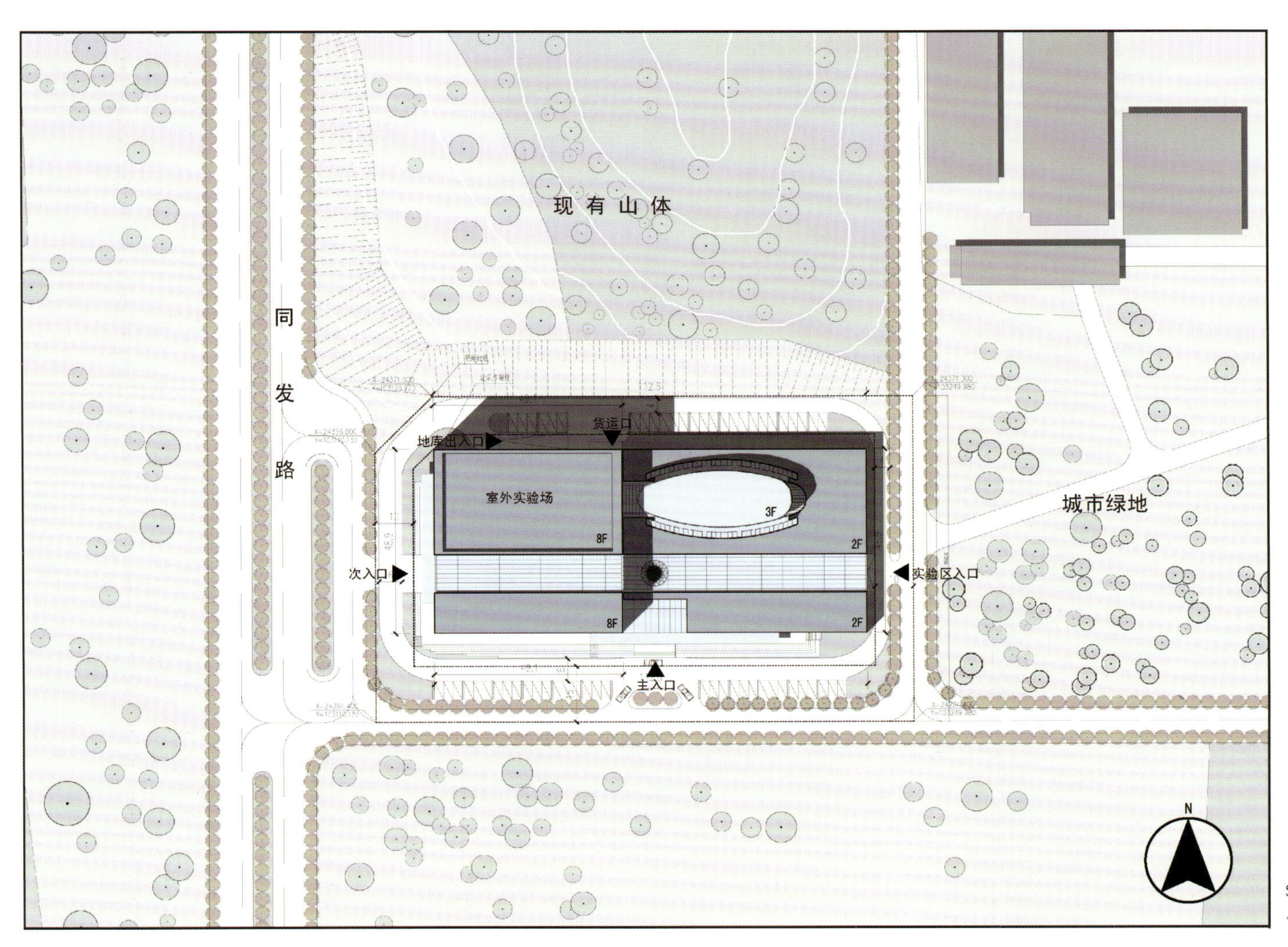

Site Plan
总平面图

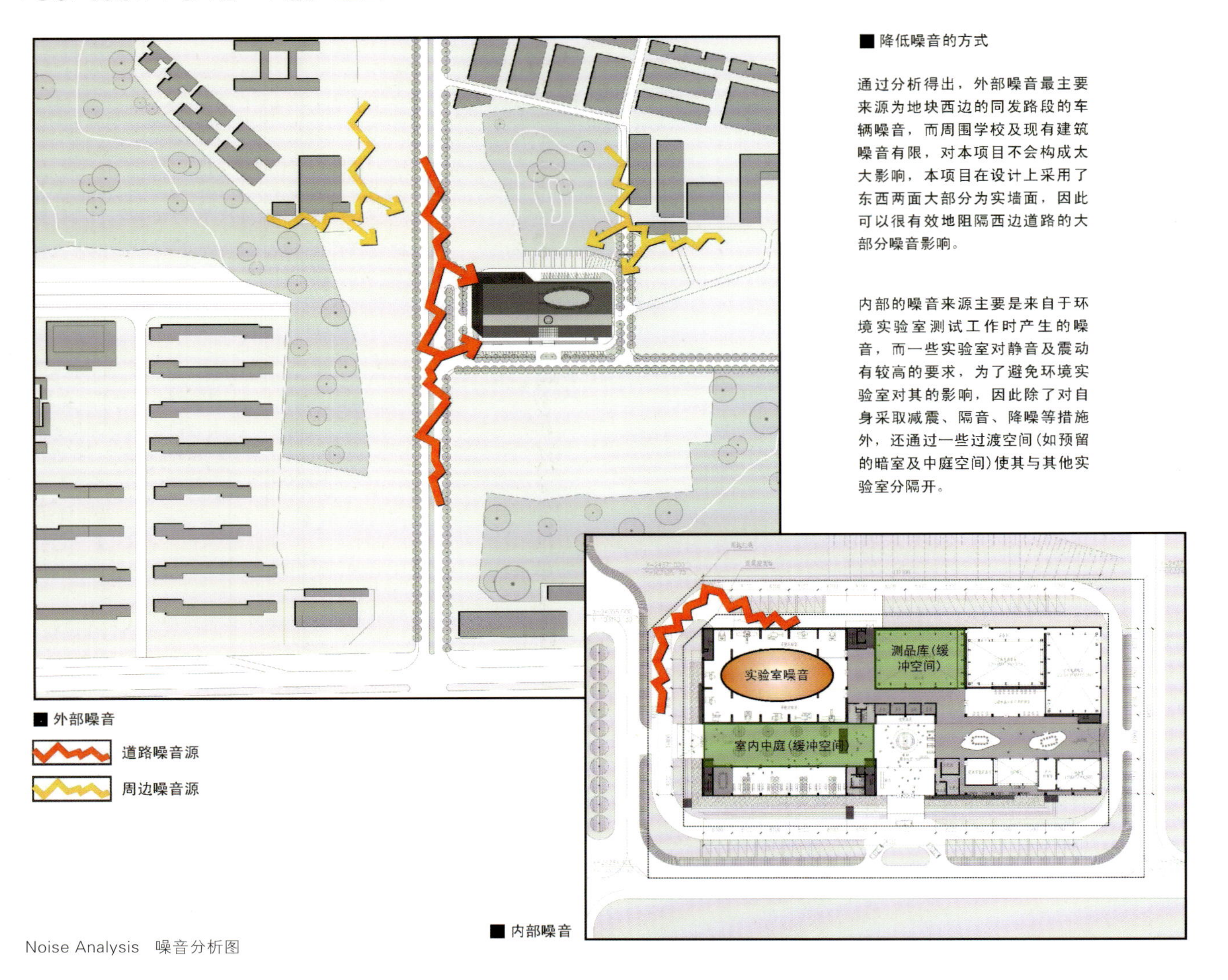

■ 降低噪音的方式

通过分析得出，外部噪音最主要来源为地块西边的同发路段的车辆噪音，而周围学校及现有建筑噪音有限，对本项目不会构成太大影响，本项目在设计上采用了东西两面大部分为实墙面，因此可以很有效地阻隔西边道路的大部分噪音影响。

内部的噪音来源主要是来自于环境实验室测试工作时产生的噪音，而一些实验室对静音及震动有较高的要求，为了避免环境实验室对其的影响，因此除了对自身采取减震、隔音、降噪等措施外，还通过一些过渡空间（如预留的暗室及中庭空间）使其与其他实验室分隔开。

Noise Analysis 噪音分析图

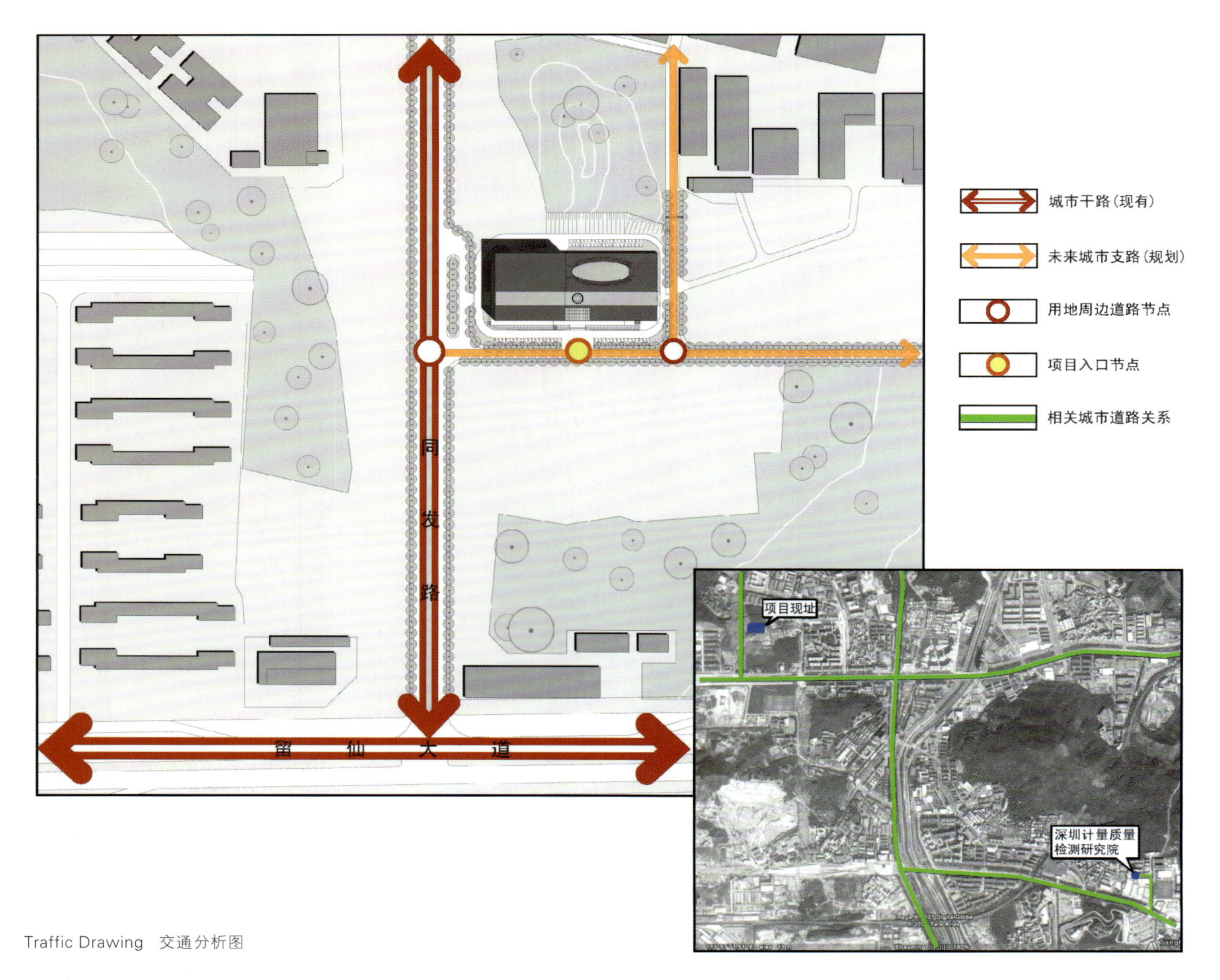

Traffic Drawing 交通分析图

■ 植被不仅仅带来绿色，还带来艺术

■ 生态停车位

■ 独特的灯光设计可以带来充满科技感的效果

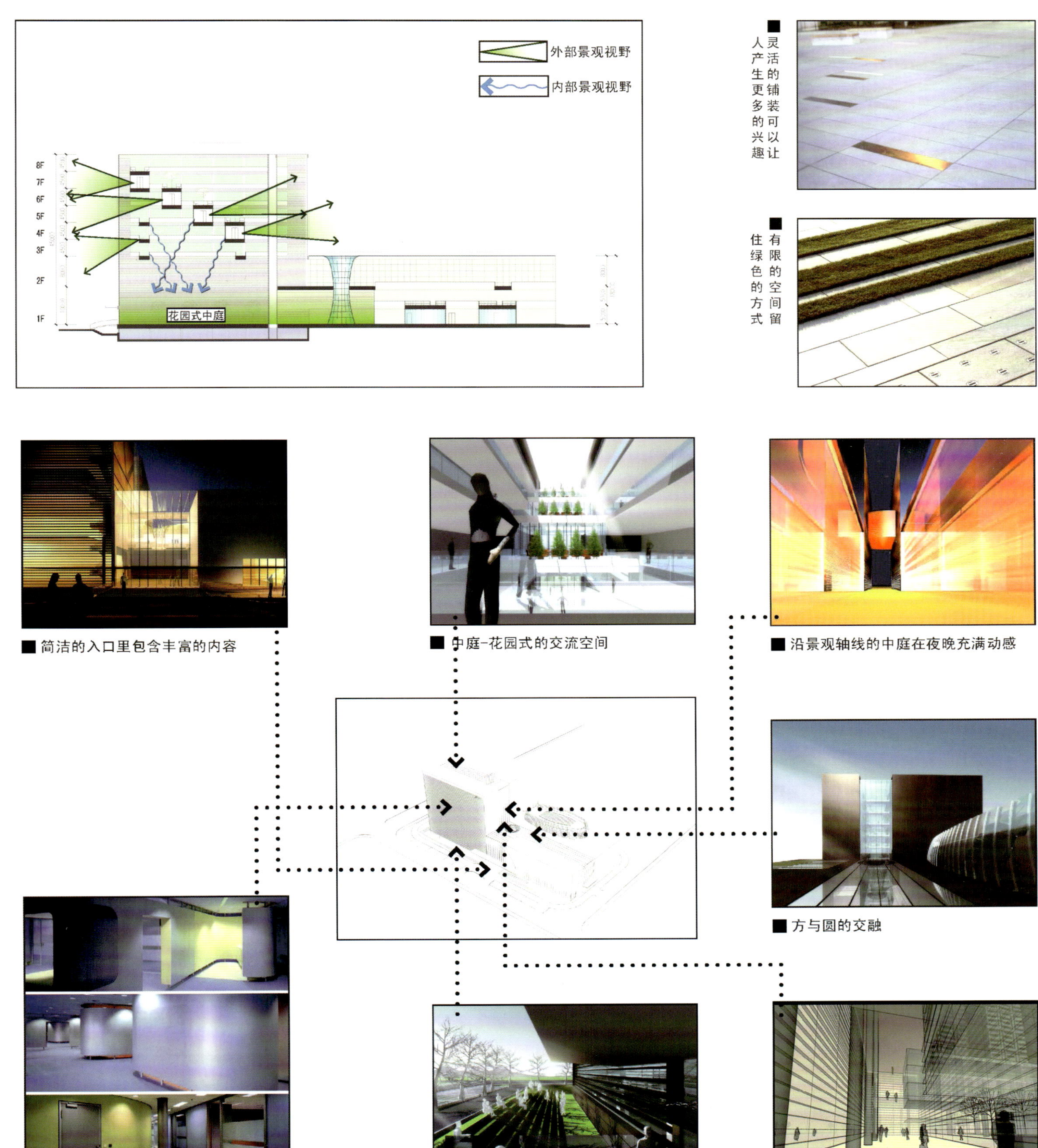

■ 灵活的铺装可以让人产生更多的兴趣

■ 有限的空间留住绿色的方式

■ 简洁的入口里包含丰富的内容

■ 中庭-花园式的交流空间

■ 沿景观轴线的中庭在夜晚充满动感

■ 方与圆的交融

■ 充满未来感的办公空间

■ 晨曦中的下沉庭院

■ 立体交通的穿插造成了空间的多变

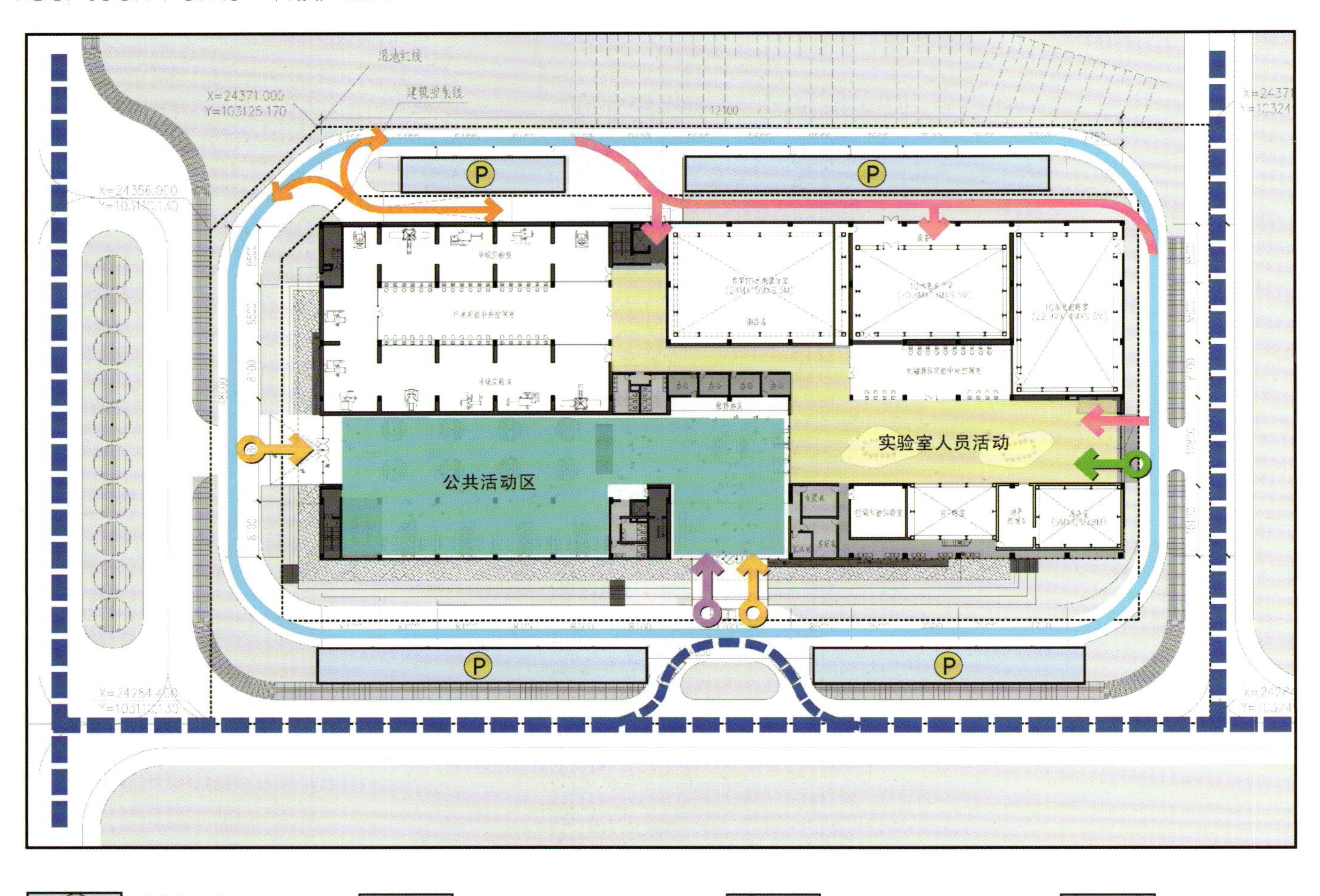

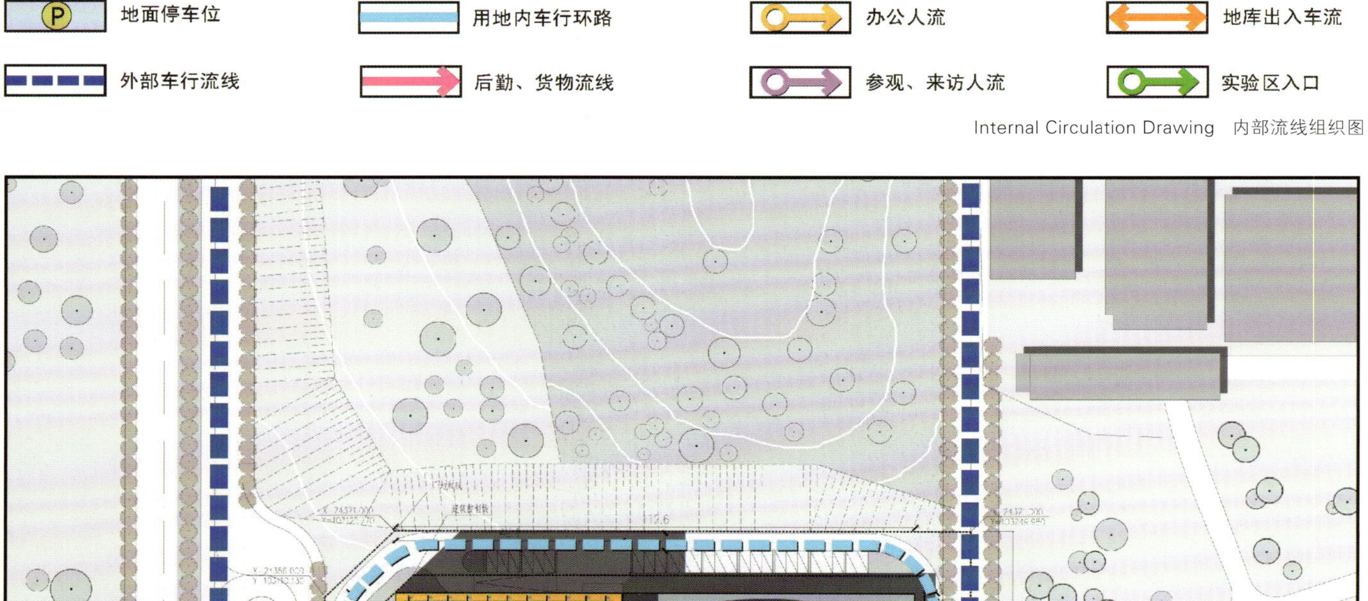

Internal Circulation Drawing 内部流线组织图

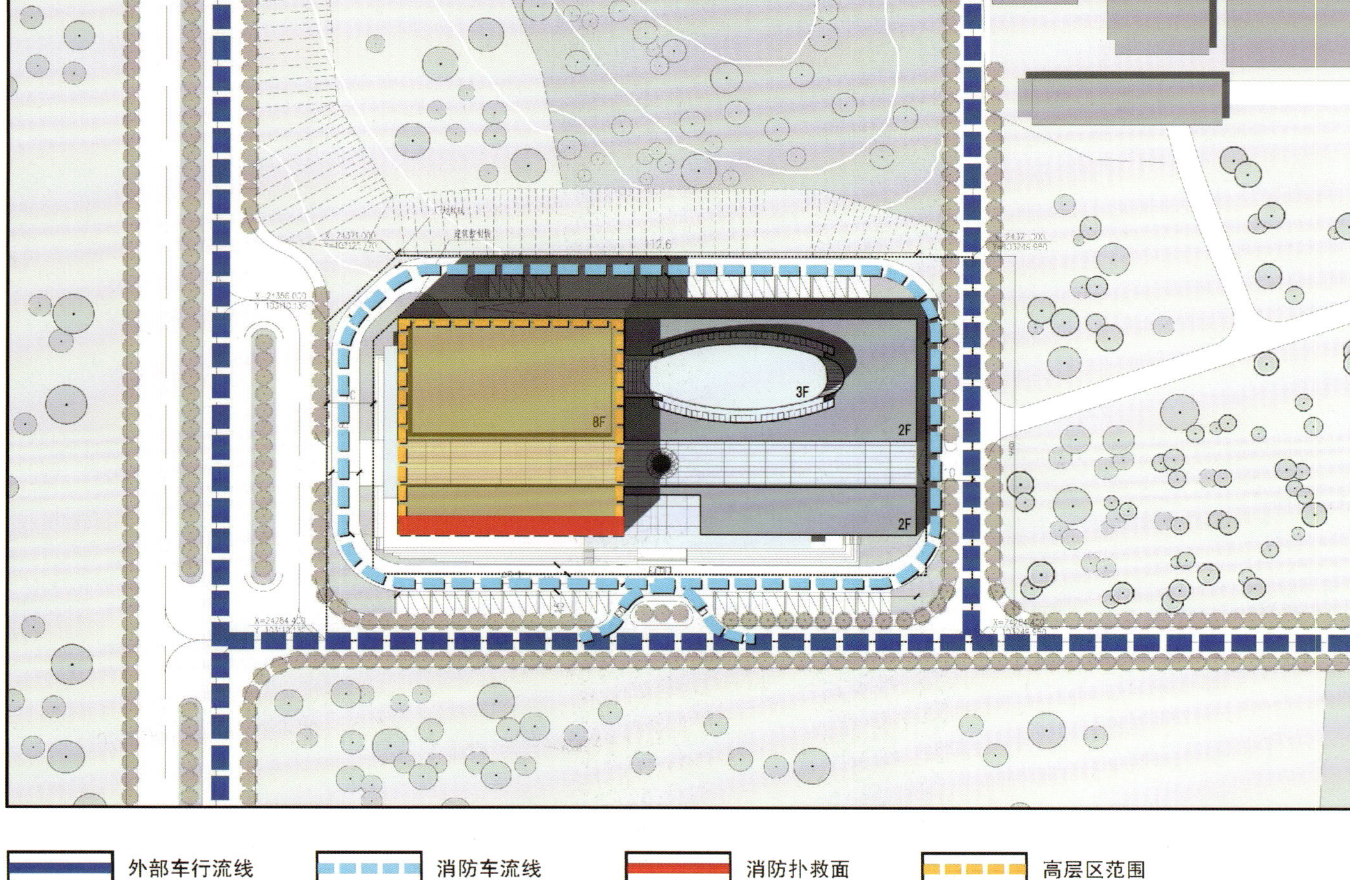

Fire Analysis Diagram 消防分析图

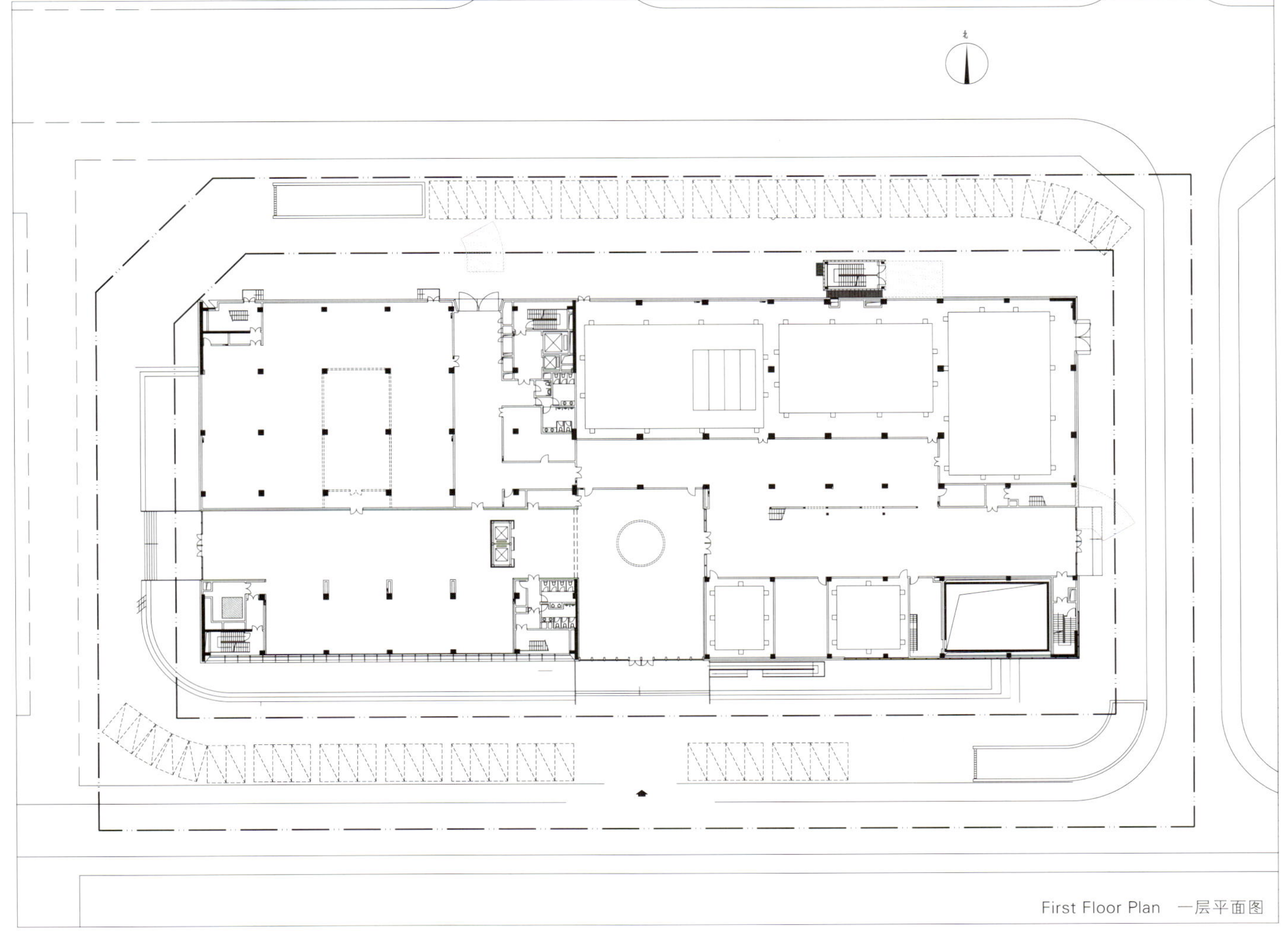

First Floor Plan 一层平面图

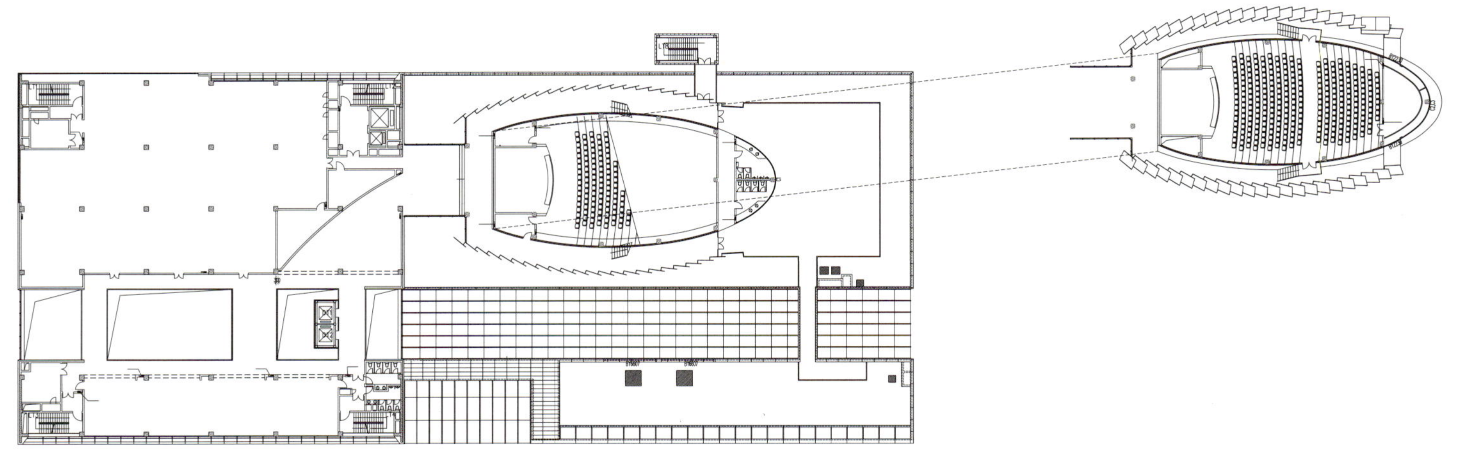

Third Floor Plan 三层平面图

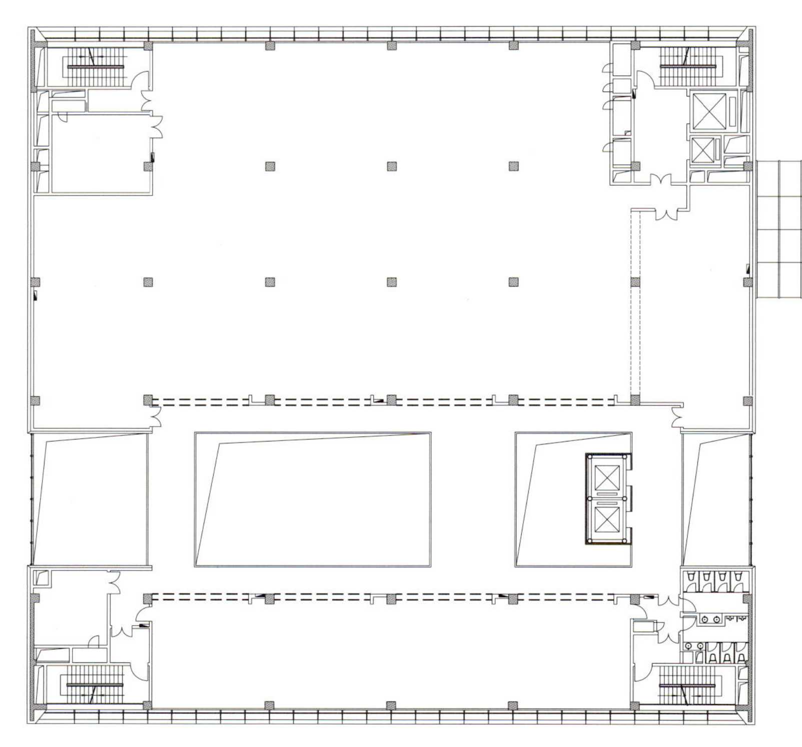

Fifth Floor Plan 五层平面图

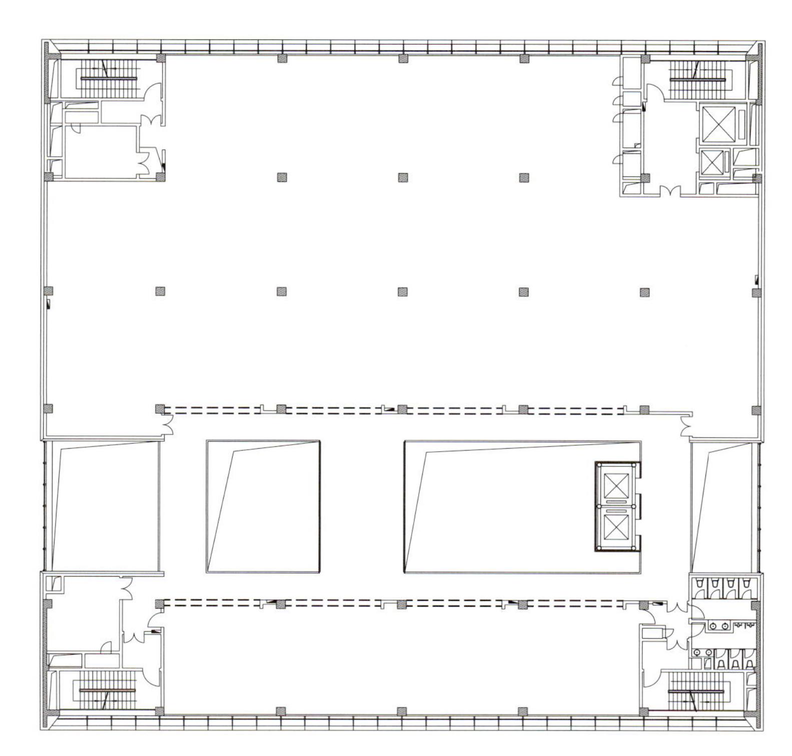

Sixth Floor Plan 六层平面图

Design Idea

Starting from the market demand, the designers have tried to build a high-quality office building for scientific research in Nanshan District, providing a modern and humanized working environment. At the same time, with reasonable and efficient overall plan, it hopes to create a modern public complex which is dynamic and functional to integrate with the surroundings.

设计理念

从关注城市需求入手，在整个构思中设计师的核心理念旨在为南山区创造一个高品质的科研办公建筑，营造既富有现代科技感又具有人性化的办公环境，同时希望通过合理有效的总体布局，创建一个与现有周边环境相协调，各功能分区间积极互动的、层次丰富的并充满活力的现代公共综合体。

Functional Layout

According to the functional layout, the building is separated into three parts: in the north is the large laboratory; in the south near to the main entrance are the service spaces and the small laboratories; the traffic links and the pubic spaces are set in the middle. And a bottom-to-top lighting atrium is designed in center to form the horizontal spatial axis. It provides soft and steady light for the interior spaces and avoids strong and direct sunlight. What's more, the atrium helps to form the micro-climate which provides natural ventilation. On this horizontal axis, the meeting room and the green corridors have created a colorful and interesting working space.

功能布局

在功能组织上，建筑划分为三部分，北侧为大空间实验室，南侧靠近主入口，为服务空间及小型实验室，中间为交通联系体及公共空间。并在中间这个部分设有通高的采光中庭，形成横向的空间轴线。中庭提供了柔和且稳定的光线，可以满足内部活动的光照要求，避免强烈的太阳光直射。而且中庭还可以形成室内的小气候，促进空气流通。在这条横向的空间轴线上，通过会议室和绿化连廊的节点设计，使得整个办公空间更加丰富和有趣。

Landscape Design

The green belt to the west of the site is well designed. And in the future, there will be urban green lands on the east and south of the site to provide great views. Inside the building, with vertical landscape design, the open and spacious atrium space will provide unique landscape effect.

景观设计

基地西边的城市绿化带比较完善，随着城市建设的进行，将在基地东面和南面形成城市绿地，将成为很好的景观。在建筑内部，通过垂直景观的设计，共享的宽阔中庭将呈现出别样的园林景观效果。

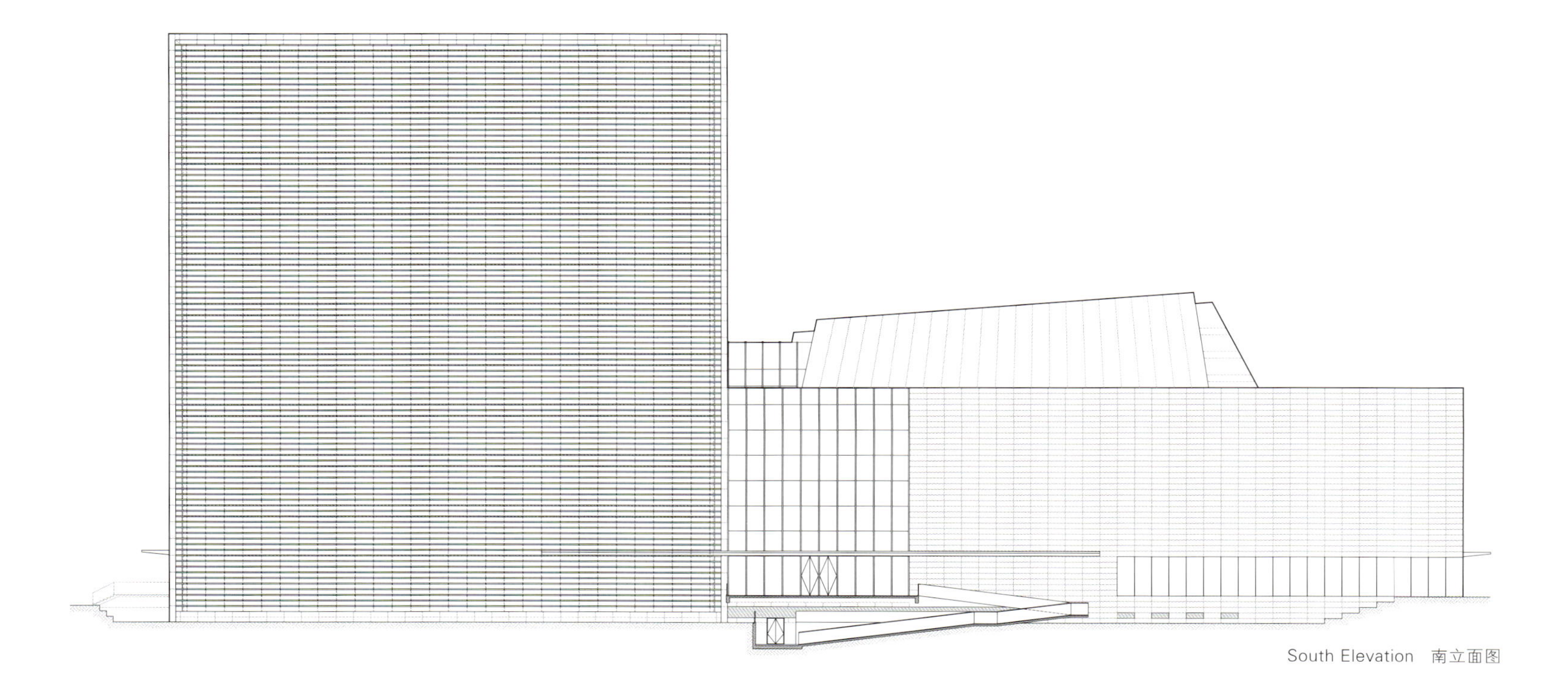

South Elevation 南立面图

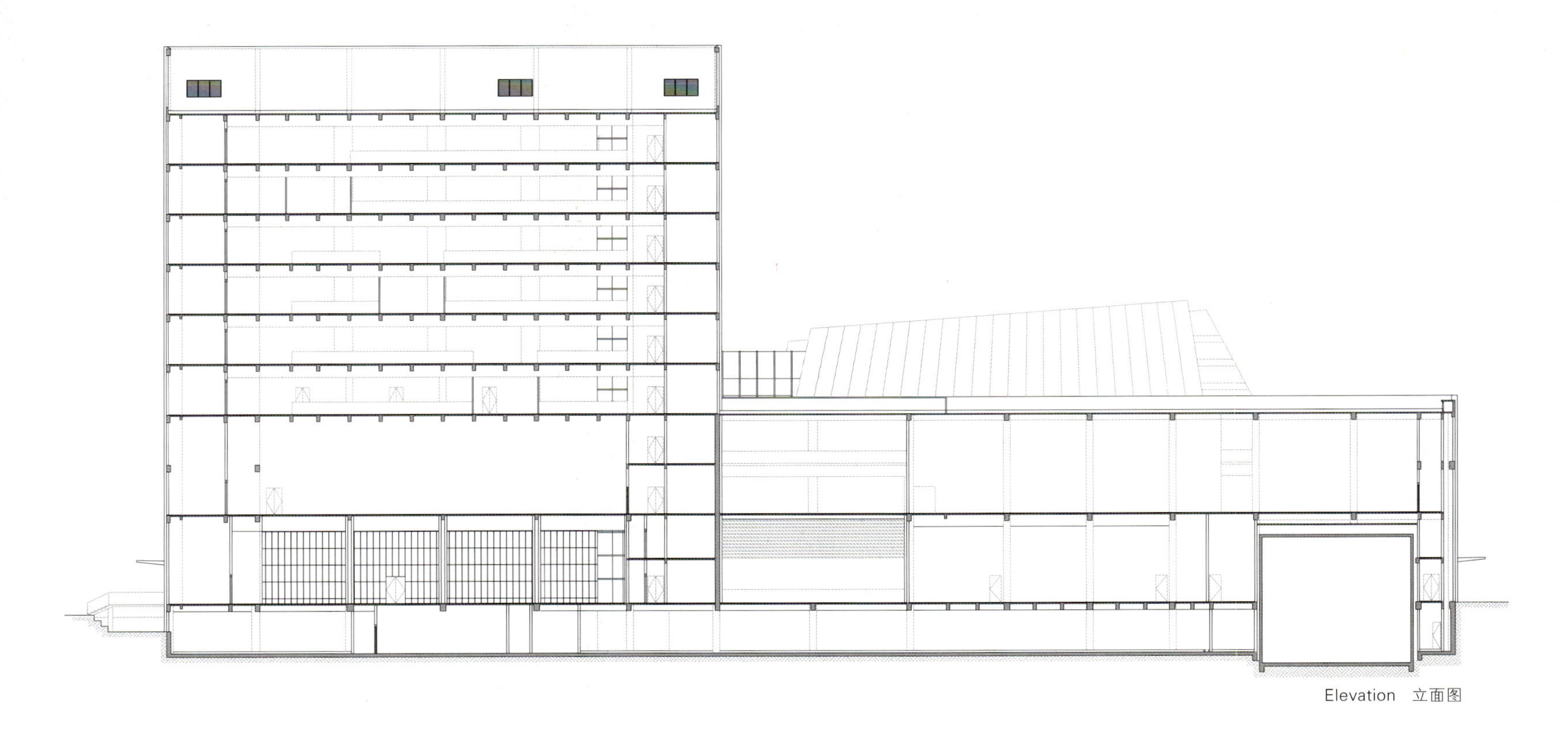

Elevation 立面图

Architectural Design

1. It makes the building full of modern sci-tech feelings by skillful interior design.

2. For the labs with special space requirements, the design enables the users to have sci-tech experience when they use them.

3. The international conference hall follows the sci-tech style to apply innovative and lively architectural form.

4. The exterior wall uses LOW-E glass to save energies efficiently. And the facade design highlights the great influence of the modern science and technology.

建筑设计

1. 通过建筑室内的处理，使得整个建筑具有超现实的科技感。

2. 对于实验室等本身具有工艺要求的房间，在其布置上，采用具有科技气氛的形式，使得建筑在其使用过程中也具有科技感。

3. 国际会议厅形式的处理，继续延续了科技感的主导，采用新颖的外形，也使得建筑形式更加活泼。

4. 外表面使用 LOW–E 玻璃，可以达到很好的节能效果。对于建筑外立面的处理，使得建筑更加体现了现代科技的影响力。

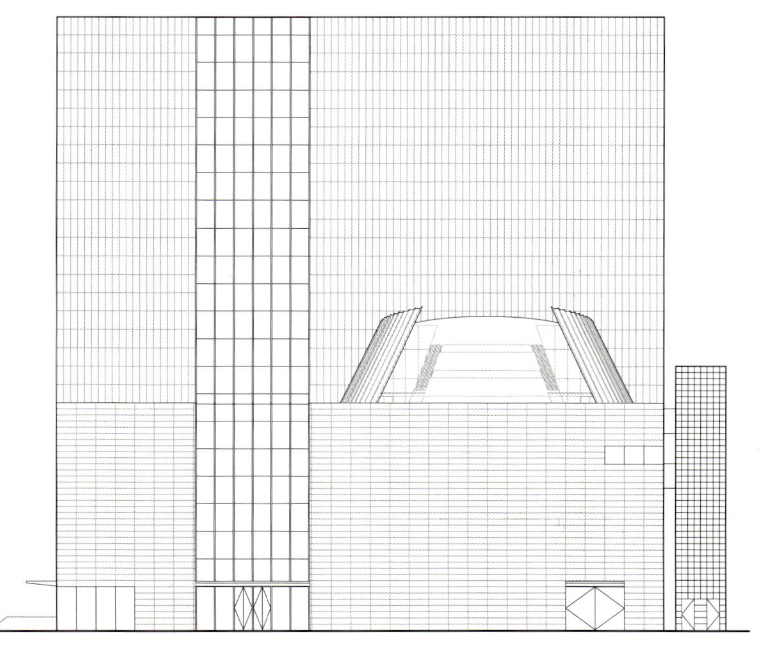

East Elevation 东立面图

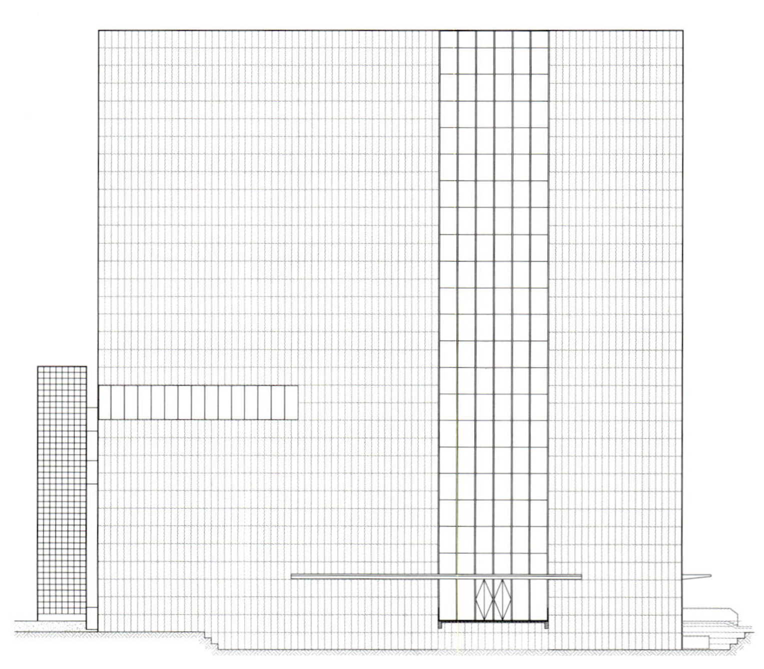

West Elevation 西立面图

深圳市计
测研究院
石岩、
Shiyan

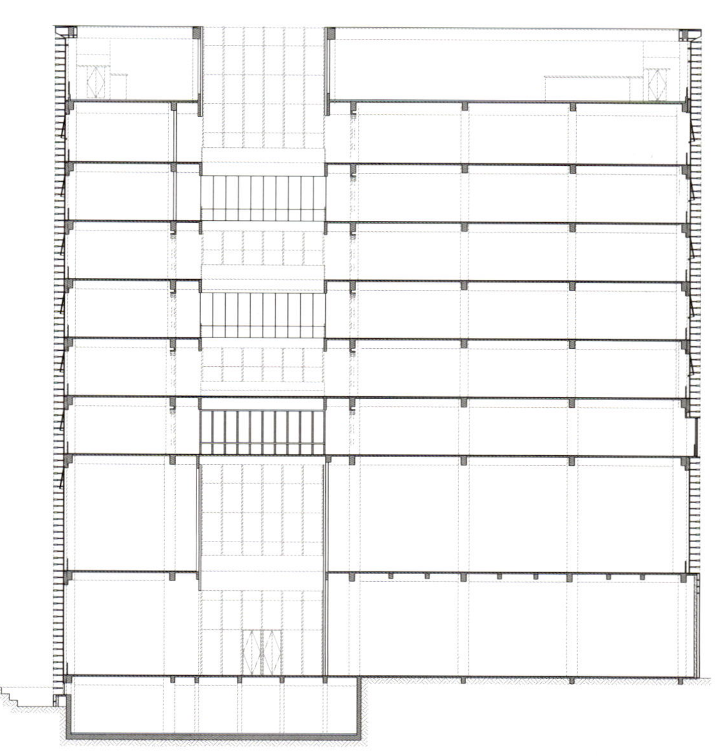

Section 1-1　1-1 剖面图

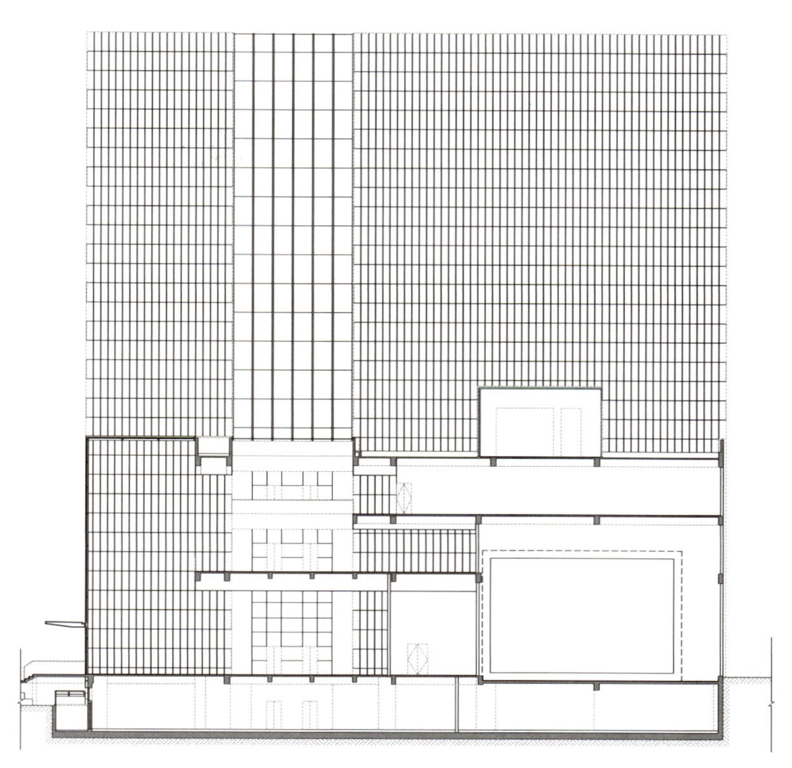

Section 2-2　2-2 剖面图

Fuzhou Strait Creative Industry Cluster

福州海峡创意产业集聚区

Features 项目亮点

The design highlights the volumes of the existing buildings and rearranges the interior spaces and the internal functions. It follows the principle of "showing respect to the history in the urban renewal design".

设计突出老厂房的原有建筑体量，合理利用原有建筑，进行适当整合，将核心业态进行有效置换与布置，突出了城市更新设计中所遵循的尊重历史的原则。

Keywords 关键词

Cultural Creativity
文化创意

Artistic Atmosphere
艺术气息

Renovation of Old Factory
旧厂房改造

Location: Cangshan District, Fuzhou, Fujian, China
Developer: Shanghai Redtown Culture Development Co., Ltd.
Architectural Design: W&R Group
Land Area: 50,096 m^2
Gross Floor Area: 57,000 m^2
Plot Ratio: 1.14
Completion Date: 2014

项目地点：中国福建省福州市仓山区
开发商：红坊文化发展有限公司
建筑设计：水石国际
用地面积：50 096 m^2
总建筑面积：57 000 m^2
容积率：1.14
竣工时间：2014 年

Overview

The project is located in Jinshan Investment Zone of Cangshan District, Fuzhou. The zone is developed as an important base for cultural industry during the 12th Five-Year-Plan period. It is the cross-strait cultural industry park and cultural cooperation center.

项目概况

项目位于福州市仓山区金山投资区内，该区是“十二五”期间政府拟打造的全国重要文化产业基地，是国家所确立建设的海峡两岸文化产业园和文化产业合作中心。

Theme and Purpose

The purpose of the design is to integrate the design sources cross the strait, and create a benchmark creative industry park in Fuzhou City and Fujian Province, which is characterized by the rebirth and industrialized development of the Chinese lacquer culture. It will promote the cultural creativity and optimize the industrial structure; upgrade the industrial level and improve the local economy; regenerate the traditional handicraft (lacquer art) and achieve the industrialized development of the traditional culture.

主题定位

整合海峡两岸设计类产业资源，打造以漆器文化的再生和产业化发展为特色的综合性设计产业园，形成福州乃至全省范围的标杆性创意产业集聚区；从而促进地区文化创意产业发展，推进产业结构调整，提升产业能级和地区经济水平；促进以漆器为主题的传统手工艺再生，实现传统文化的产业化发展。

Site Plan
总平面图

Planning

Occupying a total land area of 50,096 m², the starting area is built with the core programs including Taiwan Design Center and Fuzhou Lacquer Industry Promotion Center, which is soon becoming the hotspot of this area. The Creative Industry Cluster is integrated with buildings for three functions: cultural exhibition, business office and recreation. Among them, the cultural exhibition area will be used for promoting art designs; the Taiwan Design Center, which is for offices, research and development, and exhibition, will provide a communication platform for the designers and creative teams from Fujian and Taiwan; the recreation area will greatly upgrade the environment of the park with multiple recreational, entertainment and commercial facilities.

规划形态

项目总用地面积为 50 096 m²，先期启动区部分包含台湾设计中心和福州漆器产业促进中心等园区核心业态，利于快速形成区域热点。项目分为三类业态，即文化展示、商务办公、时尚休闲，形成综合性设计产业园。其中，文化展示打造园区艺术设计推广平台；作为商务办公的台湾设计中心集办公、研发、展示于一体，为闽台设计师和创意团队提供交流发展平台；具有设计品质的时尚休闲区改造园区整体环境，同时引入多元化的休闲娱乐型商业业态。

改造前

Architectural Renovation

Before renovation, the buildings featured monotonous forms and floor plans which could not provide multiple choices. And because the planning of the industrial buildings was in mass, the values of the buildings need to be explored. Therefore, the designers highlight the volumes of the existing buildings and redesign their interior spaces and functions. It follows the principle of "showing respect to the history in the urban renewal design".

建筑改造

改造前建筑单体造型及内部平面布置均比较单调，在考虑成本的前提下难以形成更多可供选泽的形象。传统工业厂房的规划结构、场地环境较为凌乱，原有建筑的价值有待挖掘。设计突出老厂房的原有建筑体量，合理利用原有建筑，进行适当整合，将核心业态进行有效置换与布置，突出了城市更新设计中所遵循的尊重历史的原则。

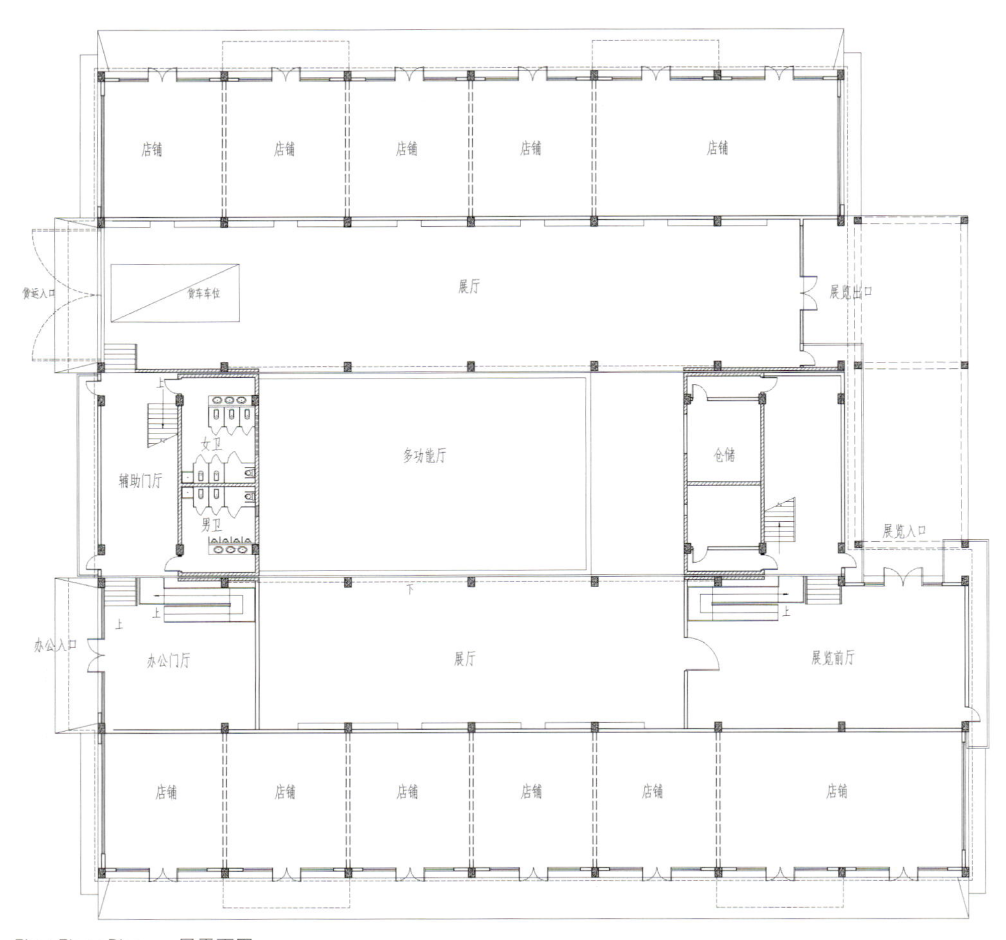

First Floor Plan 一层平面图

Facade Renewal Principle

Based on the principle of "focusing on key points and reasonably controlling the investment", it has renovated the facades in four ways with emphasis on the core area and the buildings along the roads.

立面改造原则

根据重点突出、合理投控的原则，立面改造分为四种类型进行改造。核心区域及主要沿街形象面为改造重点。

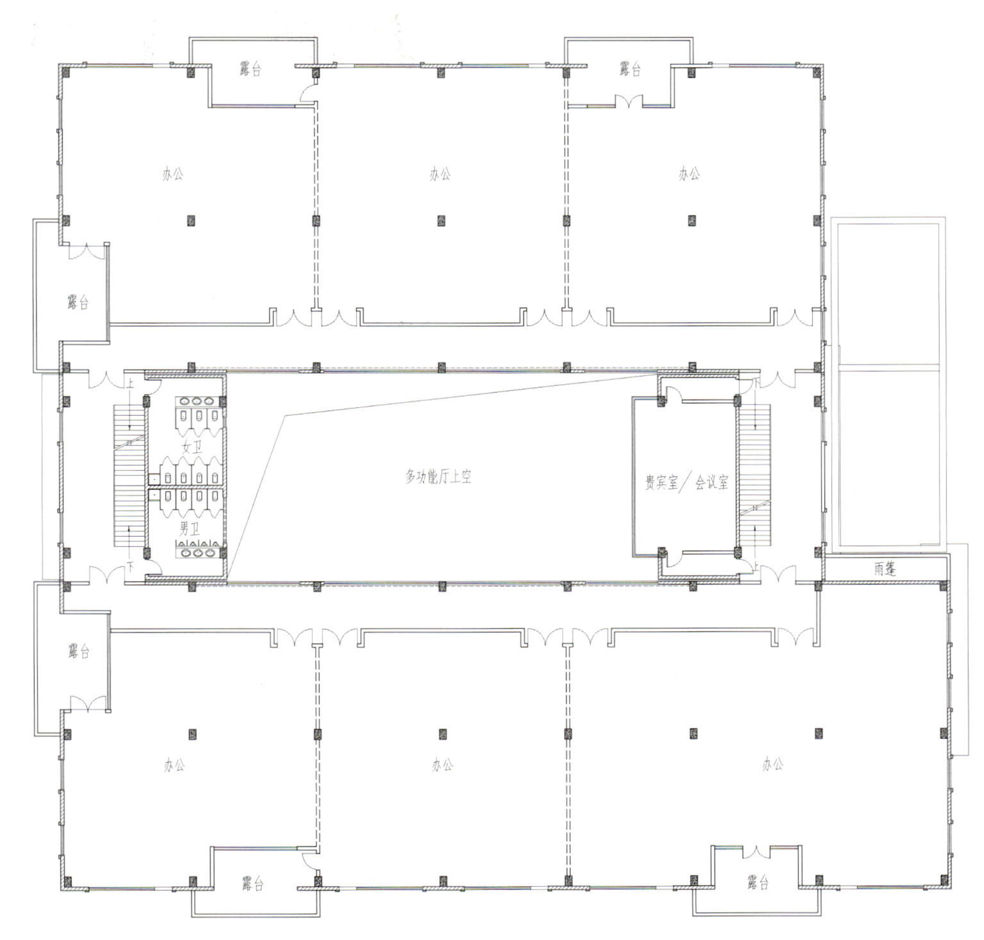

Second Floor Plan 二层平面图

6
台 湾 设 计 廊

6
台 湾 设 计 廊

South Elevation
南立面图

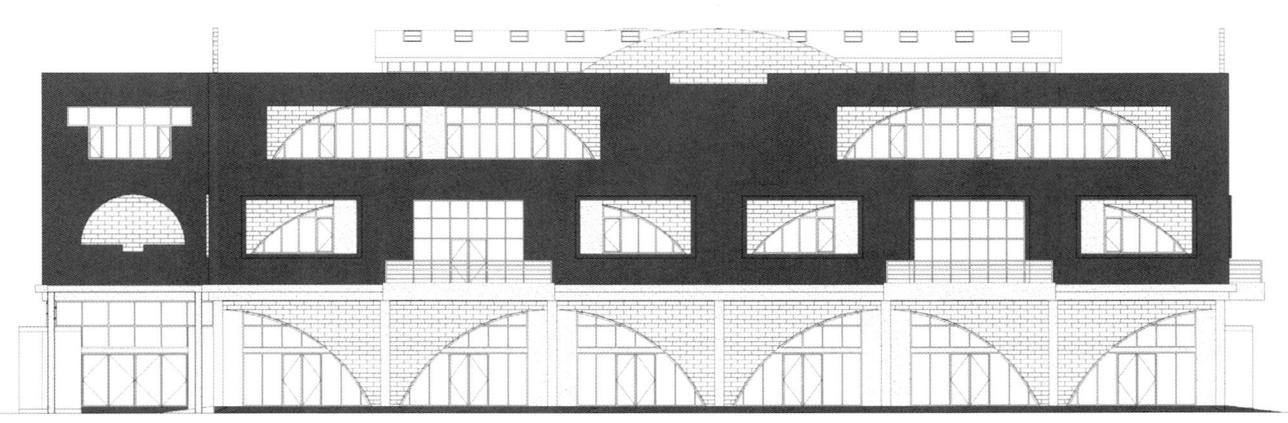

North Elevation
北立面图

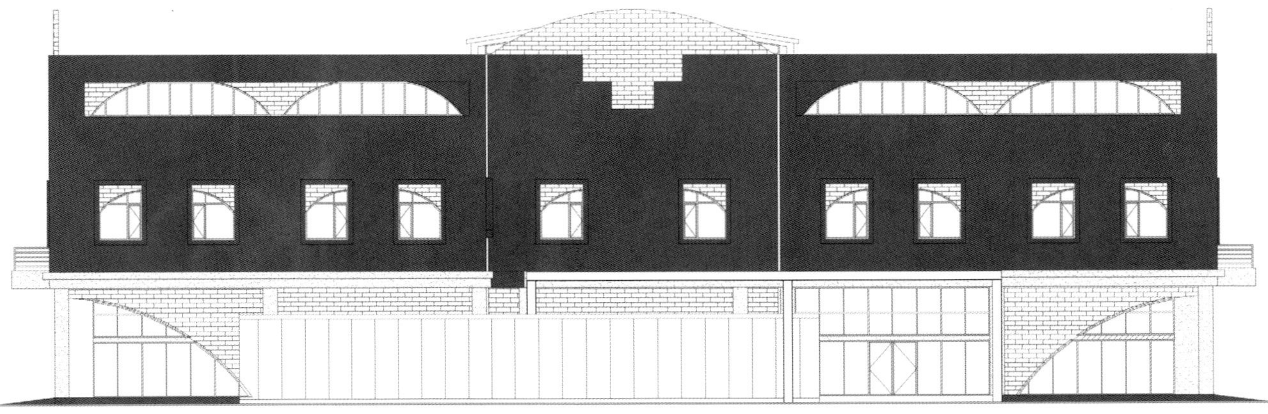

East Elevation
东立面图

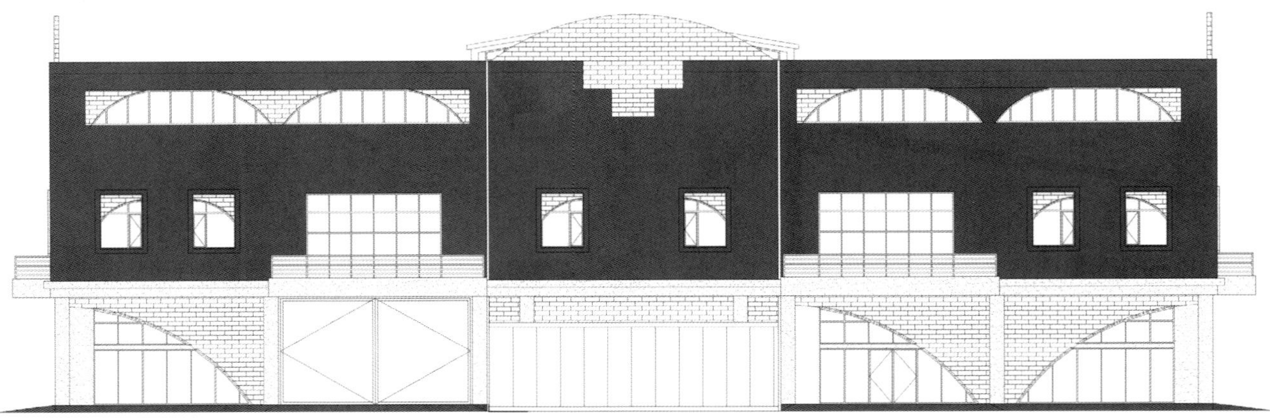

West Elevation
西立面图

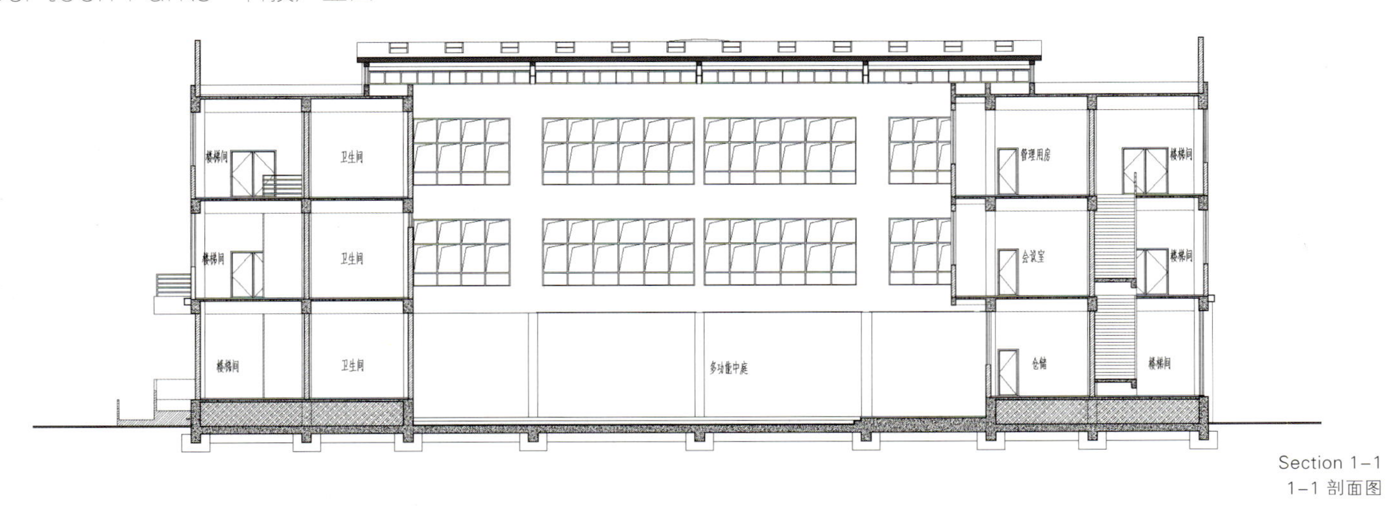

Section 1-1
1-1 剖面图

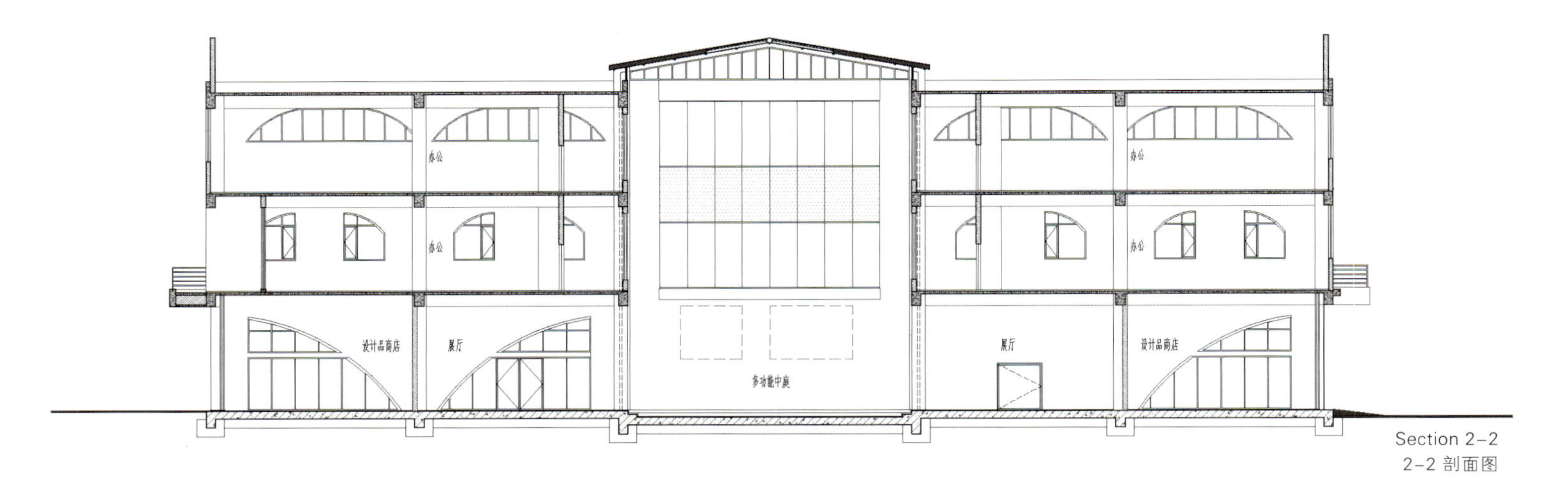

Section 2-2
2-2 剖面图

改造前

海峡创意设计中心

改造前

MIX
Cafe & Bar
MIX

改造前

Chengdu Bochuan Logistics Park

成都博川物流基地

Features 项目亮点

The architects create sharing courtyards by designing double-height ground floors, and establish the dialogue between buildings and courtyards with the landscape axis.

通过双首层的立体设计使院落共享化，通过景观轴的处理，达到建筑与建筑、建筑与院落、院落与院落的互动。

Keywords 关键词

Axis-network System

轴网系统

Three-dimensional Space

立体空间

Clear Cut

棱角分明

Location: Chengdu, Sichuan, China
Developer: Fujian Huaya Group
Planning and Design: Peddle Thorp Melbourne Pty Ltd.
Land Area: approximately 265,000 m^2
Total Floor Area: 750,000 m^2
Plot Ratio: 2.37
Status: Under Construction

项目地点：中国四川省成都市
开发商：福建华亚集团
规划设计：澳大利亚柏涛建筑设计有限公司
用地面积：约 26.5 万 m^2
总建筑面积：750 000 m^2
容积率：2.37
项目状态：在建

Background

Located in the core area of Western China, connecting the Western and Eastern, Chengdu is an important distribution center for commodities and materials. It is also designated by the State Council as the hub for science and technology, finance, commerce and trade, transportation and transmission. The advantaged location provides the basic conditions for developing modern logistics.

项目背景

成都地处中国西部地区的核心地带，是承接东部、联接西部的桥头堡，具有极强的辐射带动作用，是西南地区重要的商品物资集散地，更是国务院确定的西南地区的科技、金融、商贸中心和交通、通信枢纽，具有显著的区位优势，具备发展现代物流业的良好的基础条件。

Overview

The park is situated in the plain region of Chengdu where the terrain is flat and open for construction. The site's altitude ranges from 473 m to 485 m, which is higher in the west and lower in the east. And the land use is simple.

项目概况

项目地处成都平原地区，区域内地形平坦、地势开阔、适于建设，地面高程（成都坐标系）473 ~ 485 m，整体地势西高东低，南北向相差不大；用地性质较为简单，功能也较为单一。

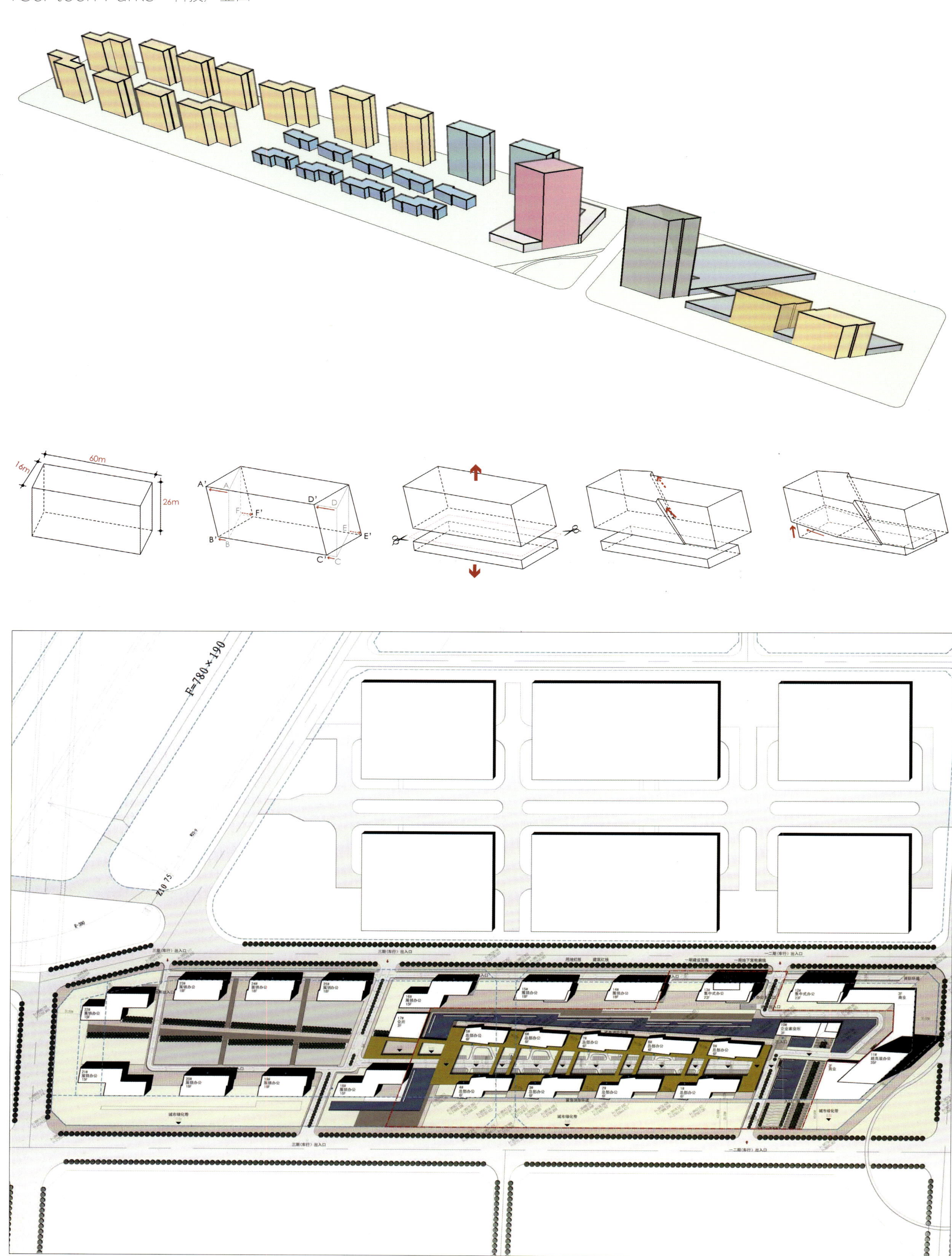
60m
16m
26m
A'
A
D'
D
F
F'
E
E'
B'
B
C'
C
F=780×190

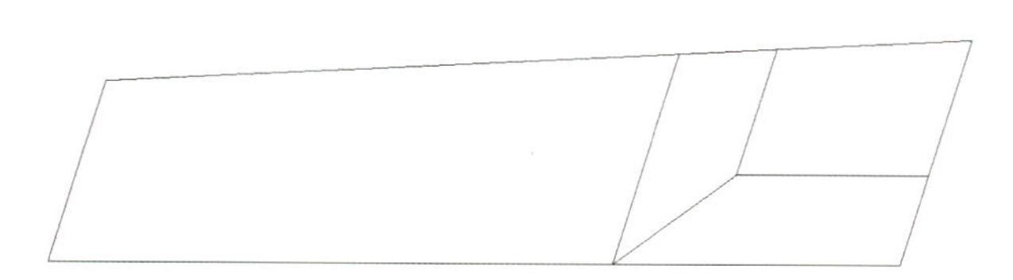

10# Roof Plan (Club)
10#（会所）屋顶平面图

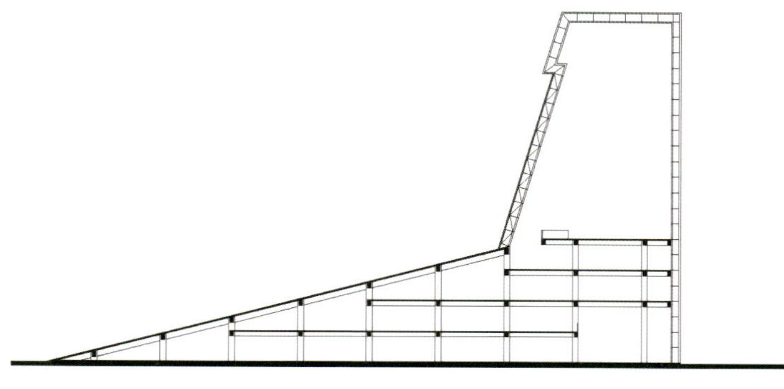

10# Section (Club)
10#（会所）剖面图

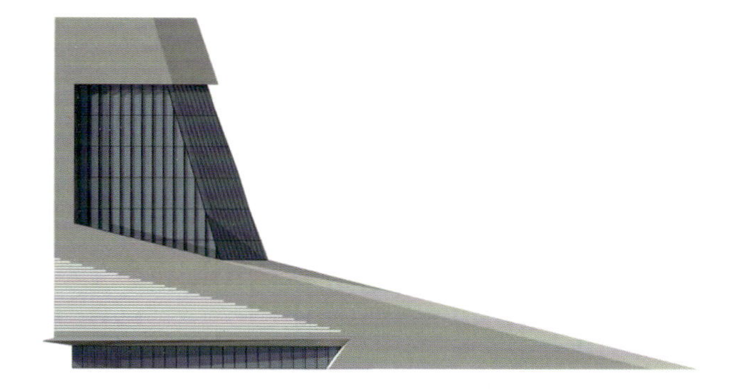

Design Idea

The logistics industry, as the support to the national regional economy, is important just like the blood vessel to our human body. So the design is inspired by the "pulsation" to stretch the existing grids and create an organic and radial pattern which forms the basic element of the buildings within the park. The lines of the "pulsation" are the interwoven rails, showing the hardness of the steel and the high efficiency of the logistics industry.

设计理念

物流行业对国家地区经济的支撑，如同血脉对人体的重要性。项目以"脉动"作为设计概念，将原始网格进行拉伸与错动，形成错落的、有机的线性放射图案，以此作为基地内建筑体的基本构成元素，脉动的线条代表的是交错的铁轨，体现了钢铁的硬朗与物流产业的高效率。

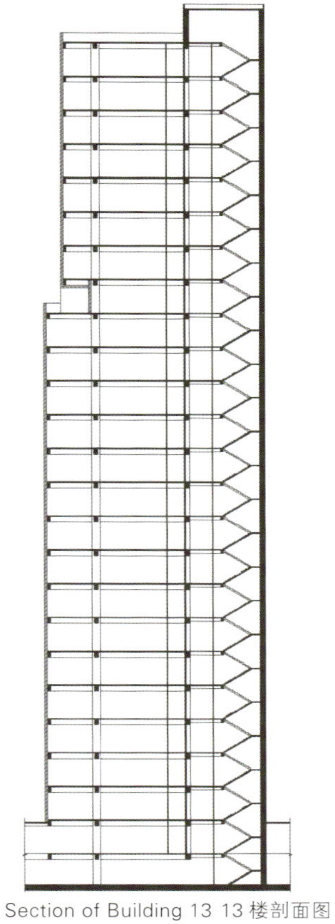

Section of Building 13 13楼剖面图

博川物流基

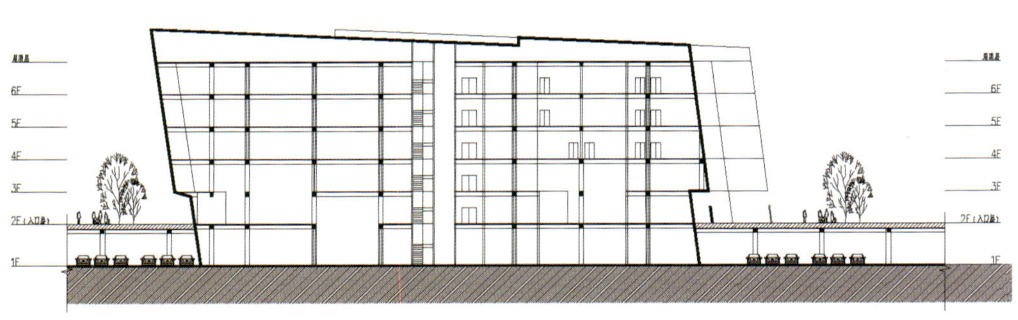

Section of Building 3 and 4 (Headquarter Buildings)
3、4# 楼（总部办公）剖面图

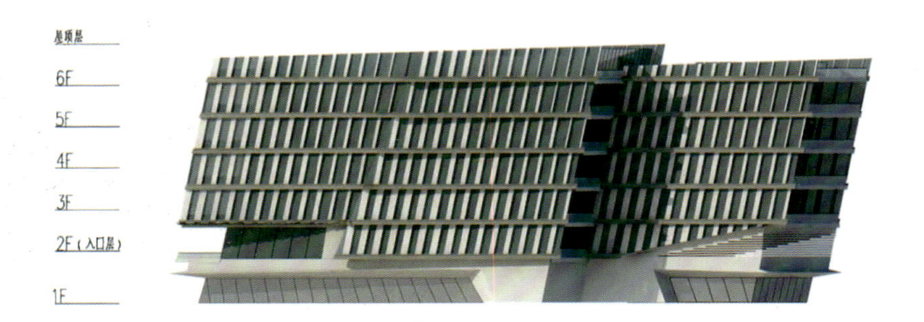

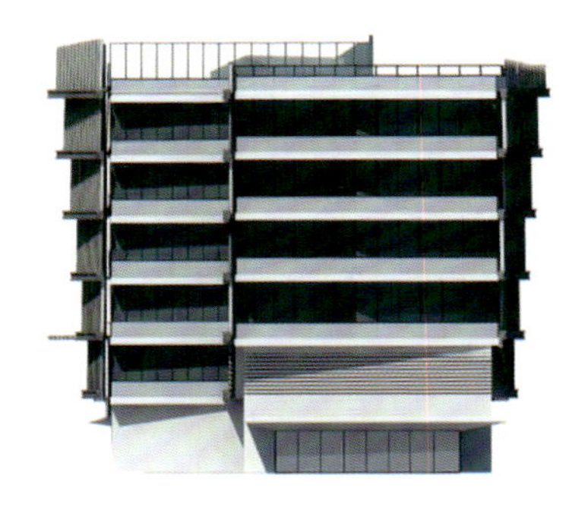

Architectural Design

The design pays attention to the interaction and communication in the modern offices. Sharing courtyards are created by designing double-height ground floors, establishing the dialogue between buildings and courtyards with the landscape axis.

All the landscapes and buildings are designed in a unified style to form an axis-net-work system.

The architects enhance the volumes of the buildings by using acute angles and clear cut, making these buildings look elegant and unique.

建筑设计

项目强调现代办公互动与交流，通过双首层的立体设计使院落共享化，通过景观轴的处理，达到建筑与建筑、建筑与院落、院落与院落之间的互动。

所有景观与建筑都由一定的轴网形成，景观与建筑统一风格，轴网统一一体化设计。

设计师通过锐角与切割的建筑设计手法加强型体感觉，使得整体外观棱角分明，独具特色。

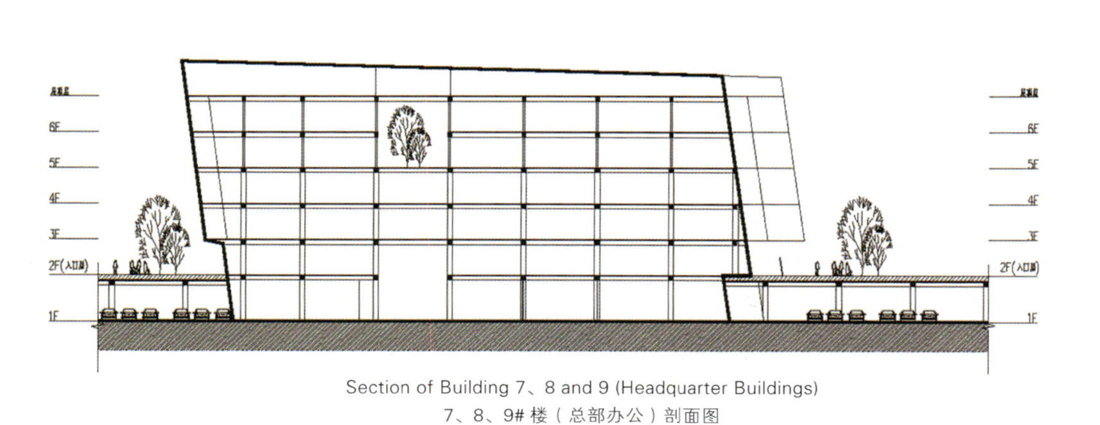

Section of Building 7、8 and 9 (Headquarter Buildings)
7、8、9# 楼（总部办公）剖面图

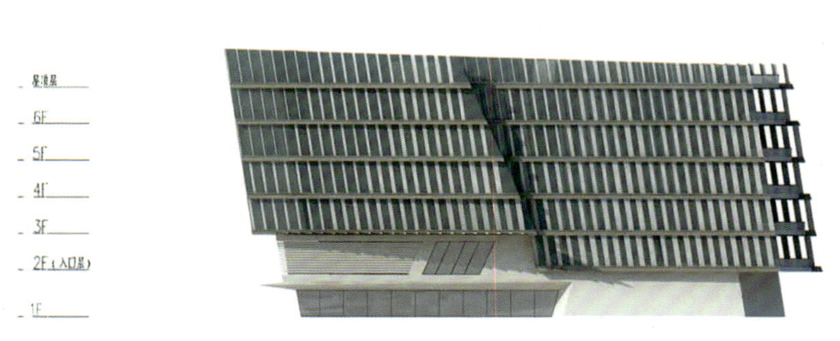

BLC

BAOSTEEL

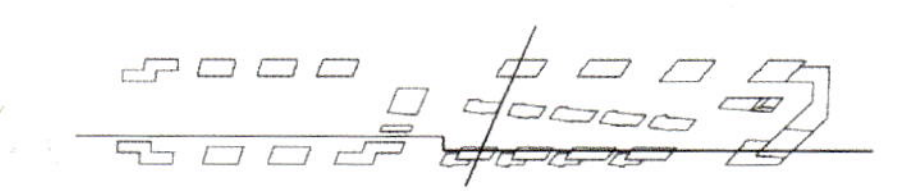

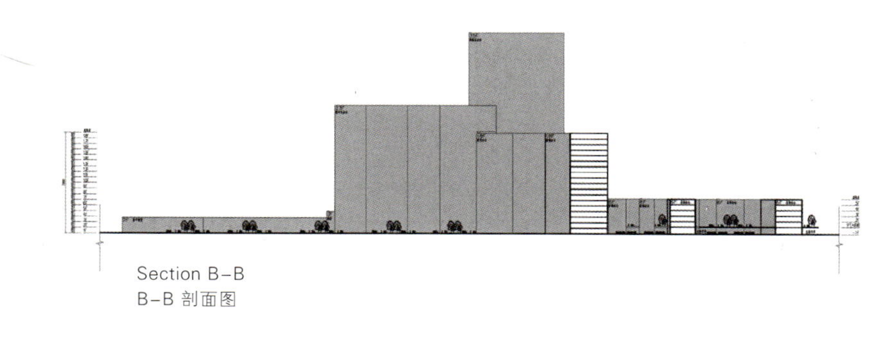

Section B-B
B-B 剖面图

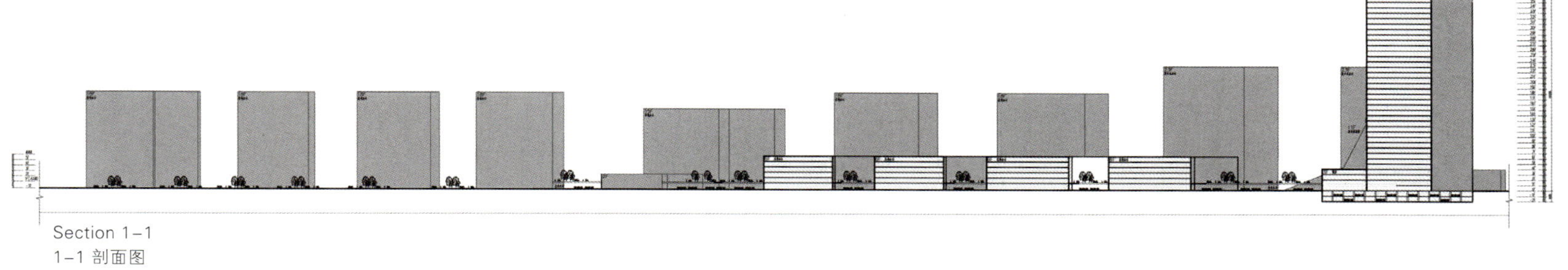

Section 1-1
1-1 剖面图

Shenzhen Bay Technology and Eco City (b-TEC)

深圳湾科技生态城

Features 项目亮点

Two kinds of urban textures (Manhattan style and Florence style) are superimposed and integrated to interpret the design idea of Shenzhen Bay Technology and Eco City (b–TEC).

项目在垂直方向叠合了曼哈顿肌理与佛罗伦萨肌理，以层面的垂直分布与有机联系，诠释了深圳湾科技生态城的设计理念。

Keywords 关键词

Superimposed Textures
肌理叠合

Varied Urban Characteristics
一城多面

Green and Ecological
绿色生态

Location: Nanshan, Shenzhen, Guangdong, China
Architectural Design: Peddle Thorp Melbourne Pty Ltd.
Land Area: 46,125 m^2
Gross Floor Area: 250,695 m^2
Plot Ratio: 5.4
Design Date: 2011
Status: not built

项目地点：中国广东省深圳市南山区
建筑设计：澳大利亚柏涛建筑设计有限公司
用地面积：46 125 m^2
总建筑面积：250 695 m^2
容积率：5.4
设计时间：2011 年
项目状态：未建

Overview

Located on the west bank of Shahe River, Nanshan District, Shenzhen, the project is to the west of Shahe Golf Court across the river and to the north of a residential area. It is planned to be built with hotel-style apartments, International Children's Center, cultural center, business apartments and offices.

项目概况

项目位于深圳南山区沙河西侧，东边是沙河高尔夫球场与之隔河对望，南边是一片住宅区，规划设计有酒店式公寓、国际儿童中心、文化活动中心、商务公寓、产业办公等多种功能场所。

Design Idea

As the development of globalization, modern cities are developed to be similar to each other. Is it possible to create a city that is distinguished with varied characteristics?

All the characteristics of a city are related to human's living. The designers have superimposed two types of urban textures: one is the traditional European cities represented by Florence, and the other is the modern American cities represented by Manhattan. The former is characterized by the radial streets to highlight the humanized scale, free style, low density and slow lifestyle. The later is characterized by the chessboard-style road system which benefits to easy arrangement of the buildings on two sizes. It represents a kinds of large-scale, high-density and fast-paced lifestyle. In this project, two kinds of textures are superimposed to form the embryonic form of Shenzhen Bay Technology and Eco City (b-TEC).

设计理念

随着全球化，千城一面成为现代城市的宿命，那么是否可以创造一城多面呢?

面可以是立面，但在城市中，更深层次上和人的生活相关的是层面。设计师在垂直方向叠合了两种城市的肌理，一种是佛罗伦萨为代表的欧洲传统城市，一种是曼哈顿为代表的美国现代都市。前者是放射状的街道形式，展现欧洲的宜人尺度，自由而中心明确，代表着小尺度、低密度、漫步、生活的舒适度、慢节奏；后者是规整的棋盘式道路网格，有利于道路两侧建筑物的布置，代表大尺度、高密度、运作的效率、快节奏。把这两种肌理叠加在一起，产生了深圳湾科技生态城的雏形。

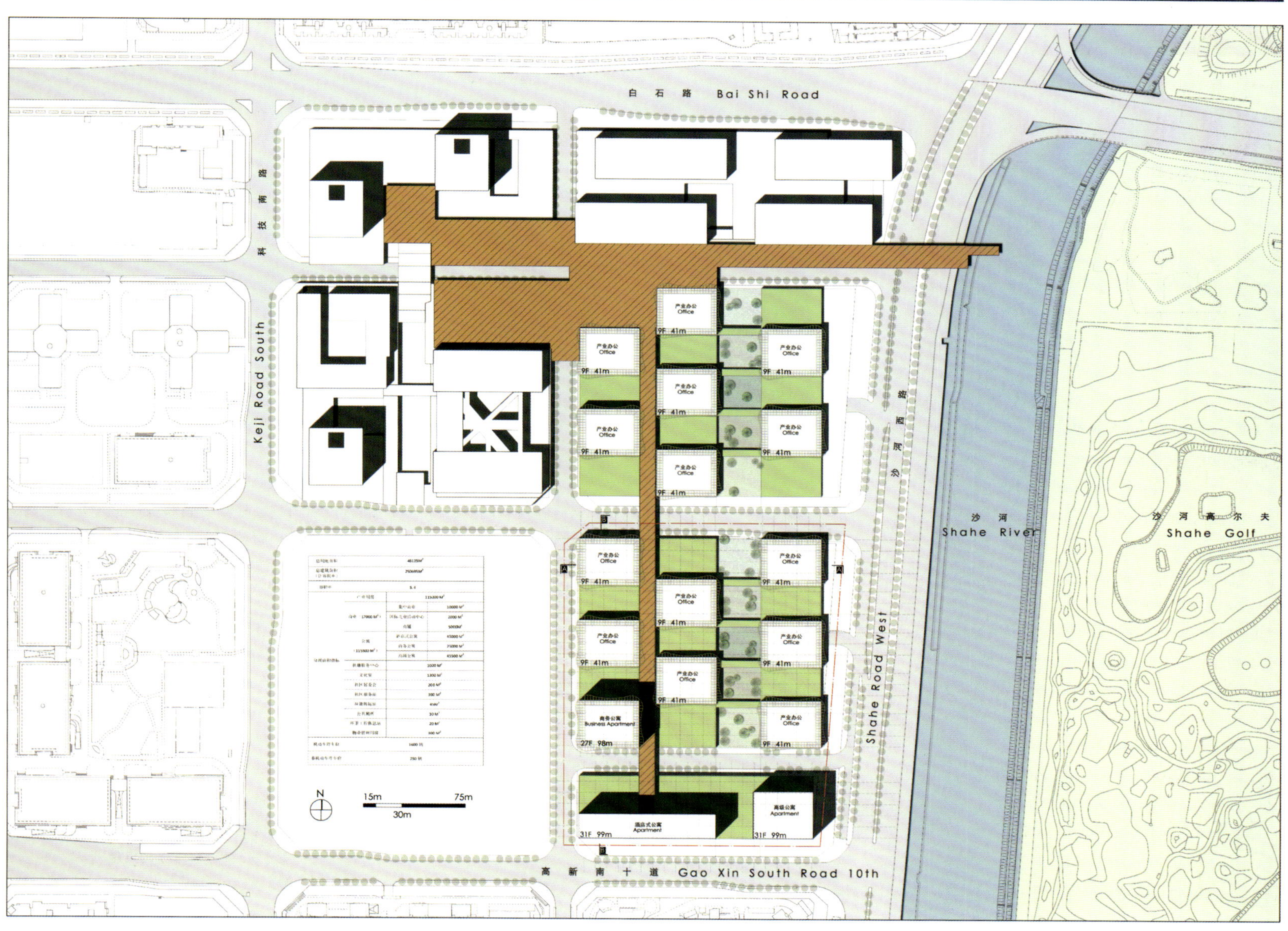

白石路 Bai Shi Road
科技南路
Keji Road South
沙河西路
Shahe Road West
沙河
Shahe River
沙河高尔夫
Shahe Golf
高新南十道 Gao Xin South Road 10th
产业办公
Office
9F 41m
商务公寓
Business Apartment
27F 98m
酒店式公寓
Apartment
31F 99m
高级公寓
Apartment
31F 99m
N
15m
30m
75m

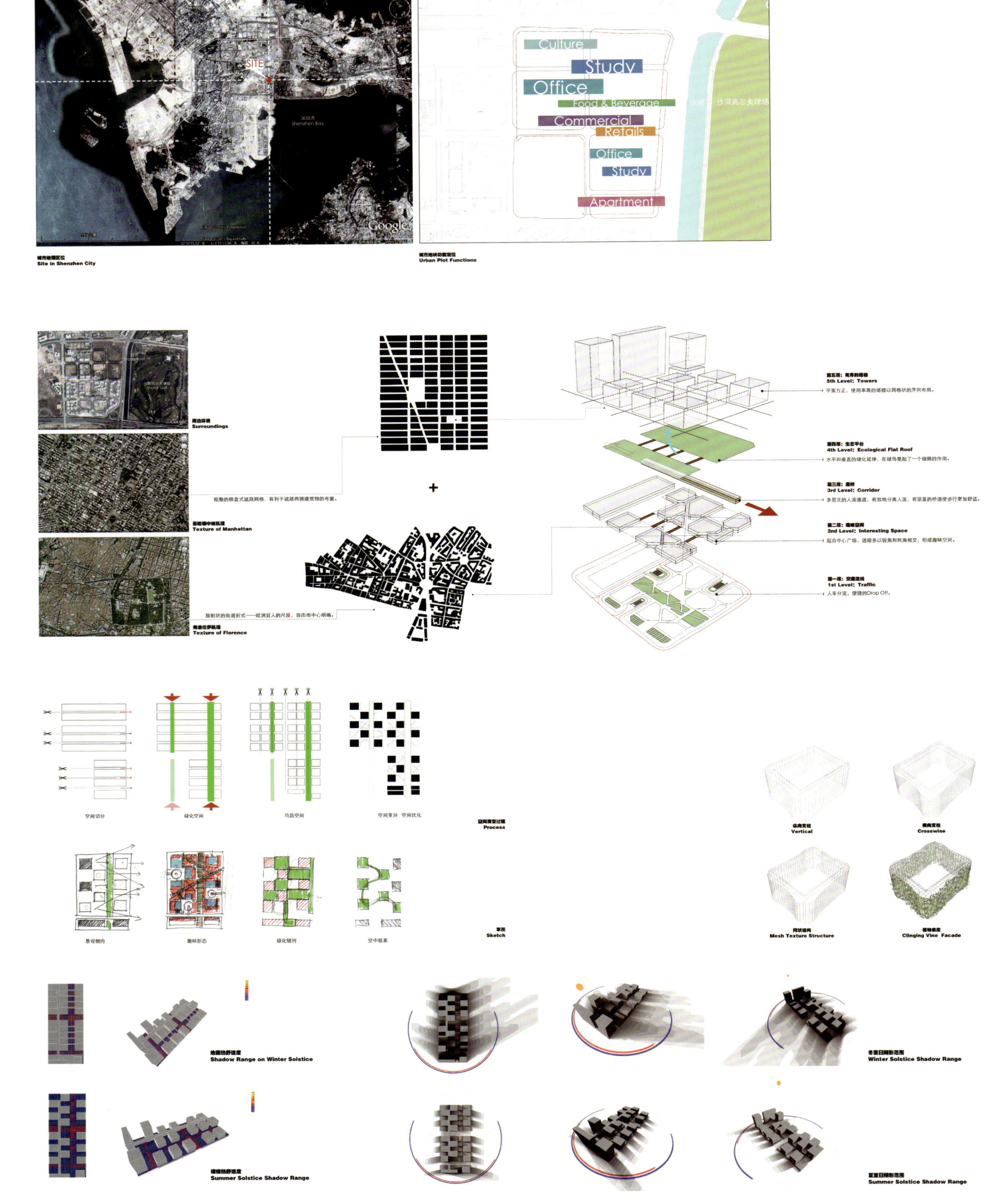

此布局方式营造的街道空间全年的热舒适度较佳。

冬至日：保证南北贯穿的空间——公共空间日照时间多，冬日让公众享受更多阳光；
夏至日：错开的布局空间营造了多面阴影，形成了夏日有效的遮阳面。

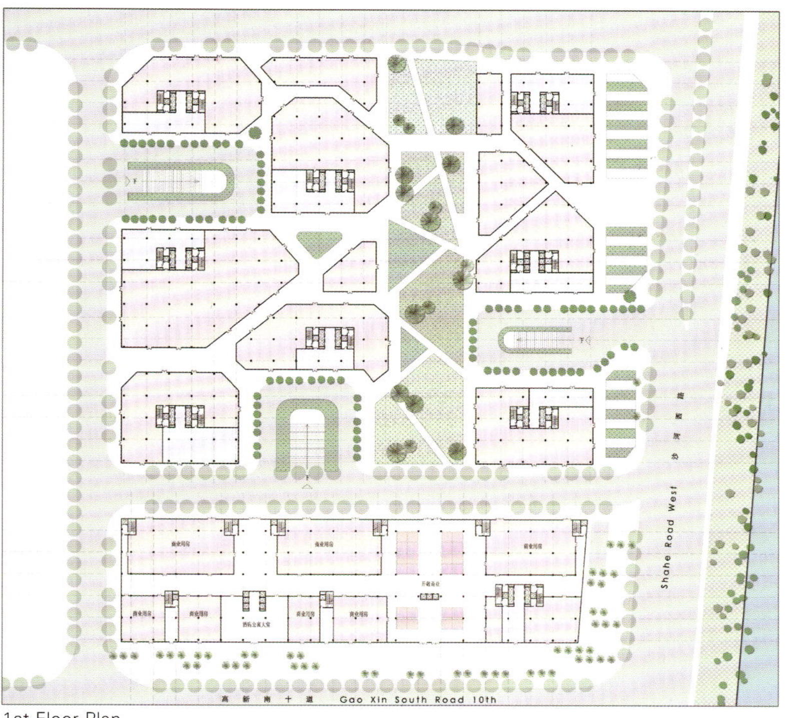

1st Floor Plan
一层平面图

2nd Floor Plan
二层平面图

Office
产业楼

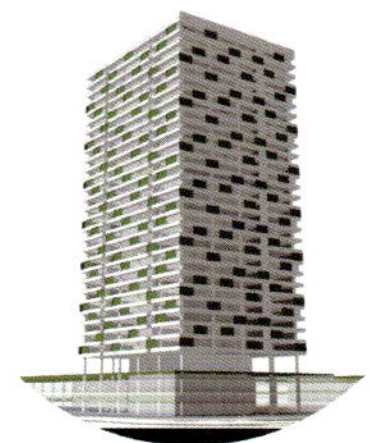

Apartment Tower
高级公寓

Business Apartment Tower
商务公寓

Apartment Tower
酒店式公寓

Office Typical Plan
产业楼标准层平面图

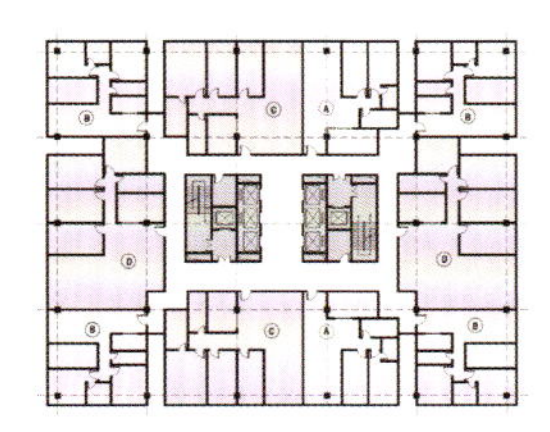

Apartment Typical Plan
高级公寓标准层平面图

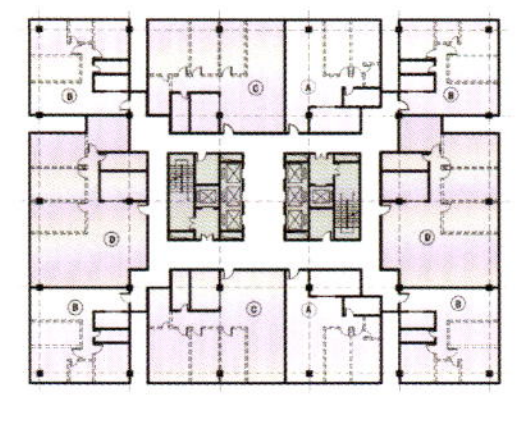

Business Apartment Typical Plan
商务公寓标准层平面图

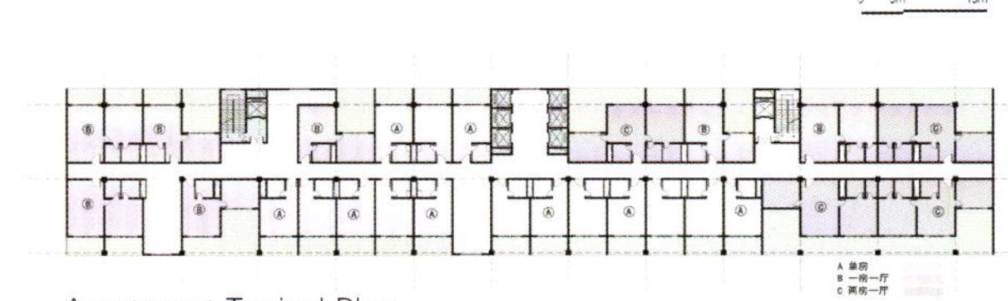

Apartment Typical Plan
酒店式公寓标准层平面图

太阳能光电

雨水收集

屋顶绿化

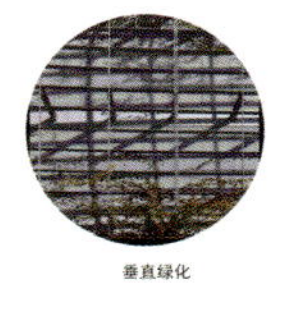

垂直绿化

节能措施
Measures of saving energy

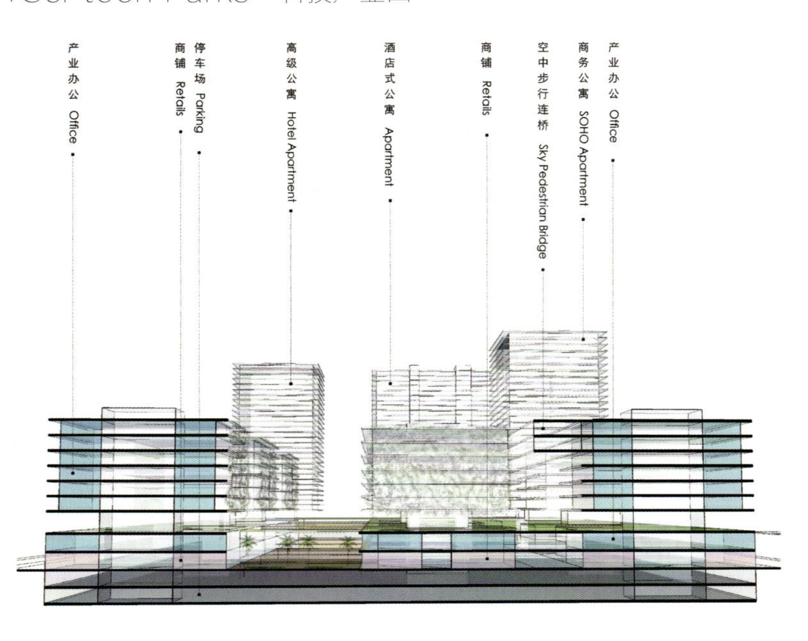

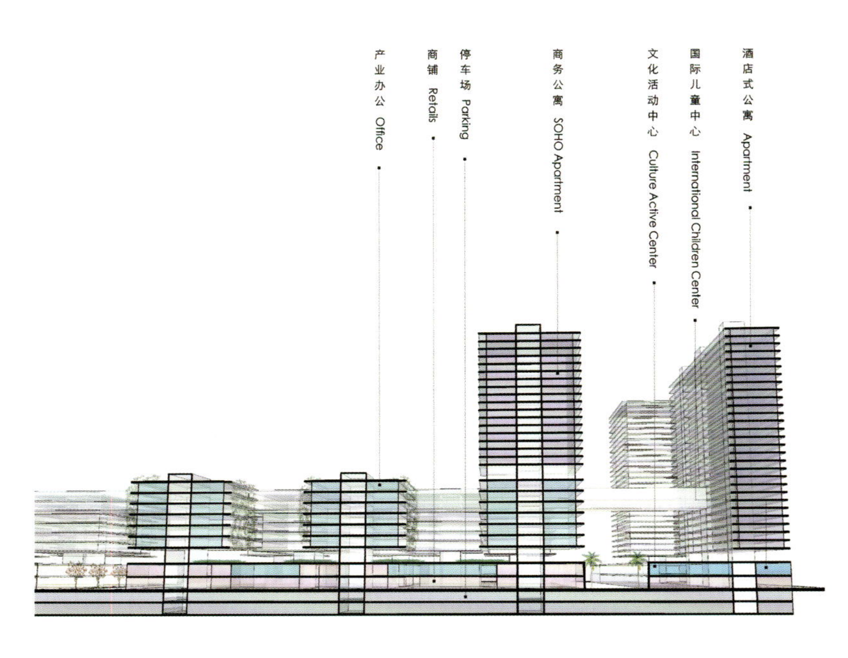

Architectural Design

The podiums are arranged in European block-style to provide catering and commercial services. The tops of the podiums are roof gardens. The towers are used for offices and apartments which are connected by vertical and horizontal traffic systems. Different people can encounter at any time on different floors to start their informal communications. It is really a new characteristic of the high-tech industry: the boundary between work and life is broken by modern network. People will get their inspirations from the informal communication because of this kind of interpenetration.

建筑设计

裙房为欧洲街区式的以餐饮为主的商业和服务业，裙房顶部为屋顶花园，塔楼部分为产业办公和公寓，并通过垂直和水平交通联系在一起，不同的人在不同时间和不同层面上相遇，这种设定的多样层面更容易激发人们间的非正式交流。尤其在当今的网络时代，工作与生活的界限被打破，是高科技产业的一个新特点。人们通过非正式交流激发彼此灵感，工作与生活的相互延伸使得价值的创造打破了一成不变的模式。

Eco Design

The project applies many energy-saving technologies and measures such as the solar photovoltaic system, the rainwater collecting system, the roof greeing, the vertical greening, etc. And the ecological design not only means the application of green technologies but also reflects in the long-lasting vitality of a complex. The design of Shenzhen Bay Technology and Eco City (b-TEC) is just interpreted by the vertical superimposition and organic connection of different aspects.

生态设计

项目采用了太阳能光电、雨水收集、屋顶绿化、垂直绿化等节能措施与方式，此外，生态不仅仅是人们对绿色科技的运用，也有另一种解读，一个聚落持续的活力是生态平衡的表现。形象的理解是层面的垂直分布与有机联系，设计师用这种聚落来诠释深圳湾科技生态城的设计。

Daylighting Design

Winter Solstice: the public spaces from south to north are connected and open, allowing people to enjoy more sunlight in winter.

Summer Solstice: staggered space design will create some shades which will be great for sun-shading in summer days.

日照设计

冬至日：保证南北贯穿的空间——公共空间日照时间多，冬日让公众享受更多阳光；

夏至日：错开的布局空间营造了多面阴影，形成了夏日有效的遮阳面。

Terra Bio Valley Conceptual Planning & Design

泰然生物谷概念性规划设计

Features 项目亮点

The plan focuses on two 25-storey towers along the High-tech Avenue which face to the urban core area in the west. The top and bottom of the buildings are cut to form unique appearance. And the skin is designed with vertical elements to shape the rising gesture.

项目重点打造高新大道面向西面城市核心区的两栋 25 层高层塔楼，在建筑的顶部和底部进行体块的切割，形成独特的外部形象，并且通过竖向印象的建筑表皮，形成向上攀升的建筑形象。

Keywords 关键词

In Harmony with Nature
自然共生

Full of Energy
富于活力

Modern Style
现代风格

Location: Wuhan East Lake High-tech Development Zone, Hubei, China
Planning and Design: Peddle Thorp Melbourne Pty Ltd.
Land Area (Industrial): 399,757.6 m²
Floor Area (Industrial): 839,400 m²
Plot Ratio (Industrial): 2.1
Land Area (Residential): 211, 415.5 m²
Floor Area (Residential): 422,800 m²
Plot Ratio (Residential): 2.0
Design Date: 2012
Status: Not Built

项目地点：中国湖北省武汉市东湖高新区
规划设计：澳大利亚柏涛建筑设计有限公司
用地面积（工业区）：399 757.6 m²
建筑面积（工业区）：839 400 m²
容积率（工业区）：2.1
用地面积（居住区）：211 415.5 m²
建筑面积（居住区）：422 800 m²
容积率（居住区）：2.0
设计时间：2012 年
项目状态：未建

Overview

The project is located in the west of Wuhan Optics Valley New District. In the east along the High-tech Avenue, there will be the administrative center of the East Lake High-tech Development Zone, the sci-tech center, the exhibition center and Huawei R & D Center. In the south, it is planned with the film and TV production base of Hubei Administration of Radio, Film and Television, the commercial center, and Jiulong Industrial Base of Wuhan Biolake. And in the southwest, the Design Capital of Hongshan District and the Inspection Center of the State Food and Drug Administration will be set.

项目概况

项目位于新规划的光谷新区西侧，东面沿高新大道依次规划有东湖高新区行政中心、科技中心、展览中心、华为研发中心，南面规划有湖北省广播电视总局影视基地、商业中心、武汉国家生物九龙产业基地等，西南规划有洪山区设计之都、国家食品监督管理局检验中心。

Planning Idea

Based on respect for nature and the surrounding environment, the architects start from the city image and urban design to do the overall plan. With the eco park in the southeast as the core, all the buildings are arranged around and are rising high from inside to outside. The multi-storey buildings adjacent to the park are designed with open ground floors to introduce the eco environment into the buildings. While the buildings standing high in the periphery open to the urban space to show the image of the Optics Valley's new district. And the high-rise residences close to Jiufeng National Forest Park are designed in point style to provide the visual corridor between the city and the park, and highlight the design idea of "harmony between architecture and environment".

Site Plan
总平面图

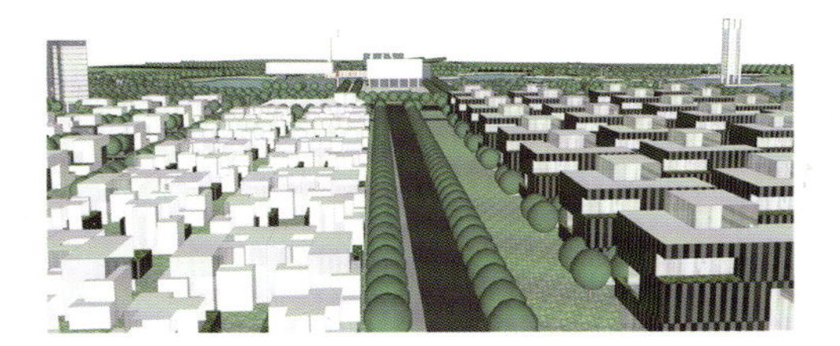

Planning Structure 规划结构

城市主干道

城市次干道

产业园区道路

出入口

Traffic 道路交通

研发孵化聚集地

中央商务聚集地

IEO总部聚集地

创新宜居社区

Function Zoning 功能分区

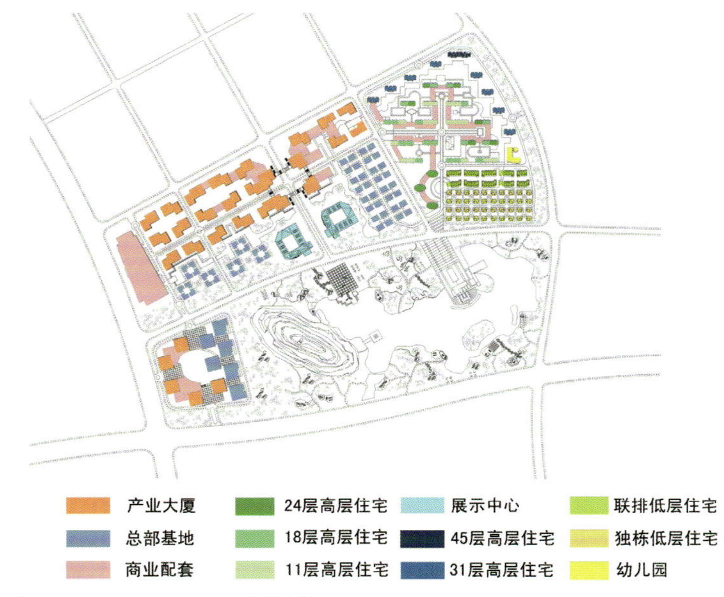

Functional Arrangement 功能布置

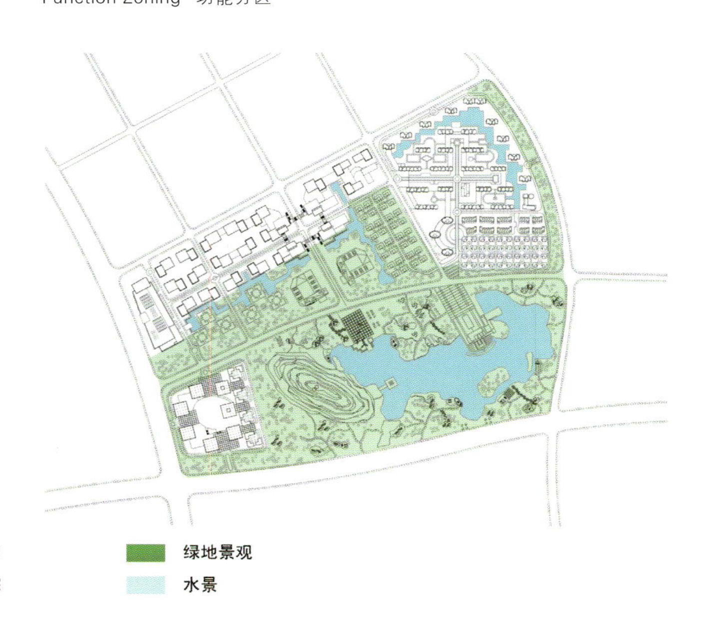

绿地景观

水景

Landscape 绿化分析

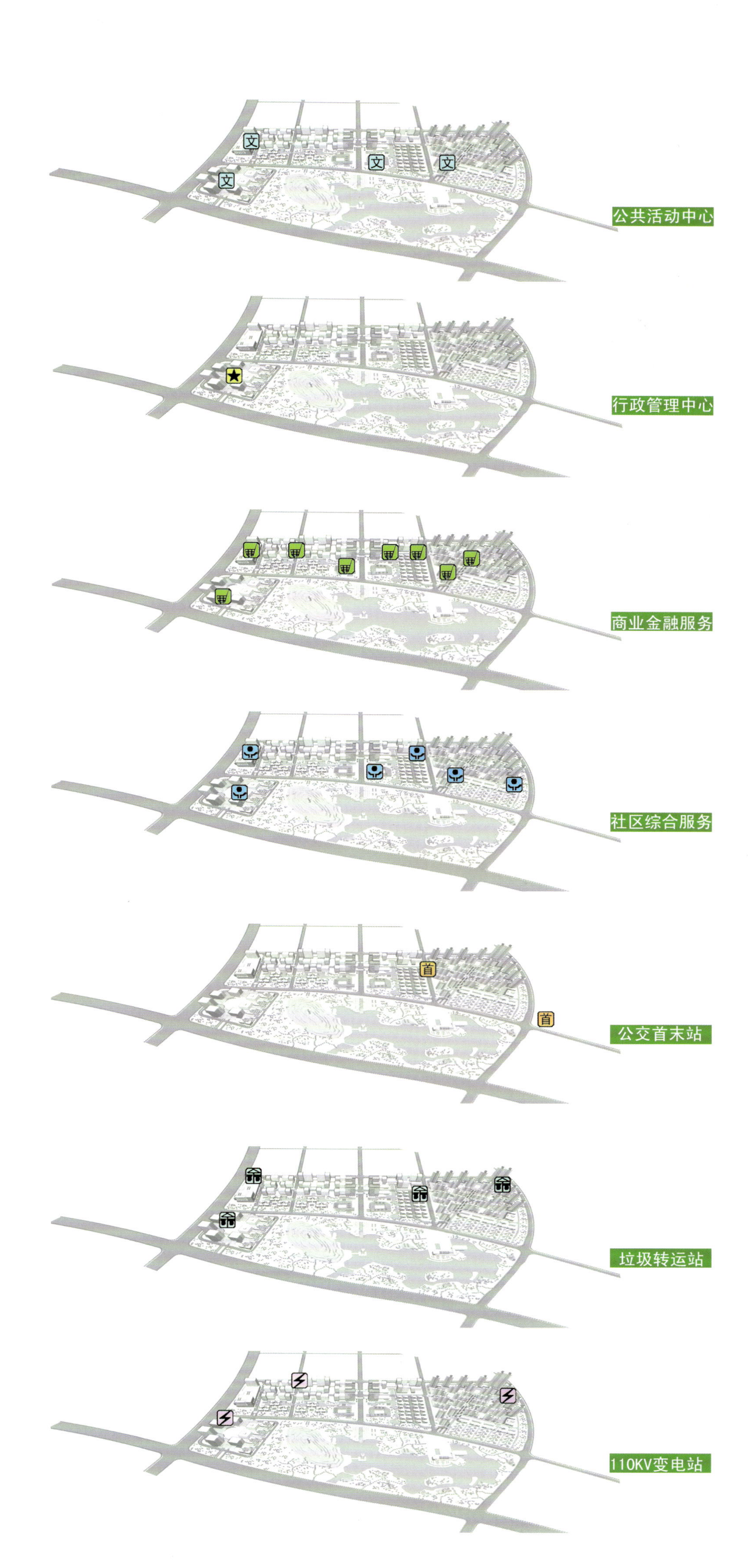

文 公共活动中心
- 共享空间
- 活动广场
- 小型图书馆
- 活动中心
- 球类棋牌活动场所
- 综合影剧院

★ 行政管理中心
- 园区管理中心
- 街道办
- 派出所
- 其他管理机构

商业金融服务中心
- 综合超市
- 餐饮
- 电信营业厅
- 银行网点
- 邮政支局

社区综合服务中心

首 公交首末站

垃圾转运站

110KV变电站

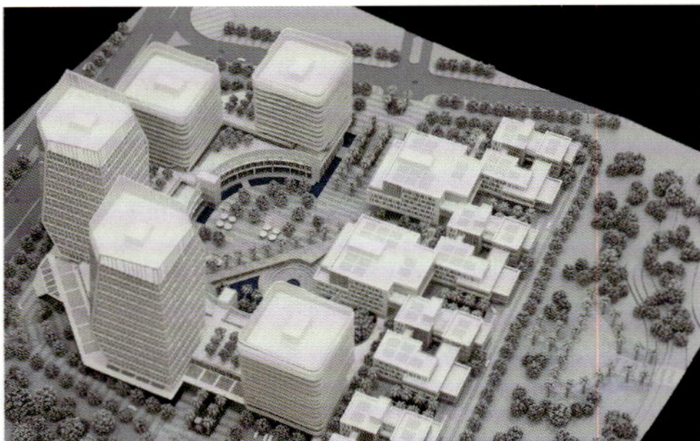

规划理念

项目充分尊重自然及周边的环境，从城市形象和城市设计的高度进行统筹规划，以东南面的生态公园为核心，建筑高度由内到外逐渐升高，内侧紧邻公园为多层建筑，通过底层架空等方式将生态环境渗透到园区内部，外侧面向城市空间，多为挺拔的高层建筑，以展示光谷新区的城市形象。临近九峰森林公园一侧的高层住宅均设计成点式建筑，最大限度地让出城市与公园之间的视觉通道，充分体现了建筑与环境共生的设计理念。

Planning Objectives

The plan tries to present a dynamic neighborhood which integrates functions like office, business, social communication and recreation together. With three elements — the innovative enterprise gathering platform, the advanced public service system and the eco residential community, it aims to create an ecological, beautiful and sustainable innovative park.

规划目标

项目试图创建一个富于活力的街区，将办公、商务、社交、休憩等各类功能复合，相互作用，通过现代城市创新企业聚集平台、先进的园区公共服务体系、生态的宜居社区这三大要素，打造生态、园林、可持续发展、现代城市创新企业聚集园区的主题。

Planning Layout

The inner road has connected different zones from east to west. And the east-west landscape axis has well combined the buildings with the landscapes to form the layout with “one belt, four axes, three core and four zones”.

规划结构

东西向的内部道路将各地块从东到西串连起来，通过东西向的景观通道，将建筑与景观有机地结合在一起，形成了一带、四轴、三核、四区的规划结构。

Functional Layout

The site is divided into four parts by the planning roads: the R&D incubator on the south of Jiufeng No.1 Road as well as the CBD, the IEO and the innovative residential community on the north of Jiufeng No.1 Road. All these functions are interpenetrated with each other to form a unified neighborhood.

功能分区

项目用地被规划道路划分成四个功能区，分别是九峰一路以南的研发孵化聚集区，九峰一路以北、自西向东的中央商务聚集区、IEO 总部聚集区、创新宜居社区，每个功能区相互渗透，使得整个社区形成一个整体。

泰然 生物谷
TERRA BIO-VALLEY

Architectural Design

Following the modern architectural style, the plan focuses on two 25-storey towers along the High-tech Avenue to make them face to the urban core area in the west. The top and bottom of the buildings are cut to form unique appearance. And the skin is designed with vertical elements to shape the rising gesture.

The buildings in the headquarters base range from 5-storey to 9-storey, and all are designed with terraces on the east to open to the eco park. It will not only archive the harmony between the architectures and the environment but also form varied architectural forms. By employing the similar skin elements like the vertical metal trim strips, the windows of different sizes and the glass curtain walls, it will form a modern-style urban complex that dialogues with the high-rises and the podiums.

建筑设计

项目延续了现代建筑的风格，重点打造高新大道面向西面城市核心区的两栋 25 层高层塔楼，在建筑的顶部和底部进行体块的切割，形成独特的外部形象，并且通过竖向印象的建筑表皮，形成向上攀升的建筑形象。

产业总部基地建筑为 5~9 层，每栋建筑东面设计了面向生态公园的露台，在充分与环境融合的同时，也形成了非常丰富的建筑形体，通过与高层建筑和裙房相似的建筑表皮，运用竖向错动的金属装饰条、大小各异的窗洞和玻璃幕墙等现代建筑材料，形成了与高层、裙房合而为一的现代风格的城市综合体。

Landscape Design

The ground floor of the buildings are open to provide views of the mountain and the green land in the east, which integrates the buildings into the surrounding environment. The central square is carefully designed with outdoor cafe, tree arrays, flower beds, lawns and fountains as well as the sunken square in the center, which will be used for displaying the enterprise culture and the communication between staff. The landscape axis runs through the sunken square leading to phase one of the Biological Innovation Zone, connecting two phases (phase one and phase two) together.

景观设计

项目通过底层架空将东侧的山景和绿化引入场地，使建筑和环境融为一体。重点打造中心广场的景观，广场设有露天茶座、树阵、花池、草坪、旱喷泉等景观小品，在中心位置设有下沉广场，用于企业文化展示和人员的交流，通过下沉广场通向生物创新一期的景观轴线，将生物创新一、二期有机地联系起来。